Lecture Notes in Mechanical Engineering

Lecture Notes in Mechanical Engineering (LNME) publishes the latest developments in Mechanical Engineering—quickly, informally and with high quality. Original research or contributions reported in proceedings and post-proceedings represents the core of LNME. Volumes published in LNME embrace all aspects, subfields and new challenges of mechanical engineering.

To submit a proposal or request further information, please contact the Springer Editor of your location:

Europe, USA, Africa: Leontina Di Cecco at Leontina.dicecco@springer.com
China: Ella Zhang at ella.zhang@cn.springernature.com
India, Rest of Asia, Australia, New Zealand: Swati Meherishi at swati.meherishi@springer.com

Topics in the series include:

- Engineering Design
- Machinery and Machine Elements
- Mechanical Structures and Stress Analysis
- Automotive Engineering
- Engine Technology
- Aerospace Technology and Astronautics
- Nanotechnology and Microengineering
- Control, Robotics, Mechatronics
- MEMS
- Theoretical and Applied Mechanics
- Dynamical Systems, Control
- Fluid Mechanics
- Engineering Thermodynamics, Heat and Mass Transfer
- Manufacturing Engineering and Smart Manufacturing
- Precision Engineering, Instrumentation, Measurement
- Materials Engineering
- Tribology and Surface Technology

Indexed by SCOPUS, EI Compendex, and INSPEC.

All books published in the series are evaluated by Web of Science for the Conference Proceedings Citation Index (CPCI).

To submit a proposal for a monograph, please check our Springer Tracts in Mechanical Engineering at https://link.springer.com/bookseries/11693.

Suvanjan Bhattacharyya · David James
Editors

Recent Trends in Sports Engineering

Select Proceedings of ICSE 2023

Springer

Editors
Suvanjan Bhattacharyya
Department of Mechanical Engineering
Birla Institute of Technology and Science
Pilani, Rajasthan, India

David James
Labosport Limited
Hucknall, UK

ISSN 2195-4356 ISSN 2195-4364 (electronic)
Lecture Notes in Mechanical Engineering
ISBN 978-981-97-8296-3 ISBN 978-981-97-8297-0 (eBook)
https://doi.org/10.1007/978-981-97-8297-0

This Springer imprint is published by the registered company Springer Nature Singapore Pte Ltd.
The registered company address is: 152 Beach Road, #21-01/04 Gateway East, Singapore 189721, Singapore

Preface

The field of sport engineering has witnessed an extraordinary evolution over the past few decades, transforming the way athletes train, compete, and achieve their peak performances. The convergence of engineering, technology, and sports science has led to ground-breaking innovations that have redefined the boundaries of athletic potential and spectator engagement. This book, *Recent Trends in Sport Engineering*, aims to capture the dynamic and rapidly advancing landscape of this inter-disciplinary field, offering a comprehensive overview of the latest developments, technologies, and research that are shaping the future of sports.

Sport engineering is no longer confined to the design of equipment and facilities. It encompasses a broad spectrum of applications, including performance analytics, biomechanics, materials science, wearable technology, and even virtual reality. These advancements are not only enhancing athletic performance and safety but also revolutionizing fan experiences and the way sports are consumed globally.

The impetus for this book arises from the need to document and disseminate the cutting-edge trends that are driving sport engineering today. As an academic and practitioner deeply immersed in this field, I have had the privilege of witnessing firsthand the transformative impact of these innovations. This book is an attempt to share that knowledge and to provide a platform for further exploration and discovery.

The chapters in this book cover a diverse array of topics, each highlighting a specific trend or breakthrough in sport engineering. From the latest in high-performance materials and advanced biomechanical analysis to the integration of artificial intelligence and big data in sports, each section is meticulously researched and presented to provide a clear and comprehensive understanding of the subject. Case studies and real-world examples are included to illustrate the practical applications of these advancements, offering readers' insights into how theory translates into practice.

Recent Trends in Sport Engineering is intended for a wide audience, including students, researchers, engineers, sports scientists, and professionals in the sporting industry. Whether you are a seasoned expert or a newcomer to the field, this book aims to serve as a valuable resource that broadens your understanding and sparks your curiosity about the future possibilities in sport engineering.

I would like to express my sincere gratitude to my colleagues, collaborators, and students whose contributions and insights have been invaluable in the creation of this book. I also extend my thanks to the athletes, coaches, and sports technologists who have shared their experiences and expertise, enriching the content with practical perspectives.

As you delve into the pages of *Recent Trends in Sport Engineering*, I hope you find the inspiration and knowledge that will propel your interest and involvement in this exciting field. May this book not only educate but also motivate you to contribute to the ongoing innovations that continue to push the boundaries of what is possible in sports.

Pilani, India — Dr. Suvanjan Bhattacharyya
Hucknall, UK — Dr. David James

Contents

About the Editors

Dr. Suvanjan Bhattacharyya is an assistant professor in the Department of Mechanical Engineering at BITS Pilani, renowned for his pioneering contributions in the field of applied thermal engineering. With a career dedicated to advancing research in energy systems, he has garnered international recognition for his work on heat transfer, fluid dynamics, and sustainable energy technologies. His prolific academic output includes over 100 high-quality SCI journal papers, underscoring his commitment to excellence in research and publication. His expertise spans a wide range of inter-disciplinary areas. His research interests include energy storage, microfluidics, and thermal management systems, where he combines experimental and numerical methods to address complex engineering challenges. Currently, he supervises 12 Ph.D. students, fostering the next generation of engineers and researchers.

Prof. David James is a leading figure in the world of sports engineering and technology. He is the CEO of the LABOSPORT Group and holds a title of visiting professor at Sheffield Hallam University. He has been working in the field of sports engineering for 20 years and has sustained a significant contribution to research and knowledge transfer through a myriad of projects with sports companies and federations. In 2018, he was awarded the title of Professor of Sports Engineering (Sheffield Hallam University Personal Chair). He is a champion for public engagement in research. More than 1 million individuals have directly connected with his public engagement activities as either participants or audiences. He has worked tirelessly to reposition science and engineering as a part of mainstream culture through major projects including exhibitions, lecture tours, public debates, theatre, school workshops and public art.

Future of Machine Learning in Sports Engineering

V. V. Prasanth and G. Nallavan

Abstract One of the emerging technologies that is used in sports engineering is machine learning. Datasets and algorithms play an important role in achieving the desired results in the field of sports engineering. We have lot of data and it plays a major role in talent identification, training and coaching, umpiring and scoring, analysis and insight generation, injury management and recovery, fan engagement, e-sports and betting, etc. The data we acquire through all of these is being trained and tested using available machine learning algorithms. The automation using artificial intelligence will be in place and it is in development, which will add more value to that data and its visualization for the audience. Qualitative research on various findings and works carried out around the world in terms of sports data and machine learning is analyzed to know the future of machine learning in sports engineering with all types of machine learning: supervised, semi-supervised, unsupervised, and reinforcement learning. Machine learning leads to more progressive development in sports engineering, mainly in track and field, football, hockey and cricket, etc, and other applications like betting and e-sports are in demand for the future. Future developments on machine learning in sports engineering are discussed in this for further developments. By developing such solutions in the future, all sports enthusiasts will get benefits. Especially, athletes and coaches will benefit. Also, other aspects like fan engagement, injury, betting, manufacturing, and testing can also be achieved.

Keywords Machine learning in sports · Sports data · Sports engineering

1 Introduction

Sports engineering is an interdisciplinary field which is evolving with the principles of STEM, which continuously pushes the boundaries of what is physically possible in the realm of sports. With the integration of machine learning into this

V. V. Prasanth (✉) · G. Nallavan
Department of Sports Technology, Tamil Nadu Physical Education and Sports University, Chennai 600127, India
e-mail: prasanth199277@gmail.com

S. Bhattacharyya and D. James (eds.), *Recent Trends in Sports Engineering*, Lecture Notes in Mechanical Engineering, https://doi.org/10.1007/978-981-97-8297-0_1

domain of sports engineering we are stepping forward toward in achieving empowerment of systems to learn from data, adapt, and make predictions or making decision based on the data that is provided for learning from talent identification, training and coaching, umpiring and scoring, analysis and insight generation, injury management and recovery, fan engagement, e-sports and betting, etc. In the future, machine learning algorithms will play a major role wherever we have sufficient data to be implemented to enhance performance of sports engineering solutions in the world of sports in all the aspects to implement various solutions by automating the processes. To know the future of machine learning algorithms in sports engineering in the aspect of talent identification, training and coaching, umpiring and scoring, analysis and insight generation, injury management and recovery, fan engagement, e-sports and betting, etc, are explored and found that how machine learning will dominate the sports realm in the upcoming years. This work helps the sports engineers to find and explore the future of machine learning in all aspects which is mentioned above in engineering the sports, which helps to identify the potential applications, advancements, and transformative effects that machine learning will have on sports. Thus, the integration of data-driven insights and the ever-evolving technologies is ready to redefine how athletes train, how teams strategize, and how fans engage with the games they love. We will also go through in detail about the impact and future of machine learning in terms of talent identification, training and coaching, umpiring and scoring, analysis and insight generation, injury management and recovery, fan engagement, e-sports and betting, etc. which gives a comprehensive understanding of the exciting opportunities and advancements in the future that are listed below:

- Improved Performance: In this, we discussed about the importance of machine learning in analyzing and optimizing athlete performance. Also, we discuss on usage of wearable devices, sensors, and other AI-based insights that helps athletes to reach their maximum potential by decreasing the injury risk.
- Advancement in Sports Analytics: In Sports Analytics, Machine Learning will play a major role is creating strategies, predictive analytics, game simulations, and real-time data processing, which offer teams and coaches to make data-driven decisions to achieve success.
- Prevention of Injury and Rehabilitation: Machine Learning extends its arms in preventing injury and helps the athlete to get rehabilitated, which can predict injury risks and accelerate recovery, ensuring athletes stay in optimal condition which is suitable.
- Fan Engagement and Entertainment: The future of sports engineering using machine learning also promises to enhance the fan experience. Using Machine Learning we can explore a lot and create personalized, immersive, and interactive experiences for sports enthusiasts, both in-stadium and through various digital platforms like mobile, television, and other entertainment media.
- Ethical and Regulatory Challenges: Machine Learning becomes integral part of sports engineering and it also concerns about ethics and regulatory. This will address issues related to data privacy, fair play, and the need for transparent guidelines in this evolving landscape.

Through the above discussion, machine learning becomes the fundamental part to get integrated with sports engineering, which shapes the future of sports in ways we are only beginning to fathom. With this work, we aim to uncover the exciting possibilities and challenges that lie ahead in the future of machine learning in sports engineering.

1.1 The Role of Machine Learning in Talent Identification

In this section, we have discussed more about the games that have already used and the potential future of machine learning in talent identification. Machine learning plays a major role in talent identification and it helps to identify cricket talents [1]. Also, to detect elite soccer players using one-class support vector machine (one-class SVM) on a dataset (N = 951) collected from 14-year-old junior players. According to the results, this approach is useful to make decision [2, 19]. To evaluate the performance of a team and identify talents, various machine learning algorithms are used in beach soccer and sepak takraw [3]. Talent identification in English football is also possible in a multidisciplinary approach [4, 6]. Machine learning algorithms are used to find the position-wise physiological, technical, tactical, and psychological attributes of football players [5]. To identify talent from the young rugby players also SVD is used with under 15 anthropometric and fitness data [7]. Identifying talents in archery is also possible with the machine learning algorithms [8]. Logistic Regression, Random Forest, and XG Boost are also used in identifying talents in athletic throwers [9]. Identifying talent is global challenge and it is still evolving with the help of machine learning algorithms which gets inputs from sensors and other wearables [10]. With the help of wearables like head-mounted camera, we can detect athlete's motion pattern and collect data for talent identification using an efficient-perspective-n-point algorithm. Also, this system creates a 3D pose model to distinguish and compare between train and test results visually [11]. Computational techniques are used in identifying talent which mainly uses the data and getting trained and tested with the efficient machine learning algorithms [12]. Result prediction, player performance assessment, player injury prediction, sports talent identification, and game strategy evaluation are also possible using sports data mining which helps the coaches and managers to take smart decisions to implement [13]. This data can be used along with the machine learning algorithms across various games with key performance indicators. Position-wise prediction for the netball players using regression approach can also be possible and it was already done [14]. Usage of PRISMA guidelines with the ICT-based technologies like CNN, AI, Hawkeye, and Computer Vision is used to identify cricket domain-specific skills like bowling, batting, and wicket keeping. Along with this, other methods like SVM are also used [15, 17]. Machine learning changed the way we think, analyze, strategize, and execute the game in real time with the help of data and insights provided with the help of algorithms [16]. Various technologies like IoT, fog, and cloud computing are used to get data from experts, coaches, players, and institutes using a mobile framework which can be accessible, scalable, cost-effective, secure, platform/location independent, and fast in operations are used in cricket talent identification [18]. Classification and prediction of low and high potential archers is also possible with the help of KNN algorithms and logistic regression [20]. Also, the cricket Talent Identification method (TiD) which is using 28 parameters and 9 different future selection techniques that are mutual information, information gain ratio, correlation, chi square, univariate root mean square error (RMSE), receiver operating characteristic (ROC) with decision tree classifier, reliefF, boruta and oneRin twenty-twenty cricket [21]. In the realm of sports engineering, the integration of machine learning has emerged as a game-changer. The games that are discussed above are limited and the synergy between machine learning and sports engineering in talent identification is somewhat limitless which will be helpful in finding talents in all the games with the help of required parameters and techniques.

1.2 The Role of Machine Learning in Training and Coaching

The very first initial stage of the sports is training and coaching in this machine learning plays major role in getting more insights to the trainers and coaches. AI along with sensors provides data from various weights used in weight training and also in leg press machine performing 3–5 sets of 10–12 repetitions and also we can monitor other training workloads during the session [22, 28, 32]. Machine learning techniques have great potential in improving the players' performance by providing personalized training plans, which are automated and adaptive. Also, this framework for personalized training integrated with the performance indicators such as training load measures, physiological constraints, and behavior change feature's goal setting and self-monitoring [23]. Online coaching platforms also use the machine learning techniques to design, implement, and evaluate the performance of the sports people and their data are summarized and provided to the coaches in ranking manner which all did remotely [24]. Daily step count of the employees or general people can be calculated with the help of machine learning algorithms to make sure the target is achieved on every day. The best one is Random Forest Algorithm which is providing the best results in terms of daily step count and it is very useful in automating personalized training or coaching in the future [25]. Smart assistant for professional volleyball training which is using machine learning techniques called SAETA is used to monitor both players and their environment, which produces real-time data that is analyzed by different modules of a decision engine [26]. Smart Sports Training (SST) like deep learning, computer intelligence, and data mining is used across various games like soccer, running, tennis, basketball, swimming, cycling, triathlon, fitness (gym), table tennis, volley ball, rowing, shooting, and other games which uses combination of various IoT devices with machine learning algorithms [27]. Archery training and coaching is also implemented with machine learning techniques to identify the area of improvement and personalized plans for the same [8]. Deep learning is one of the in-depth concepts of machine learning which helps the basketball players to recognize their posture action and analysis systems are proposed which used the video of the game play to correct the errors with the feedback generated [29]. Also, used in soccer to analyze and investigate the relationship between the training load and wellness index. By targeting the training load for a week, wellness index is predicted and that enhances the adoptions, readiness and reduces the potential drops in performance-related poor wellness scores [30]. Some of the machine learning algorithms are used to analyze and predict the human health to know whether the particular athlete is a perfect fit or not for a particular game with the help of mobile device which gathers data from the users, which is named as MHfit [31]. Dynamic posture monitoring algorithm is used to identify posture or movement of the human body while the volleyball and football players are playing. Also, this allows them to get their mistakes corrected timely and improve strength and training efficiency [33, 34]. Artificial intelligence is one of the unavoidable integrations with sports which is used in all the aspects of training which includes utilization of AI algorithms, ML, DL, and Computer Vision in Sports and its related apps to personalize the training programs to analyze, assess, change, and optimize training methods [35].

1.3 The Role of Machine Learning in Umpiring and Scoring

Automation is everywhere and that can't be limited within particular activities likewise the automation of umpiring and scoring is made in many games. In cricket, automated run outs and no ball decision systems are framed with the help of two machine learning techniques SVM and CNN which will overcome the human error to provide fair decision [36, 53]. Also, detects third umpire's decision and provides the facility of automated scoring which gets updated in an automated manner using the CNN technique, DNN, and KNN [39–42, 47]. Machine learning helps to automatically identify the cricket umpire postures for the events like six, no ball, out, wide, none using CNN and standard classifiers [37, 48]. Object detection with computer vision and motion tracking using machine learning algorithms will optimize the results of ball tracking which will help the umpire in making decision about the game play in cricket [38, 41, 42, 47]. Machine learning algorithms are widely used in table tennis to track the ball movement and provide insights and helps in making decisions while doing umpiring [43]. Decision-making in tennis is one of the important activities for the umpire to know the ball and player location and it should be systemized with the advanced algorithms to provide accurate results [44]. Multi-view camera setup and multi-agent system (MAS) framework are used to detect and track the table tennis and create 3D representation of the flight, trajectory of the ball [45]. AI officiating is one of the emerging fields that provides accuracy and efficiency in various sports and games which will be now automated by providing decisions accurately with the help of datasets trained into the system [46]. Histogram-Oriented Gradient (HOG) and Non-Linear Support Vector Machine are used to classify the gesture action and recognize the movement to identify the activity of the umpire [49]. Summarization of important events or creating highlights is also possible with the help of advanced machine learning algorithms by categorizing the events of the match [50]. Also, human gestures and video annotation using accelerometers are made possible with the help of wearing the wrist bands on the human who is performing gesture [51]. With that gestures from the umpires the score board can be updated automatically [52, 56]. Home aid advantage of major cricket nations is possible with the help of classification and regression tree and the other logistic regression approaches [54]. We have lot of proven methods to claim that the technological solutions provide good results that the human match officials [55]. Prediction of match result using the machine learning algorithms is also possible with the required dataset for doing such activities for various games like football, cricket, etc. [57, 58, 60]. Automation of sports commentary using machine learning is also possible by training the correct dataset with the algorithm and then it can be implemented for automation [59]. With the large amount of data from NHL predicting expert system using big data and machine learning is providing lot of inputs to get the prediction of the games [61, 62].

1.4 The Role of Machine Learning in Analysis and Insight Generation

Analysis and insight generation is one of the important areas where the players get more knowledge about their play and correct their mistake by taking certain measures and training to overcome the feedback which was given. Data-driven analyses for quantitative analysis of basketball and football are done to extract features or rules from data and to visualize the same. Also, location-wise data is also tracked using machine learning [63, 64, 68, 72]. From the last decades, the way of consumption and analyses in sports is changed by Artificial Intelligence (AI) using machine learning paradigm to enhance the performance through sports analytics [16, 65, 68]. There are lot of data available in sports and researchers are being united to visualize and streamline the data with the help of machine learning, computer vision, and artificial intelligence [66]. Machine learning is used to determine the on-field playing positions of a group of football players by characterizing them into 52 non-spatiotemporal descriptors and dimensionality reduction techniques were used [67]. AlpacaML is an iOS application which is connected with sensors and allows to check the movement of the young athletes which is using machine learning algorithms [69, 74]. There are few applications also available to design and implement machine learning and statistical analysis with data mining in sports [70]. Sports news also be categorized using machine learning techniques which leads to the usage of NLP experts in recent years in the sports field [71]. Usage of various technologies which work on data based are possible easily with sports because, sports included with lot of data [73, 75]. Through match analysis using various technology available we can predict the results of the team sports too [57]. Cricket batsmen with 16 attributes and bowlers with 18 attributes are taken and framework Hortonworks Hue 2.2.0 is sued to analyze the performance [76]. And deep learning algorithms and AI plays a major role in capturing, tracking, and visualizing the data to the stakeholders in various games like soccer, football, and other sports events [77, 78]. With the help of data what we have collected we can use the same for analysis and proceed to get the predictions and the same can be used for betting with accuracy in all the games with data [79, 80].

1.5 The Role of Machine Learning in Injury Management and Recovery

Every sports person needs to focus on being injury free which is the important factor to achieve success. By taking necessary actions like warm up and cool down along with this technology also plays a major role in avoiding and predicting injury and also it helps us to get recovered if in case it happens [81]. By analyzing the scientific quantitative and qualitative data of the elite level athletes we can get more data about prevention of injury, formulation, and implementing rehabilitation programs are possible [82, 83]. Athlete's sports injury prediction is acting as a best model using machine learning by analyzing various physical fitness data like exercise volume, load, body metabolism, and other physical indicators, which is very helpful in prediction and makes necessary training optimization and takes prevention measures on the same [84]. This approach is not limited to football alone and it applies to all the elite games like futsal that will allow the coaches, physical trainers, and medical practitioners to take decision for prevention of injury [85]. Specifically in soccer, possibility of injury is very high when compared to other games. There are some multidimensional approaches availabe to forecast the injury in professional soccer. With the help of gps measurements, we can collect lot of data from the same and it can be used to train and predict using the machine learning models or algorithms by training with the professional soccer player data [32, 101]. Injury prediction in football can done by taking anthropometric, motor coordination, and physical performance measures. By continuously monitored by the coaches and medical staffs with the preseason measurements of anthropometric data of height, weight, sitting height, and physical fitness data like strength, flexibility, speed, agility, and endurance can be performed. With this data the machine learning algorithm can forecast the injured players in the hold-out test which is having 85% precision, 85% recall (sensitivity), and 85% accuracy. Further to this, the injury can be classified as acute with 78% precision, 78% recall, and 78% accuracy [86, 87]. Sport-related concussion (SRC) is monitored with the help of machine learning. In which CatBoost is the better on to perform recovery time prediction on both male and female [88, 95]. Trauma prediction of athletes is also possible with the help of artificial intelligence, through which the healthier life can be achieved for an athlete [89]. Also, electroencephalogram is used along with machine learning algorithms to predict trauma injury [105]. Machine learning algorithms are integrated with the sensors like IMU for action recognition in real time [90]. Injury prediction is possible and is classified into two categories using Naïve Bayes model as severe and non-severe by comparing the actual to the predicted ones and results recorded from the sports medicine background [91]. Monitoring the exercise data in real time, predicting the risk of physical recovery and injury is possible with the wearable devices in real time for athletes [92]. Anthropometric measurement with blood parameters and urine analysis is performed to predict and monitor the injury level with metabolic, anabolic, and catabolic activities with machine learning [93]. Not only machine learning we can also use deep learning algorithms like OCNN (optimized CNN), SCM (self-coding method of convolution), SAR (self-adjustment resizing algorithm), and CNN in predicting the injury with the help of the data [94]. SVM and Random Forest from machine learning play a major role in sports medicine by detecting the knee joint injury from the athletes by FMS (functional movement screen) and BMI calculation [96, 104]. Also, deep learning plays important role in prediction of head injury in pediatrics but the main challenges we are facing are limited data or small sample size, noisy features, and self-reported symptoms are there still. It will be streamlined in the future and the data availability will be higher [97]. In MLB, the data are used for predicting injuries that might occur in the next season with 84 ML algorithms which were developed in 2020 while the same can be done with the other games in the future [98]. Identification of anterior cruciate ligament (ACL) injuries is common with all the athletes. To analyze this several machine learning models were tested and feature selection is implemented with the learning model LOSO-CV (leave-one-subject-out cross-validation) to estimate training and test errors. For this the motions sensors are applied on lower limbs for the post-ACL subjects in rugby. Using this method, the clinicians and sports scientists can get RTS assessment of injured athletes and in rehab programs [99]. MRI imaging is used to monitor the activity of central nervous system and respiratory functions during the preparation activities, which is a transitional phase from the static to moving stage that is analyzed with the help of deep learning algorithms [100]. Accuracy in prediction of football injury or any injury is possible by combining both screening and monitoring the data using ML techniques with gradient boosting with ROSE upsampling within a leave-one-out cross-validation which is used for data analysis [102]. In basketball, the physical fitness and injury risk are identified and then the data for each player can be standardized using Z-scores. To focus negative impact of class imbalance on model performance, synthetic minority over-sampling techniques and edited nearest neighbor algorithms are used to sample the training set with lower limb data [103].

1.6 The Role of Machine Learning in Fan Engagement

In the world of sports, fan engagement is one of the important areas we have to focus a lot with the help of machine learning algorithms. In this, the sentiment analysis and content categorization play a major role in creating fan engagement. An analytic framework is provided to sports organizations to effectively use AI and social media, which consist of multiple stages like data collection, data preparation, sentiment mining, and content categorization. This can be applied to any game and their fans to create personalized content, in which they are interested [106]. Opinions, feelings, emotion, and subjectivity are called as sentiment which is analyzed from social media like Twitter suing ML algorithms like MultinominalNB, SVM, Linear SVM, and decision trees to get the fans engaged with the games like football, baseball, etc. [108, 111, 112]. Financial outcomes are increased nowadays with the help of fan engagement through different social media or TV broadcasting [109, 115]. As per the examination, it is proved that Augmented Reality (AR) and Virtual Reality (VR) are associated with sensors and other IoT to provide real-time, high-fidelity data for performance monitoring and injury prevention with the help of AI and ML techniques [65, 107]. Various technologies are implemented to have the fans engaged in the sports events happening across the world, the technologies are ball trajectory for table tennis, and bounce distribution, badminton shuttlecock tracking and trajectory, augmented reality, virtual reality, real-time 3D broadcasting, in-stadium, video stream pushing using 5G network with Mobile Edge Computing along with AI-based analytics on the sports during live stream will engage them more and more [110, 117]. Machine learning algorithms like PCA (principal component analysis)-based k-means algorithms are used to read the aggression level of the fan's when there is doping, match fixing, black whistles, and football hooliganism are occurring during the famous or popular matches in the world of sports [113]. During COVID-19 the impact on fans becomes very high due to the usage of cutting-edge technologies like Over-the-Top (OTT), Non-Fungible Tokens (NFTs), block chain, and AI and VR with information technology in enhancing fan's engagement specifically in India [114, 116]. Technologies like Computer Vision and Deep Learning play a crucial role in predicting the players' performance and generate game stats from the videos based on the effective training dataset we provide with unique dynamics and camera angles with DeepSportradar-v2, which is a multi-sport suite of computer vision that addresses the high-quality datasets for different sports [118]. Various studies like examining fan engagement in green game sporting event related to environment are done to understand the fans' perspective in keep game green using value-belief-norm (VBN) theory. With this the sports fans are getting more knowledge through environmental education in maintaining the green game practices [119]. Fan engagement gives a large space to be occupied by the digital marketing and sponsorship in major football league around the world which engages the fans to get updated with the valuable insights and news of their favorite teams [120]. All the needs of the fans are being currently engaged by the latest sports technologies like wearable, Internet of Things (IoT) applications, media, and communications [121]. Deep fan is about the cultural and ritual theory which contribute to the understanding of mass mediated sports experience through advertising. Media sports fans and sporting event are closely parallel for the mythic celebration [122]. In simple words, AI and machine learning are not limited to player performance analysis, injury prevention, game strategy development, etc. which also extend its arms to fan engagement in creating personalized contents like advertisements, marketing, and other contents to the fans of the various games [123].

1.7 The Role of Machine Learning in E-Sports and Betting

E-sports and betting is one of the fast-growing fields that is growing along with the machine learning usage in the field of sports. These e-sports matches can be predicted with the help of the support vector machine algorithm and Gaussian processes with the help of the large dataset of professional matches to predict the winning or losing possibility in this field [124]. Adoption of a block chain-based framework in e-sports industry is investigated with the help of Ajzen's theory of planned behavior (TPB) in Reddit platform using the RStudio package Reddit Extractor for scraping and analysis of sentiments [125]. Team fight Tactics (TFT) prediction using machine learning is possible in e-sports. Algorithms like logistic regression, SVM rbf (radius basis function), KNN, Gaussian Naïve Bayes, Decision Tree, Random Forest, and DFNN are used to study the usage in e-sports comparing to traditional sports. In that Random Forest gives the highest accuracy of 0.914 [126, 130, 133]. Using the pre-game data of league of legends and game application programming interface (API), various game data like skills of players, synergies between players of the same team, or ability to beat another player classification algorithms can aim to provide best model accuracy above 0.70 using the data [127]. Also, logistic regression, decision, and neural network concepts are used to predict the match results of league of legends and also using other methods and models in machine learning [136, 141]. By collecting tweets about the traditional gambling and e-sports gambling scrutiny of compliance with all sections of the Committee of Advertising practice (CAP) code is conducted along with machine learning algorithms. Out of which 10 tranditional and 68 e-sports tweets are original from 250 tweets of each, respectively [128]. Along with the evolution of technology, sports evolved and bookmarkers also. With digitalization these houses are available to access anywhere online. Due to this digitalization the sports betting becomes more profitable with e-Sports like Counter-Strike: Global Offensive. The data available for this game is collected and bused for training the machine learning algorithms that consist of 2500 matches which results in profiting from the Counter-Strike: Global Offensive bets in long term [129]. E-sports Trajectory and Actions (ESTA) dataset is collected from the server logs for Counter-Strike game and it's used to deploy in machine learning models to predict the winning data using player-specific information [131]. With Counter-Strike dataset of 40,000 matches are obtained and analyzed to predict using six machine learning models out of six models' Elo rating performed with 64% which is the best one and followed by random forest with 63% accuracy [138]. E-sports can be played in solo player or multiplayer in the environment where there is no physical stadium infrastructure is not required. Using these technological advancements some of the companies like Gamer Sensei and DOJO Madness have raised significant amounts of funding since 2016 and creating a platform to improve player skills and connect with mentors and trainers by giving game-specific analytical data and related tools [132]. Match prediction as a service, which provides match result prediction based on laid down algorithm using Breima bagging method and random subspaces methods. A service that is developed in e-sports discipline "Dota 2" with machine learning methods [134,

144]. By training a Weka LR algorithm and Microsoft LightGBM algorithm with the Mixed-InGame and Pro-InGame data, we can use two Weka feature selector CfsSubsetEval with BestFirstSearch to compare algorithm accuracies in real-time data exploration [145]. E-sports and betting in the e-sports games play a major role in usage of sports tech and innovation nowadays [135]. Statistically reliable prediction of the combination of game-related preferences and socio-demographic factors help us to determine occasional and frequent gamers in semi-peripheral country using SVM and Random Forest [137]. Automation is achieved with AI and AI is enabling usage of technologies like big data, virtual reality, AI chips and GPUs, online gaming, and cloud platforms while playing video games and e-sports. Using this technology, we can generate procedural content, player experience modeling, antisocial behavior detection, win prediction, player telemetry analytics, intelligent tutoring, and training. With the help of this the following case studies consist of DeepMind Alpha Go, Alpha Star, and Microsoft HoloLens [139]. To improve the marketing campaign for e-sports, Facebook, TikTok, and Instagram are the central platforms using the new technology application like AI, Machine Learning, and automation to succeed in a short period of time [140]. Performance analytics in e-sports is one of the fields which is growing faster, which uses technologies for visualization and creation of dashboards for the same using popular programming languages and software applications with machine learning and AI for e-sports analytics by collecting the data from various matches. Those who are doing well in these skills can excel in e-sports analytics jobs in the current era [142]. Fuzzy system uses the real-world data and determines a player's percentage chance of winning a duel in online first-person shooter video game like Counter-Strike Global Offensive. Also, these intelligent analytical tools provide useful information or suggestion like strength and weaknesses of the e-sports player [143].

2 Results and Discussion

From this work, we found that various findings support that machine learning is one of the major parts that is already occupied and going to occupy unavoidable role in sports. As per the findings we have gone through above, it is very clear that machine learning and its development will develop sporting activity around the world in all the games.

Figure 1 shows the machine learning and its applications in various activities that are made with the help of machine learning in terms of sports engineering to add more value to the end users. The data that is presented above are taken for study purpose and this might change over a period of time due to evaluation and usage of technology in other games. Talent identification is one of the important and basic areas where the athletes need to identify themselves that which game they are suitable and which game will make them excel in the future according to various criteria. We have reviewed around 21 works in this section and it clearly shows the usage of algorithms in talent identification and also it will be useful for sure with lot of data in the future. On the next section, the usage of machine learning and its algorithms are verified in past and present around 14 works in this to find the future potential in training and coaching by checking with various algorithms and games. We knew that, umpiring and scoring is related with each other and this is done with the help of machine learning algorithm to provide fair decision for the event or action that happened in any game which is getting systemized. Upon checking on previous work we should work more in this area to optimize the solutions to make the audience or fans to accept the decisions without any debate about the machine decisions by implementing rules and regulations with advanced technology that is combined with machine learning. However, this is not only limited to umpiring and scoring, but also extends to analysis and insight generation for the improvement of player's performance to indicate them and keep informed with the strength and weakness about them and their opponents too. Injury occurs in most of the games specifically in football athletes who are facing lot of injury which can be predicted and prevented with the help of sensors and other wearable devices which collect the data and forecast the injury. Even though the injury happened technological solutions help athletes to recover and get them back on track with the appropriate treatment for the injuries. Another important factor to generate revenue to the sporting industry is fan engagement for every game which depends on the interest area what the fans are focusing. These are being identified in various games and they are kept updated with the marketing and interest-specific content to the fans with the help of machine learning algorithms with the previous track record or data. Especially, e-sports and betting is the fastest growing around the world which enables the players to use most of the advanced technologies in their daily routine and it has been a part of their activity in playing the game online whether it may be single or multiplayer. Using the same we have lot of benefits that favor in predicting the performance of normal sports and e-sports prediction earned by using betting platforms. On the whole umpiring and scoring is leading the table in implementation of machine learning solutions and followed by injury management and recovery is taking place. Next to that there are lot of works going on in e-sports and betting (fantasy) and fan engagement equally and, on the other hand, analysis and insight generation are growing following the other verticals. Next to that training and coaching are taking place and at the last we have talent identification. Through which every vertical needs development and still the basic elements of sports are talent identification and next is training and coaching which has to be streamlined in the future with advanced developments.

3 Conclusion and Future Work

Specifically in sports engineering innovation and transformation is taking part with the implementation of machine learning algorithms in reshaping athlete performance, analytics, training, coaching, umpiring, scoring, fan engagement, talent identification, e-sports, and betting with ethical landscapes of sports. Through the findings this work highlights that there are exciting possibilities in the future in all the

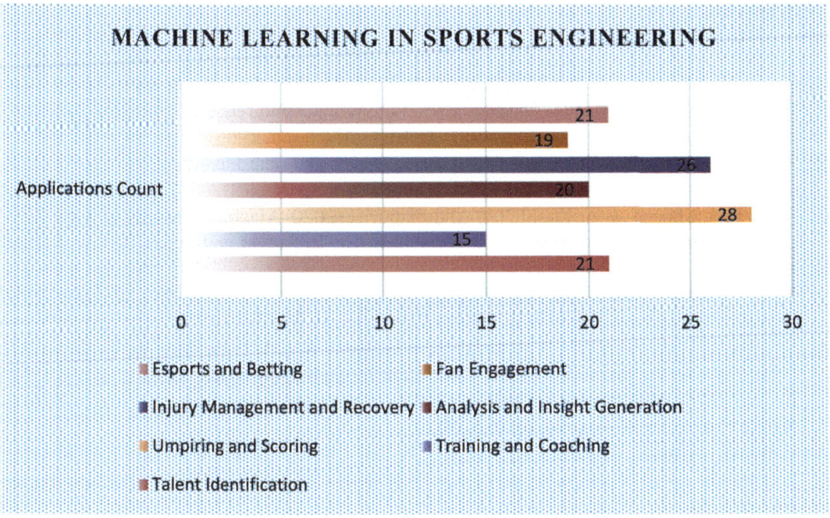

Fig. 1 Machine learning and its applications in terms of sports engineering in past and present scenarios

areas which are discussed above. According to the findings, it is clear that the machine learning will enhance performance of the athletes by providing personalized training, injury prediction, and other skill improvement techniques. Also, data-driven decision-making tools and real-time insights are provided to them in terms of analytics using machine learning. It helps in athletes' safety and lifelong career by predicting, preventing injury and faster recovery from injury. By providing personalized content, immersive experiences and interactive engagement with the help of machine learning sports industry is getting benefited. However, data privacy and fair play are to be addressed to maintain authenticity of fan experience. Due to rise of machine learning there is a need for proper and clear guidelines, transparency, athletes consent, and awareness campaigns to explore the ethical and regulatory challenges in sports engineering successfully. To achieve full advantage of transformative power of machine learning in sports, we need to ensure athletes well-being, fair competition, and data privacy in the future. Also, opens the door to numerous avenues for future research and development like automation in organizing the sports events or culture in all the aspects discussed above. Some key areas for future work include ethical frameworks, transparency, and accountability to gain trust among athletes, coaches and fans, AI Bias and Fairness, Legal and regulatory frameworks to be done whenever needed, advanced fan engagement by adapting emerging technologies like AR/VR, and real-time performance optimization which enables athletes to make necessary strategies and tactics during the game. Machine learning in sports engineering for the future is a dynamic and evolving field and working in the above concerns will maximize the positive impact while addressing potential pitfalls and challenges.

References

1. Reyaz N, Ahamad G, Khan NJ, Naseem M (2022) Machine learning in sports talent identification: a systematic review. In: 2022 2nd international conference on emerging frontiers in electrical and electronic technologies (ICEFEET). IEEE, pp. 1–6
2. Jauhiainen S, Äyrämö S, Forsman H, Kauppi JP (2019) Talent identification in soccer using a one-class support vector machine. Int J Comput Sci Sport 18(3)
3. Musa RM, Majeed APA, Kosni NA, Abdullah MR (2020) Machine learning in team sports: performance analysis and talent identification in Beach Soccer and Sepak-takraw. Springer Nature
4. Kelly AL, Calvo AL, Dos Santos SDL, Jiménez Sáiz SL (2022) Special issue "talent identification and development in youth sports". Sports 10(12):189
5. Sujatha R, Maheswari BU, Kavitha D, Kandaswamy K (2022) Developing a talent identification model for predicting player position in football using machine learning algorithms. In: International conference on advances in data science and computing technologies. Springer Nature Singapore, Singapore, pp 285–293
6. Kelly AL, Williams CA, Cook R, Sáiz SLJ, Wilson MR (2022) A multidisciplinary investigation into the talent development processes at an english football academy: a machine learning approach. Sports 10(10):159
7. Till K, Jones BL, Cobley S, Morley D, O'Hara J, Chapman C, Beggs CB (2016) Identifying talent in youth sport: a novel methodology using higher-dimensional analysis. PLoS ONE 11(5):e0155047
8. Musa R M, Taha Z, Majeed APA, Abdullah MR (2019) Machine learning in sports: identifying potential archers. Springer Singapore
9. Fernández E, Izquierdo JM, Zarauz A, Redondo JC, Prediction of sports talent in young throwers using machine learning predicción Del Talento Deportivo En Jóvenes Lanzadores Utilizando machine learning
10. Leite N, Calvo AL, Cumming S, Gonçalves B, Calleja-Gonzalez J (2021) Talent identification and development in sports performance. Front Sports Active Living 3:729167
11. Almasi M, Ghaeinian SA, Samiee S, Fathi H (2020) Investigating the application of human motion recognition for athletics talent identification using the head-mounted camera. In: 2020 international conference on inventive computation technologies (ICICT). IEEE, pp 54–59
12. Khan NJ, Ahamad G, Naseem M, Sohail SS (2023) Computational efficiency in sports talent identification-a systematic review. Int J Appl Decis Sci 16(3):358–384
13. Keshtkar Langaroudi M, Yamaghani M (2019) Sports result prediction based on machine learning and computational intelligence approaches: a survey. J Adv Comput Engin Technol 5(1):27–36
14. Jasni NH, Mustapha A, Tenah SS, Mostafa SA, Razali N (2022) Prediction of player position for talent identification in association netball: a regression-based approach. Int J Adv Intell Inf 8(1)
15. Reyaz N, Ahamad G, Naseem M, Ali J, Rahmani KI (2023) Information communication and technology in sports: a meticulous review. Front Sports Active Living 5
16. Chmait N, Westerbeek H (2021) Artificial intelligence and machine learning in sport research: an introduction for non-data scientists. Front Sports Active Living 3:363
17. Reyaz N, Ahamad G, Khan NJ, Naseem M, Ali J (2024) SVMCTI: support vector machine based cricket talent ıdentification model. Int J Inf Technol 1–14
18. Khan NJ, Ahamad G, Naseem M (2022) An IoT/FOG based framework for sports talent identification in COVID-19 like situations. Int J Inf Technol 14(5):2513–2521
19. Bergkamp TL, Niessen ASM, den Hartigh RJ, Frencken WG, Meijer RR (2019) Methodological issues in soccer talent identification research. Sports Med 49:1317–1335
20. Muazu Musa R, Abdul Majeed PP, A, Taha Z, Chang SW, Ab Nasir AF, Abdullah MR (2019) A machine learning approach of predicting high potential archers by means of physical fitness indicators. PLoS ONE 14(1):e0209638

21. Khan NJ, Ahamad G, Reyaz N, Naseem M (2023) Optimal feature selection for cricket talent identification. Int J Adv Technol Engin Explor 10(98):67
22. Novatchkov H, Baca A (2013) Artificial intelligence in sports on the example of weight training. J Sports Sci Med 12(1):27
23. Zahran L, El Beltagy M, Saleh M (2020) A conceptual framework for the generation of adaptive training plans in sports coaching. In: Proceedings of the international conference on advanced intelligent systems and informatics 2019. Springer International Publishing, pp 673–684
24. Iguider W (2022) Machine learning models for sports remote coaching platforms
25. Dijkhuis TB, Blaauw FJ, Van Ittersum MW, Velthuijsen H, Aiello M (2018) Personalized physical activity coaching: a machine learning approach. Sensors 18(2):623
26. Vales-Alonso J, Chaves-Diéguez D, López-Matencio P, Alcaraz JJ, Parrado-García FJ, Gonzalez-Castano FJ (2015) SAETA: A smart coaching assistant for professional volleyball training. IEEE Trans Syst, Man, Cybern: Syst 45(8):1138–1150
27. Rajšp A, Fister I Jr (2020) A systematic literature review of intelligent data analysis methods for smart sport training. Appl Sci 10(9):3013
28. Novatchkov H, Baca A (2012) Machine learning methods for the automatic evaluation of exercises on sensor-equipped weight training machines. Proc Engin 34:562–567
29. Yao P (2021) Real-time analysis of basketball sports data based on deep learning. Complexity 2021:1–11
30. Perri E, Simonelli C, Rossi A, Trecroci A, Alberti G, Iaia FM (2021) Relationship between wellness index and internal training load in soccer: application of a machine learning model. Int J Sports Physiol Perform 16(5):695–703
31. Miah J, Mamun M, Rahman MM, Mahmud MI, Ahmad S, Nasir MHB (2022) Mhfit: Mobile health data for predicting athletics fitness using machine learning models. In: 2022 2nd international seminar on machine learning, optimization, and data science (ISMODE). IEEE, pp 584–589
32. Rossi A, Pappalardo L, Cintia P, Iaia FM, Fernández J, Medina D (2018) Effective injury forecasting in soccer with GPS training data and machine learning. PLoS ONE 13(7):e0201264
33. Sun Z, Sun P (2022) Attitude monitoring algorithm for volleyball sports training based on machine learning in the context of artificial intelligence. Secur Commun Netw
34. Herold M, Goes F, Nopp S, Bauer P, Thompson C, Meyer T (2019) Machine learning in men's professional football: Current applications and future directions for improving attacking play. Int J Sports Sci Coach 14(6):798–817
35. Bodemer O (2023) Enhancing individual sports training through artificial intelligence: a comprehensive review. Authorea Preprints
36. Iyer GN, Bala VS, Sohan B, Dharmesh R, Raman V (2020) Automated third umpire decision making in cricket using machine learning techniques. In: 2020 4th international conference on intelligent computing and control systems (ICICCS). IEEE, pp 1216–1221
37. Samaraweera WJ, Premaratne SC, Dharmaratne AT (2020) Deep learning for classification of cricket umpire postures. In: Neural information processing: 27th international conference, ICONIP 2020, Bangkok, Thailand, November 18–22, 2020, Proceedings, Part V 27. Springer International Publishing, pp 563–570
38. Arora U, Verma S, Sahni S, Sharma T (2017) Cricket umpire assistance and ball tracking system using a single smartphone camera. PeerJ Preprints 5:e3402v1
39. Kowsher M, Alam MA, Uddin MJ, Ahmed F, Ullah MW, Islam MR (2019) Detecting third umpire decisions and automated scoring system of cricket. In: 2019 international conference on computer, communication, chemical, materials and electronic engineering (IC4ME2). IEEE, pp 1–8
40. John DT, Kumar KS, Nair VT, Visakh P, Poorna BR (2019) Cricket scoreboard automation using umpire gestures. Int J Res Engin, Sci Manag 2(7)
41. Nandyal S, Kattimani SL (2022) Cricket event recognition and classification from umpire action gestures using convolutional neural network. Int J Adv Comput Sci Appl 13(6)

42. Wyawahare M, Dhanawade A, Dharyekar S, Dhole A, Dhopade M (2022) Automating scorecard and commentary based on umpire gesture recognition. In: International Conference on Advancements in Smart Computing and Information Security. Springer Nature Switzerland, Cham, pp 158–168
43. Ji YF, Wang CL, Shi ZH, Ren J, Zhu L (2017) Table tennis service umpiring system based on video identification—for height and angle of the throwing ball. In: Wearable sensors and robots: proceedings of international conference on wearable sensors and robots 2015. Springer Singapore, pp 495–509
44. Kovalchik SA, Sackmann J, Reid M (2017) Player, official or machine?: uses of the challenge system in professional tennis. Int J Perform Anal Sport 17(6):961–969
45. Wong P, Myint H, Dooley L, Hopgood A (2023) A multi-view automatic table tennis umpiring framework. In: Proceedings of the institution of mechanical engineers, part p: journal of sports engineering and technology, 17543371231171433
46. Ajmal K, Arun N (2023) Use of artificial intelligence and machine learning in sports. ФІЗИЧНА КУЛЬТУРА В УНІВЕРСИТЕТСЬКІЙ ОСВІТІ: СВІТОВА ПРАКТИКА ТА СУЧАСНІ ТРЕНДИ, 219
47. Samaraweera WJ (2019) Detection and classification of umpire events in cricket videos (Doctoral dissertation)
48. Shingrakhia H, Patel H (2022) SGRNN-AM and HRF-DBN: a hybrid machine learning model for cricket video summarization. Vis Comput 38(7):2285–2301
49. Nandyal S, Kattimani SL (2021) Umpire gesture detection and recognition using HOG and non-linear support vector machine (NL-SVM) classification of deep features in cricket videos. J Phys: Conf Ser 2070(1):012148. IOP Publishing
50. Vasudevan V, Gounder MS (2023) A systematic review on machine learning-based sports video summarization techniques. Smart Comput Vis 1–34
51. Chambers GS, Venkatesh S, West GA, Bui HH (2002) Hierarchical recognition of intentional human gestures for sports video annotation. In: 2002 international conference on pattern recognition, vol 2. IEEE, pp 1082–1085
52. Nair VK, Jose RR, Anil PB, Tom M, PL L (2020) Automation of cricket scoreboard by recognizing umpire gestures. Int J Innovat Sci Modern Engin 6:1–7
53. Ravi A, Venugopal H, Paul S, Tizhoosh HR (2018) A dataset and preliminary results for umpire pose detection using SVM classification of deep features. In: 2018 IEEE symposium series on computational intelligence (SSCI). IEEE, pp 1396–1402
54. Jayalath KP (2018) A machine learning approach to analyze ODI cricket predictors. J Sports Anal 4(1):73–84
55. Collins H (2010) The philosophy of umpiring and the introduction of decision-aid technology. J Philos Sport 37(2):135–146
56. Shahjalal MA, Ahmad Z, Rayan R, Alam L (2017) An approach to automate the scorecard in cricket with computer vision and machine learning. In: 2017 3rd international conference on electrical information and communication technology (EICT). IEEE, pp 1–6
57. Bunker R, Susnjak T (2022) The application of machine learning techniques for predicting match results in team sport: a review. J Artif Intell Res 73:1285–1322
58. Horvat T, Job J (2020) The use of machine learning in sport outcome prediction: a review. Wiley Interdisciplin Rev: Data Mining Knowl Discov 10(5):e1380
59. Lee G, Bulitko V, Ludvig E (2012) Sports commentary recommendation system (SCoReS): machine learning for automated narrative. In: Proceedings of the AAAI conference on artificial intelligence and interactive digital entertainment, vol. 8, No. 1, pp 32–37
60. Hucaljuk J, Rakipović A (2011) Predicting football scores using machine learning techniques. In: 2011 proceedings of the 34th international convention MIPRO. IEEE, pp 1623–1627
61. Bunker RP, Thabtah F (2019) A machine learning framework for sport result prediction. Appl Comput Inf 15(1):27–33
62. Gu W, Foster K, Shang J, Wei L (2019) A game-predicting expert system using big data and machine learning. Expert Syst Appl 130:293–305

63. Fujii K (2021) Data-driven analysis for understanding team sports behaviors. J Robot Mechatron 33(3):505–514
64. Brooks JD (2018) Using machine learning to derive insights from sports location data (Doctoral dissertation, Massachusetts Institute of Technology)
65. Chase C (2020) The data revolution: cloud computing, artificial intelligence, and machine learning in the future of sports. In: 21st century sports: how technologies will change sports in the digital age, pp 175–189
66. Brefeld U, Davis J, Lames M, Little JJ (2022) Machine learning in sports (Dagstuhl Seminar 21411). In: Dagstuhl reports, vol 11, No 9. Schloss Dagstuhl-Leibniz-Zentrum für Informatik
67. García-Aliaga A, Marquina M, Coteron J, Rodriguez-Gonzalez A, Luengo-Sanchez S (2021) In-game behaviour analysis of football players using machine learning techniques based on player statistics. Int J Sports Sci Coach 16(1):148–157
68. Jayal A, McRobert A, Oatley G, O'Donoghue P (2018) Sports analytics: analysis, visualisation and decision making in sports performance. Routledge
69. Zimmermann-Niefield A, Turner M, Murphy B, Kane SK, Shapiro RB (2019) Youth learning machine learning through building models of athletic moves. In: Proceedings of the 18th ACM international conference on interaction design and children, pp 121–132
70. Dooley C (2017) Data visualisation and machine learning web application with potential use in sports data analytics (Doctoral dissertation, Master's thesis, OE Gaillimh)
71. Barua A, Sharif O, Hoque MM (2021) Multi-class sports news categorization using machine learning techniques: resource creation and evaluation. Procedia Comput Sci 193:112–121
72. Sarlis V, Tjortjis C (2020) Sports analytics—Evaluation of basketball players and team performance. Inf Syst 93:101562
73. Patel D, Shah D, Shah M (2020) The intertwine of brain and body: a quantitative analysis on how big data influences the system of sports. Ann Data Sci 7:1–16
74. Ghosh I, Ramasamy Ramamurthy S, Chakma A, Roy N (2023) Sports analytics review: artificial intelligence applications, emerging technologies, and algorithmic perspective. Wiley Interdisciplin Rev: Data Mining Knowl Discov e1496
75. Pavitt J, Braines D, Tomsett R (2021) Cognitive analysis in sports: aupporting match analysis and scouting through artificial intelligence. Appl AI Lett 2(1):e21
76. Kaur A, Kaur R, Jagdev G (2021) Analyzing and exploring the impact of big data analytics in sports sector. SN Comput Sci 2(3):184
77. Mgaya GB, Liu H, Zhang B (2021) A survey on applications of modern deep learning techniques in team sports analytics. In: Proceedings of the 12th international conference on soft computing and pattern recognition (SoCPaR 2020) 12. Springer International Publishing, pp 434–443
78. Araújo D, Couceiro M, Seifert L, Sarmento H, Davids K (2021) Artificial intelligence in sport performance analysis. Routledge
79. Wilkens S (2021) Sports prediction and betting models in the machine learning age: the case of tennis. J Sports Anal 7(2):99–117
80. Nguyen NH, Nguyen DTA, Ma B, Hu J (2022) The application of machine learning and deep learning in sport: predicting NBA players' performance and popularity. J Inf Telecommun 6(2):217–235
81. Lövdal SS, Den Hartigh RJ, Azzopardi G (2021) Injury prediction in competitive runners with machine learning. Int J Sports Physiol Perform 16(10):1522–1531
82. Liu Z, Zhang J, Wu D (2022) Sports injury identification method based on machine learning model. In: Computational intelligence and neuroscience, 2022
83. Rossi A, Pappalardo L, Cintia P (2021) A narrative review for a machine learning application in sports: an example based on injury forecasting in soccer. Sports 10(1):5
84. He K (2021) Prediction model of juvenile football players' sports injury based on text classification technology of machine learning. Mob Inf Syst 2021:1–10
85. Ruiz-Pérez I, López-Valenciano A, Hernández-Sánchez S, Puerta-Callejón JM, De Ste Croix M, Sainz de Baranda P, Ayala F (2021) A field-based approach to determine soft tissue injury risk in elite futsal using novel machine learning techniques. Front Psychol 12:610210

86. Rommers N, Rössler R, Verhagen E, Vandecasteele F, Verstockt S, Vaeyens R, Witvrouw E (2020) A machine learning approach to assess injury risk in elite youth football players. Med Sci Sports Exerc 52(8):1745–1751
87. Majumdar A, Bakirov R, Hodges D, Scott S, Rees T (2022) Machine learning for understanding and predicting injuries in football. Sports Med-Open 8(1):1–10
88. Chu Y, Knell G, Brayton RP, Burkhart SO, Jiang X, Shams S (2022) Machine learning to predict sports-related concussion recovery using clinical data. Ann Phys Rehabil Med 65(4):101626
89. Kakavas G, Malliaropoulos N, Pruna R, Maffulli N (2020) Artificial intelligence: a tool for sports trauma prediction. Injury 51:S63–S65
90. Sigurdson H, Chan JH (2021) Machine learning applications to sports injury: a review. icSPORTS, 157–168
91. Madanan M, Venugopal A, Velayudhan N (2020) Using machine learning techniques in sports medicine to predict injuries and provide recommendation to orthopaedic treatments after surgery
92. Cui J, Du H, Wu X (2023) Data analysis of physical recovery and injury prevention in sports teaching based on wearable devices. Prevent Med 107589
93. Petrovsky DV, Pustovoyt VI, Nikolsky KS, Malsagova KA, Kopylov AT, Stepanov AA, Kaysheva AL (2022) Tracking health, performance and recovery in athletes using machine learning. Sports 10(10):160
94. Song H, Montenegro-Marin CE, krishnamoorthy, S. (2021) Secure prediction and assessment of sports injuries using deep learning based convolutional neural network. J Ambient Intell Humaniz Comput 12:3399–3410
95. Tjønndal A, Røsten S (2022) Safeguarding athletes against head injuries through advances in technology: a scoping review of the uses of machine learning in the management of sports-related concussion. Front Sports Active Living 4:837643
96. Liu H (2023) Value evaluation of knee joint sports injury detection model-aided diagnosis based on machine learning. Front Phys 11:1166275
97. Richards NL (2018) Machine learning for the prevention and prognosis of pediatric head injury in sport (Doctoral dissertation, University of Cincinnati)
98. Karnuta JM, Luu BC, Haeberle HS, Saluan PM, Frangiamore SJ, Stearns KL, Ramkumar PN (2020) Machine learning outperforms regression analysis to predict next-season Major League Baseball player injuries: epidemiology and validation of 13,982 player-years from performance and injury profile trends, 2000–2017. Orthop J Sports Med 8(11):2325967120963046
99. Tedesco S, Crowe C, Ryan A, Sica M, Scheurer S, Clifford AM, O'Flynn B (2020) Motion sensors-based machine learning approach for the identification of anterior cruciate ligament gait patterns in on-the-field activities in rugby players. Sensors 20(11):3029
100. Ba H (2020) Medical sports rehabilitation deep learning system of sports injury based on MRI image analysis. J Med Imaging Health Inf 10(5):1091–1097
101. Majumdar A, Bakirov R, Hodges D, Scott S, Rees T (2022) Machine learning for understanding and predicting injuries in soccer. Sports Med-Open 8(1)
102. Hecksteden A, Schmartz GP, Egyptien Y, Aus der Fünten K, Keller A, Meyer T (2023) Forecasting football injuries by combining screening, monitoring and machine learning. Sci Med Football 7(3):214–228
103. Huang Y, Li C, Bai Z, Wang Y, Ye X, Gui Y, Lu Q (2023) The impact of sport-specific physical fitness change patterns on lower limb non-contact injury risk in youth female basketball players: a pilot study based on field testing and machine learning. Front Physiol 14:1182755
104. Vasavada K, Shankar DS, Bi AS, Moran J, Petrera M, Kahan J, Alaia MJ (2022) Predictors using machine learning of complete peroneal nerve palsy recovery after multiligamentous knee injury: a multicenter retrospective cohort study. Orthop J Sports Med 10(9):23259671221121410
105. Bazarian JJ, Elbin RJ, Casa DJ, Hotz GA, Neville C, Lopez RM, Covassin T (2021) Validation of a machine learning brain electrical activity–based index to aid in diagnosing concussion among athletes. JAMA Netw Open 4(2):e2037349–e2037349

106. Zadeh H, A. (2021) Quantifying fan engagement in sports using text analytics. J Data, Inf Manag 3(3):197–208
107. El-Maghrabi Y, Sharif M (2022) Game changers or game predictors? Big data analytics in sports for performance enhancement and fan engagement. J Contemp Healthcare Anal 6(6):19–39
108. Bhangle RS, Sornalakshmi K (2018) Twitter sentimental analysis on fan engagement. In: Advances in big data and cloud computing, pp 27–39. Springer Singapore
109. Previati A (2020) The importance of fan engagement and fan management in sports. St. John's University (New York)
110. Wu CW, Shieh MD, Lien JJJ, Yang JF, Chu WT, Huang TH, Jheng CS (2022) Enhancing fan engagement in a 5G stadium with AI-based technologies and live streaming. IEEE Syst J 16(4):6590–6601
111. Fan M, Billings A, Zhu X, Yu P (2020) Twitter-based BIRGing: big data analysis of English national team fans during the 2018 FIFA World Cup. Commun Sport 8(3):317–345
112. Yadav J, Misra M, Rana NP, Singh K (2022) Exploring the synergy between nano-influencers and sports community: behavior mapping through machine learning. Inf Technol People 35(7):1829–1854
113. Guo H, Liu B, Yang Z (2022) Machine learning-based emotion factor analysis of sport fan community. In: Security and communication networks, 2022
114. Mahajan K, Pal A, Desai A (2023) Revolutionizing fan engagement: adopting trends and technologies in the vibrant Indian sports landscape. Int J Manag Thinking 1(2):116–135
115. Kennedy H, Kunkel T, Funk DC (2021) Using predictive analytics to measure effectiveness of social media engagement: a digital measurement perspective. Sport Mark Q 30(4):265–277
116. Yadav J, Yadav R, Sahore N, Mendiratta A (2023) Digital social engagements and knowledge sharing among sports fans: role of interaction, identification, and interface. Technol Forecast Soc Chang 195:122792
117. Jheng S, Enhancing fan engagement in a 5G stadium with AI-based technologies and live streaming
118. Istasse M, Somers V, Elancheliyan P, De J, Zambrano D (2023) DeepSportradar-v2: a multi-sport computer vision dataset for sport understandings. In: Proceedings of the 6th international workshop on multimedia content analysis in sports, pp 23–29
119. Casper JM, Pfahl ME, McCullough B (2014) Intercollegiate sport and the environment: examining fan engagement based on athletics department sustainability efforts. J Issues Intercollegiate Athlet 7(1):65–91
120. Barve YD (2023) Assessing the impacts of digital marketing and sponsorship on the football industry as a business and their implications on fan engagement (Doctoral dissertation, Dublin, National College of Ireland)
121. Schlegel MU, Hill C (2023) The reach of sports technologies. 21st century sports: how technologies will change sports in the digital age. Springer International Publishing, Cham, pp 105–125
122. Real MR, Mechikoff RA (1992) Deep fan: Mythic identification, technology, and advertising in spectator sports. Sociol Sport J 9(4):323–339
123. Yan C (2022) The impact of artificial intelligence on sports. Int J Sports, Yoga Phys Activ 1(1):8–14. ISSN: 3005–5083
124. Bailey K (2020) Statistical learning for Esports match prediction
125. Yadav J, Misra M, Rana NP, Singh K, Goundar S (2022) Netizens' behavior towards a blockchain-based esports framework: a TPB and machine learning integrated approach. Int J Sports Mark Spons 23(4):665–683
126. Casian A, Zannato M (2020) Predicting team fight tactics results with machine learning techniques
127. Hitar-Garcia JA, Moran-Fernandez L, Bolon-Canedo V (2022) Machine learning methods for predicting league of legends game outcome. IEEE Trans Games
128. Nairn A, Rossi R (2021) Keep Up! The big problem of regulating social media advertising for esports gambling. Gaming Law Rev 25(10):435–443

129. Coelho CMM (2023) OddAssist-Um sistema de recomendação de apostas em eSports (Doctoral dissertation)
130. Xenopoulos P (2023) Data mining for esports (Doctoral dissertation, New York University Tandon School of Engineering)
131. Xenopoulos P, Silva C (2022) ESTA: an esports trajectory and action dataset. arXiv preprint arXiv:2209.09861
132. Abbas BK, Jasim AI, Nsaif WS (2019) A comparative study of the growth of electronic sports in the world and the important global e-sports achievements. Int J Comput Sci Mob Comput 1(8):144–153
133. Hrytsenko AK, Hrytsenko KH (2022) Predicting the results of Esports matches by means of machine learning (Doctoral dissertation, Sumy State University)
134. Stepanov YA, Grigorev SA, Burmin LN (2021). DOTA 2 betting prediction service preview
135. Allen SV, Hopkins WG (2017) Big data and more at the 2017 sports analytics conference, Australia. Sportscience 21:23–27
136. Kuzmenko OV, Hrytsenko KH, Yarovenko HM, Kushnerov OS, Hrytsenko AK (2021) Predictive modeling of the outcomes of cyber sport matches using data mining technologies
137. Gosztonyi M (2023) Who are the gamers? Profiling adult gamers using machine learning approaches. Telemat Informat Rep 11:100074
138. Švec O (2022) Predicting counter-strike game outcomes with machine learning
139. Chan L, Hogaboam L, Cao R (2022) Artificial intelligence in video games and eSports. Applied artificial intelligence in business: concepts and cases. Springer International Publishing, Cham, pp 335–352
140. Le DQ (2022) Application of digitalization and technology in developing digital marketing for the e-sports industry in Vietnam
141. Gilles A (2023) Statistical analysis and machine learning to improve league championship series teams
142. Rubleske J, Fletcher T, Westerfeld B (2020) E-sports analytics: a primer and resource for student research projects and lesson plans. J Instruct Pedagog 23
143. Deane S, Khuman AS (2021) The application of fuzzy logic in determining outcomes of esports events. Fuzzy logic: recent applications and developments. Springer International Publishing, Cham, pp 235–252
144. Summerville A, Cook M, Steenhuisen B (2016) Draft-analysis of the ancients: predicting draft picks in dota 2 using machine learning. In: Proceedings of the AAAI conference on artificial intelligence and interactive digital entertainment, vol 12, No 2, pp 100–106
145. Hodge VJ, Devlin S, Sephton N, Block F, Cowling PI, Drachen A (2019) Win prediction in multiplayer esports: live professional match prediction. IEEE Trans Games 13(4):368–379

Survey of Technologies Toward Goal-Line Decision-Making

V. V. Prasanth and G. Nallavan

Abstract Goal-line decisions are more important decisions that should be made by the team of officials to provide a decision about the outcome of the action or event that takes place during the attempt to score, which will reward their efforts fairly both the side. The technologies that are associated with decision-making are hawk eye, video assistant referee or instant review system or video replay system, electronic timing system, electronic scoring system, and photo finish which is using some of the key technologies like Smart Balls, RFID, Internet of Things, Artificial Intelligence and Computer Vision, Wearable Technology, and other sensors included in this solutions based on the need for various games and still growing. This work is carried out to look into the technologies involved in various games in this modern times. In this work, we are going to concentrate on all the games that are using goal-line decision-making technologies listed above. The qualitative research of various findings is analyzed, and the technological importance and roles are identified. We will also have a clear picture of how the game is becoming fairer with the help of various technologies and providing good and accurate decisions to the players around the world. The real effort of the players will be appreciated with the accurate data that is generated using the available technologies by the referee and organizers to the teams and fans. Nowadays, we have lot of technologies that can add value to the existing system.

Keywords Goal-line decision · Goal-line technology · Technologies in sport

1 Introduction

There is always a need for technological advancement in all the fields. Among those fields, sports is one of the unavoidable areas through which the full of fair decisions to be given to the players and teams adopting the rules and regulations of the game along

V. V. Prasanth (✉) · G. Nallavan
Department of Sports Technology, Tamil Nadu Physical Education and Sports University, Chennai 600127, India
e-mail: prasanth199277@gmail.com

S. Bhattacharyya and D. James (eds.), *Recent Trends in Sports Engineering*, Lecture Notes in Mechanical Engineering, https://doi.org/10.1007/978-981-97-8297-0_2

with the technologies available. The need of accurate and fair officiating should be there in all the games. To provide such decisions accuracy is always concerned while using technology in terms of goal-line decision-making to provide fair decisions. Nowadays, we have lot of technologies emerged as a game changer and fair decision provider which assists the officials to make decisions in critical situations. This work is carried out to look into the goal-line decision-making technologies involved in various games in this modern times and also to find the ways where the improvement is needed for the future research and development. This work embarks to explore the technologies that are used in goal-line decision-making in a diverse array of sports, tracing their development, assessing their impact, and addressing the challenges inherent to their implementation across various games. Through this work, we have discussed the technologies that are used in various games and their development across similar games and their growth in sports industry.

1.1 System of Goal-Line Decision-Making

This is an overview of the role of various devices and methods that are used in goal-line decision-making which is helpful for the officials to get accurate results or provide fair decisions to the team. This system is included with various camera systems, sensors, and other input devices to determine and provide accurate information to the officials by processing the data into the computer algorithms and communicates to the officials to follow the rules more fair enough and provide decisions to the players or teams. Here in the very first stage we are using the high-speed cameras that are placed in various strategized locations around the stadium or goal line to capture the movement of the ball or puck from all the possible dimensions. These cameras are capturing a series of images continuously and it enables the system to analyze the trajectory and position of the ball. Triangulation systems are used to detect the precise location of the ball which can calculate the exact 3D coordinates of the ball and provide accuracy while making decisions [1]. Triangulation can be described in terms of a function τ such that

$$x \sim \tau\left(y_1', y_2', C_1, C_2\right)$$

where y_1', y_2' the homogeneous coordinates of detected image points and C_1, C_2 are the matrices of camera, and x (3D point) is representing the resulting 3D point homogeneously [2, 3]. The sign $\sim$ implies that τ is only required to produce vector which is equal to x up to a multiplication by a non-zero scalar since homogeneous vectors are involved [2, 3]. Sensors are of various types with different functionalities available in the market, using which we can get additional data like impact force or ball spin, which is used for secondary analysis or to enhance the overall understanding of the situation. For processing and doing all the activity, we have discussed above we need computers and algorithms to process the data that are collected from the cameras and sensors, which will give the outcome by processing the algorithm with the collected data and determine whether the ball, puck, or person crossed the goal line or not to provide real-time and correct decision. Also, this provides factors like ball deformation, spin, and any potential obstructions that might affect the decision. Once the processing is done with the algorithms the system needs to communicate to the officials and other stakeholders in real time. Also, the decisions are informed to the officials with the help of watch-like device worn by the referee or through large displays around the stadium which ensures the correct decision is provided promptly during the game. On all of the above, the system should follow the workflow of the refereeing while providing decision-making process by adopting the rules and regulations of various games, respectively. Through this integration the natural flow of the game will not get affected which ensures match officials can confidently rely on the technology to make the accurate decisions [4].

2 Games and Goal-Line Decision-Making Technologies

We have categorized sports into various types based on the participant count and skill set that is involved in that games. To explore the technology that is involved in every category and every game we are having sports activities like team sports (soccer, basketball, baseball, softball, volleyball, rugby, American football, field hockey, ice hockey, and lacrosse), individual sports (tennis, golf, table tennis/ping pong, badminton, boxing, mixed martial arts, wrestling, track and field/athletics, cross-country racing, and gymnastics), racquet sports (tennis, badminton, squash, and racquetball), water sports (swimming, diving, water polo, rowing, canoeing, kayaking, sailing, and surfing), winter sports (skiing (Alpine, cross-country, freestyle), snowboarding, ice skating (figure skating, speed skating), bobsleigh, curling, and ice hockey), motor sports (Formula 1 racing, NASCAR racing, MotoGP (motorcycle racing), rally racing, and drag racing), target sports (archery, shooting (rifle, pistol, shotgun), and darts), extreme sports (skateboarding, BMX (bicycle motocross), snowboarding, surfing, parkour, and rock climbing), martial arts, combat sports, and other miscellaneous sports (horse racing, equestrian sports, cricket, polo, triathlon, pentathlon). On conducting survey among various works that are conducted previously with the help of online survey and paper survey. We are discussing on the impact of technology in goal-line decision-making among various games and their applications in real time.

2.1 Technologies Involved in Team Sports

In this work, we have taken various team sports those are: soccer (football), basketball, baseball, softball, volleyball, rugby, American football, field hockey, ice hockey, and lacrosse. Now, we will discuss about the every game and the technologies that are involved in these games to make decisions on goal line. Soccer is one of the famous games played around the world which is using goal-line monitoring system to provide fair decision and avoid human errors [5–7]. In soccer, we are using Hawk-Eye Goal-Line Technology which is using cameras to track the ball's position and determine whether it has crossed the goal line [8–10]. Before implementation of goal-line technology in soccer so many tests and evaluations have been done to implement in real time [11]. GoalControl 4D: similar to hawk eye, this technology also uses cameras to detect position of ball in 3D space and assist the referees to make accurate decisions [12, 13]. There is an another technology in soccer that also supports goals which are majorly used for reviewing the event occurred in the game and this also provides review or replay to ensure the accuracy. Another two important technologies that take place in most of the games are RFID and Instant Replay System which is VAR in Soccer [14–17]. Also, Soccer is using other technologies called EPTS (Electronic Performance and Tracking System) which tracks the players' performance in higher level of games [18]. In basketball, we are using Instant Replay System to review specific situations which depend on the league or tournament, but goal-line technology is not commonly used. Case Study: Houston Rockets is a professional basketball team competing in NBA and they have implemented cutting-edge technologies for player performance improvement, injury prevention, and other team strategies. In baseball, we have an Instant Replay System (Challenge System) that allows teams to challenge certain calls, related to check whether a home run was hit or if a ball was fair or foul. Softball also uses the IRS (Challenge System) which is similar to baseball to review specific calls, ensuring accuracy in decisions. Volleyball also uses video replays to challenge a specific situation during a match which is included with ball in/out decisions. Rugby is the game which is focusing on goal line more after soccer. In Rugby, we are using Television Match Official (TMO) which uses the footage to assist the on-field referee in making decisions which includes goal line more [13, 19]. American Football is also using Instant Replay System, in which coaches do challenge for specific calls that includes touchdowns and goal-line plays and also helpful for the officials in reviewing the certain plays made by the team during the match. In field hockey also we are using Video Referral System and in soccer it is named as VAR (Video Assistant Referee) that allows teams to review the play and decisions which also includes goals and goal-line situations [20–23]. Ice hockey also uses Video Review System, this is used by the coaches to challenge certain calls that are included with goals and video view. Lacrosse is the game which is governed by various bodies and the goal-line technology used for decision-making is not standardized. However, this game is also using video review system during leagues or tournaments. There are other games like cricket which is using DRS for ball tracking which assists the umpire [24, 25]. In handball, we are using video referee system which is helping to find the ball tracking with the help of high-speed camera placed around the play area [26]. Polo is a team sport but the players are ranked individually, in which we are using electronic scoreboards and GPS tracking system is used to track horse and player movements along with the video review system [27, 28]. Wood ball is the game for which there are some works going on to work with an accurate system to check the goal crossing [29].

2.2 Technologies Involved in Individual and Racquet Sports

In this work, we have taken various individual sports those are: tennis, golf, table tennis (ping pong), badminton, boxing, mixed martial arts (MMA), wrestling, track and field (athletics), cross-country running, and gymnastics. Now, we will discuss about the every game and the technology that are involved in these games. Tennis is a game which is using Hawk-Eye technology and FoxTenn for tracking the ball trajectory and to check whether it is in or out [30]. Golf—we are using video review system to check the specific incidents or rule violations but not specifically for goal-line decisions. Table tennis (ping pong) is using instant replay system which again depends on the level of play, which is used to review certain points or situations. In badminton also we are using instant replay system that is similar to table tennis that is used in higher levels of the play to review the events that occur during the game. In boxing, we are using instant replay system for reviewing the decision or some specific situations mainly in professional boxing matches. Mixed Martial Arts (MMA) is also using IRS to review certain incidents like illegal blows or other rule violations but there is no need or use of goal-line decisions. In wrestling, we are using instant review system to review scoring decisions and other critical moments during a match. Here is the most important or famous event which is track and field (athletics) we are using photo finish technology during the sprint events finish line to determine the order of finish to extremely close races. Also, cycling uses the same technology in the finish line in crucial situations. Cross-country running is using the timing system which is electronic timing systems that are commonly used for accurate measurement of the time of each runner during cross-country races. Gymnastics also uses IRS in some competitions to review the routines or specific elements for judgment making. Apart from the above ten games we have another category called racquet sports in which we have tennis, badminton, which is already discussed, likewise squash and racquetball is also considered to be the individual games which also use IRS for reviewing specific situations or line calls. In individual games, we also have skateboarding, BMX (bicycle motocross), horse racing, equestrian sports, triathlon, and pentathlon. In skateboard and BMX, we are using video recording and analysis for documenting and reviewing the tricks. In scoring, the judges use the factors like difficulty, execution, and style and using photo finish will extend hands to the referee for making correct decision. In horse racing and equestrian, the electronic timing system is used for measurement of race times and electronic scoring system based on precision and style, respectively. Horse racing also uses photo finish by installing high-speed cameras to capture images at the finish line to determine the winner. In triathlon and pentathlon, we are using electronic timing system which is used for accurate measure of race times in each round along with video review system and photo finish in the finish line of every event in critical situations. Case Study: Sarah Williams a rising star in tennis with the help of coaching team she implemented cutting-edge technologies in all aspects like smarts racquets, video analysis, VR, and biomechanical assessment using motion capture technology with high-speed cameras to gain an edge over the opponents.

2.3 Technologies Involved in Water Sports

In this section, we have taken various sports which are swimming, diving, water polo, rowing, canoeing, kayaking, sailing, and surfing. Now, we will discuss about the every game and the technology that are involved in these games. In swimming, we are using electronic timing system for the accurate measure of the race times and electronic touchpads for sensing the pressure at the pool edges. In diving, judges need to assess the dives which is based on execution, form, and difficulty for that we are using video replay in major competitions. In water polo, there is no technology used for goal-line decisions. However, referees are relying on observations and provide real-time decisions with video referral system. Rowing is also one of the famous sports that uses electronic timing systems for measuring the race times more of like swimming. Electronic timing system is also being used in canoeing and kayaking for tracking race times during training. In sailing, we are using GPS technology for tracking and monitoring the sailboat positions at the time of racing which is happening along with photo finish or electronic timing system. Case Study: US Olympic swimming team is consistently producing world-class athletes with the help of underwater cameras with analysis software, performance tracking wearable, hydrodynamic simulation software, and VR Training environments.

2.4 Technologies Involved in Winter Sports

In this section, we have taken various winter sports which are: skiing (Alpine, cross-country, freestyle), snowboarding, ice skating (figure skating, speed skating), bobsleigh, curling, and ice hockey. Now, we will discuss about the every game and the technology that are involved in these games. In skiing, we have three types they are Alpine, cross-country, and freestyle. To measure the accurate race times, we are using electronic timing systems and photo finish, and along with this we are also using video replay for reviewing performance in some situations. In snowboarding also we are using electronic timing system for race times. Like diving the judges examine their performance with the help of style, difficulty, and execution. Ice skating is also having two types which are figure skating and speed skating. In these games, the key performance indicators are technical and artistic elements using which the judges assess their performance along with the electronic timing systems which are used for measuring race times in speed skating. Bobsleigh is using photoelectric cells and electronic timing systems for accurate measurement of racing times. Also, coaches using the same during the training time. Curling is an event where the electronic hog line technology is used for detecting if a stone is released before the hog line and also video replay is used for specific situations. Ice hockey is using goal judging technology which is used to check the goal with the help of high-speed cameras around the goalpost. Case Study: Swiss National Alpine Ski Team focusing on performance optimization and injury prevention by integrating cutting-edge technologies into their training and preparation processes which includes ski performance sensors, biomechanical analysis systems, VR simulation, and weather and snow condition analysis tools.

2.5 *Technologies Involved in Motor Sports*

In this section, we have taken various motor sports which are: Formula 1 racing, NASCAR racing, MotoGP (motorcycle racing), rally racing, and drag racing. Now, we will discuss about the every game and the technology that are involved in these games. In racing, photo finish is a common system that is employed in all the games. In Formula 1, the electronic timing systems are used for precise measurement of lap times with the help of transponders on cars along with car telemetry which provides various aspects of vehicle performance in the monitor and also video review system during races. NASCAR racing is using electronic timing systems to measure lap times and overall race times using transponders along with car telemetry and video replays. MotoGP is also using electronic timing system that provides precise measurement of the lap times and also using on-bike telemetry for data analysis and performance optimization with the help of electronic aids. Rally racing is using electronic timing systems and onboard computer and telemetry along with the video replay system. Drag racing is again using the sensors at the start and finish lines for the measurement of elapsed times which is electronic timing systems along with the video replay for determining the winners and review the race [31]. Case Study: One of the well renowned team is Mercedes-AMG Petronas Formula one and they are consistently competing and winning multiple championship in highest level by implementing cutting-edge technologies like telemetry systems, simulation software, data analytics and machine learning, and additive manufacturing (3D printing).

2.6 *Technologies Involved in Target Sports*

In this section, we have taken various target sports which are: archery, shooting (rifle, pistol, and shotgun), and darts. Now, we will discuss about the every game and the technology that are involved in these games. In archery, we are using electronic scoring systems and high-speed cameras for reviewing trajectories and scoring calls which are close. In shooting, we have three types they are rifle, pistol, and shotgun which are using electronic target systems that are equipped with the sensors to detect and score the shots along with the automated scoring systems based on the shot placement using the electronic scoring software. In darts, we are using electronic dartboards that come with electronic scoring features that track and score automatically. Some of the systems also use cameras or sensors to track dart placements for scoring. Case Study: South Korean National Archery Team is a renowned and dominance team in archery, they are consistently producing world-class athletes with the help of cutting-edge technologies like biomechanical analysis, high-speed cameras with video analysis software, bow tuning equipment, mental training apps, and biofeedback devices.

3 Data Analysis and Visualization

We have major technologies and then categorized the goal-line technologies into five categories as per the above survey conducted among various online and paper surveys. The categories are 1. photo finish, 2. hawk eye, 3. electronic timing system, 4. instant review system or video replay system, and 5. electronic scoring system. The games which are using photo finish are: athletics/track and field, cycling (track and road), horse racing, swimming (using electronic touchpads also), skiing (Alpine and cross-country), speed skating, bobsleigh, skeleton, and luge, rowing, biathlon, triathlon, canoeing/kayaking, sailing, bike racing, greyhound racing, car racing, boat racing (dragon boat racing), inline skating races, harness racing, camel racing, marathon and road races, tobogganing, roller derby, dog sledding races, inline hockey, and motor-boat racing. The games which are using Hawk-Eye Technology are tennis, cricket (DRS), squash, badminton, table tennis, volleyball, snooker, football (soccer), and rugby [32]. The games which are using electronic timing system are soccer, field hockey, ice hockey, basketball, handball, water polo, volleyball, rugby, and American football. The games which are using instant or video review system are soccer, cricket, tennis, basketball, American football, ice hockey, baseball, rugby, volley-ball, handball, water polo, badminton, field hockey, golf, boxing, MMA, snooker, bowling, darts, curling, bobsleigh, skeleton and luge, surfing, skateboarding, gymnastics, and equestrian. The games which are using electronic scoring systems are tennis, table tennis, badminton, volleyball, basketball, soccer, American football, baseball, cricket, golf, hockey, rugby, water polo, bowling, e-sports, archery, shooting, cycling (track). Figure 1 shows the number of games which is using which technology in making decision during goal-line decision-making.

4 Results and Discussion

Our aim is to find the technologies that are being used in various games. As per the findings in the data and visualization part, it is clear that photo finish is being used among most of the games. Followed by IRS or video review system being used to provide fair decision during the game including weightlifting [25]. After IRS, we are having electronic scoring system which is being followed in 18 games and the remaining 2 technologies hawk-eye and electronic timing systems are being used in 9 games each. This is due to the cost of the implementation of these systems being very high. The data that is presented are dynamic that might change over a period of time due to evaluation and usage of technology in other games. Only larger or

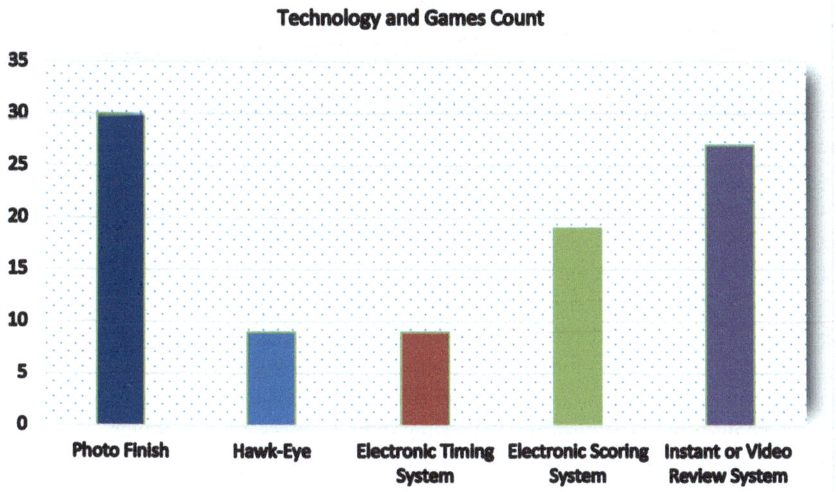

Fig. 1 Game-wise usage of various goal-line decision-making technologies that are being implemented in real time

high growth teams, organizations or sports bodies can accommodate this kind of technologies. The alternates for this kind of technologies should be created in the future to make the availability of technology to all the level of players in providing accurate results or decision to them. Meanwhile whatever technologies come in place should be cost saving and time saving and adopt the rules and regulations of the game. Through this work and findings, we can clearly say that the technology in goal-line decision-making plays a major role in taking decisions and assisting umpires on field with the real time and necessary data with accuracy. Evolution in decision-making across various sports gives us greater accuracy, fairness, and fan engagement during the match. Implementing technologies should be done according to the distinctive features and challenges by adopting the rules and regulations also in a way the solution should support the game and keep the fans engaged all the way during the game. Due to this implementation of technologies we can overcome controversies and enhance the overall experience. As we are in a world, where e-sports is growing tremendously and those are also having in game replay systems and to make the highlights and photo finish is also being used in racing games as in real world. Technology is going to play a major role in sports officiating and other activities in the sports world with the innovative implementation among various sports which will add more excitement, integrity, and transparency of the games [33].

5 Conclusion and Future Work

Initially, by creating simulation and analysis of the playing environment with the technologies we can help the athletes to develop their performance and safety in the training. As per this work, we were looking into five major technologies that are discussed in this section and apart from this we have biomechanical analysis and other training applications which are creating major changes in the athletes' performance to the peak. As the technologies have emerged as a game changer in various sports, by enhancing accuracy and fairness in determining whether a goal or target has been achieved. Since various technologies like electronic timing system, electronic scoring system, photo finish, hawk-eye goal-line technology have evolved, there is always a need to modernize the existing technologies by adding the solution like robotics and other features with more accurate values for the automated outcome and to create a new solution for the games which doesn't have proper solution in the level-wise games [34]. Nowadays, we are using this kind of solutions in higher level matches alone. It needs to be scaled down in a low price range which can be affordable for the beginners also. For making this kind of solution in this era, we need to adapt to the technologies like artificial intelligence, machine learning, computer vision, multi-sensor integration, and augmented reality [35, 36]. This kind of solutions will enhance the fairness of decisions and reduces error with the help of historical data that is used for training the system. As per the results, we should aim to work in design and develop an affordable solution that every level of team or player should use to achieve fairness in their game.

References

1. Mussabayev R, Kalimoldayev M, Amirgaliyev Y, Tairova A, Mussabayev T (2018) Calculation of 3D coordinates of a point on the basis of a stereoscopic system. Open Eng 8(1):109–117. https://doi.org/10.1515/eng-2018-0016
2. https://en.wikipedia.org/wiki/Triangulation_(computer_vision)
3. Hartley R, Zisserman A (2003) Multiple view geometry in computer vision. Cambridge University Press. ISBN 978–0–521–54051–3
4. Brown RV (1990) Assessment uncertainty technology for making and defending risky decisions. J Behav Decis Mak 3(3):213–228
5. Spagnolo P, Leo M, Mazzeo P, Nitti M, Stella E, Distante A (2013) Non-invasive soccer goal line technology: a real case study. In: Proceedings of the IEEE conference on computer vision and pattern recognition workshops. pp 1011–1018
6. Nlandu T (2012) The fallacies of the assumptions behind the arguments for goal-line technology in soccer. Sport, Ethics Philos 6(4):451–466
7. Le J, Kurtz J (2016) Critical essay assignment: the implications of technological advancement in soccer
8. Surujlal J, Jordaan DB (2013) Goal line technology in soccer: are referees ready for technology in decision making?: technology and innovation. Afr J Phys Health Educ, Recreat Danc 19(2):245–257
9. Winand M, Fergusson C (2018) More decision-aid technology in sport? an analysis of football supporters' perceptions on goal-line technology. Soccer & Soc 19(7):966–985

10. Ryall E (2012) Are there any good arguments against goal-line technology? Sport, Ethics Philos 6(4):439–450
11. Spagnolo P, Mazzeo PL, Leo M, Nitti M, Stella E, Distante A (2015) On-field testing and evaluation of a goal-line technology system. Computer vision in sports. Springer International Publishing, Cham, pp 67–90
12. Ahmadi Z, Sobhani N (2014) Arbitration management with using artificial intelligence technology (the sample: goal-line technology in football). Model Artif Intell 2:48–58
13. Collins H (2019) Applying philosophy to refereeing and umpiring technology. Philos 4(2):21
14. Ghosh S, Sasmal S, Bhui S, Dutta S, Mukherjee S, Majumder A, Ganguly B (2019) Radio frequency identification based goal line technology for quick decision making in a football match. In: 2019 devices for integrated circuit (DevIC). IEEE, pp 441–445
15. Kolbinger O, Linke D, Link D, Lames M (2015) Do we need goal line technology in soccer or could video proof be a more suitable choice: a cost-benefit-analysis of goal line technology in soccer and thoughts about an introduction of video proof. In: Sports science research and technology support: second international congress, icSPORTS 2014, Rome, Italy, October 24–26, 2014, revised selected papers 2. Springer International Publishing, pp 107–118
16. Ugondo PI, Tsokwa M (2019) Interpreting video assistant referee and goal-line technology communication: the pitch-based referees perspectives. Int J Trend Sci Res Dev 3(4):1058–1062
17. Karanasios S, Upreti B, Iannacci F (2023) When is a goal a goal? addressing equivocality with technology
18. Pino-Ortega J, Rico-González M (2021) Standardization of electronic performance and tracking systems. In: The use of applied technology in team sport. Routledge, pp 3–10
19. Swales P, Cooper J, Catlow A (2014) Technically… it was a GoalRef
20. Vergani C The impact of technology on decision-making in sports: the case of video assistant referee
21. Ajjampur S, Shridhar V, Shashikala KP (2021) Goal line technology using line axis detection. In: 2021 international conference on disruptive technologies for multi-disciplinary research and applications (CENTCON), vol 1. IEEE, pp 304–307
22. Mani A Goal-line technology in soccer: discussion and evaluation of systems
23. Spitz J, Wagemans J, Memmert D, Williams AM, Helsen WF (2021) Video assistant referees (VAR): the impact of technology on decision making in association football referees. J Sports Sci 39(2):147–153
24. Shivakumar R (2018) What technology says about decision-making: evidence from cricket's decision review system (DRS). J Sports Econ 19(3):315–331
25. Michalopoulos N, Katsini C, Raptis GE (2019) Using video-based technology in powerlifting sport to support referees' decision making. In: Extended abstracts of the 2019 CHI conference on human factors in computing systems. pp 1–6
26. Morillo JP, Reigal RE, Hernández-Mendo A, Montaña A, Morales-Sánchez V (2017) Decision-making by handball referees: design of an ad hoc observation instrument and polar coordinate analysis. Front Psychol 8:1842
27. Leveaux R, Messerschmitt M (2015) The changing shape of sport through information technologies. In: Proceedings of the 26th international business information management association conference-innovation management and sustainable economic competitive advantage: from regional development to global growth, IBIMA 2015
28. Leveaux R (2017) An examination of information technologies changing the shape of sport. The Communications of the IBIMA
29. Chandrasegaran J, Mustafa WA, Jamlos MA, Idrus SZS (2020) Design woodball line detection and monitoring system: a preliminary study. In: IOP conference series: materials science and engineering, vol 917, no 1. IOP Publishing, p 012069
30. Kolbinger O, Lames M (2017) Scientific approaches to technological officiating aids in game sports. Curr Issues Sport Sci (CISS)
31. Permana Y, Soegiyanto S, Kusuma DWY (2021) A development of sensor aids for goalkeeping violations over the goal line on first point penalty kicks on futsal. J Phys Educ Sport 10(4):385–393

32. Singh Bal B, Dureja G (2012) Hawk eye: a logical innovative technology use in sports for effective decision making. Sport Sci Rev 21
33. Kotak D, Nayak K, Kurup L (2014) Virtual reality-a helping hand for referees. Int J Curr Eng Technol
34. Haobin S, Lin Z, Wei P, Shichao W (2015) Robot soccer confrontation decision-making technology based on MOGM: multi-objective game model. J Intell & Fuzzy Syst 28(2):713–724
35. Desbordes M (2015) Emerging technologies and sports events. Sport, Bus Manag: Int J 5(2):175–190
36. Petrović LT, Milovanović D, Desbordes M (2015) Emerging technologies and sports events: Innovative information and communication solutions. Sport, Bus Manag: Int J 5(2):175–190

BallTraj: Fast-Moving Ball Trajectory Tracking Using an Encoder–Decoder Network

Aditya Somani and Aruna Malapati

Abstract Ball trajectory data is a crucial piece of information that must be accurately tracked for tennis professionals. High-end systems like hawk eye are crucial for precise tennis ball trajectory tracking. However, their high cost restricts their accessibility. As a more cost-effective alternative, a single camera can be used, but it poses challenges due to the tennis ball's small size and rapid movement. In this study, we propose a solution to the challenges faced by low-resource tennis analysis systems that utilize a single camera. Specifically, we introduce a heat map-based encoder–decoder network with skip connections, which learns the trajectory patterns of a tennis ball, tracks its high-speed movement, and produces dependable data. Additionally, we present a synthetic data generation technique that enhances model accuracy by a certain percentage through the generation of relevant data. The proposed model was trained using over 15 k frames and another 13 k + generated frames from various lighting conditions, camera angles, and court surfaces. The model's performance was assessed using frames captured in different conditions, yielding a precision of 98.2% and a recall of 91.4%.

Keywords Ball tracking · Encoder–decoder network · Synthetic data

1 Introduction

Post-match or practice analysis is a critical component of athlete training and development in tennis. Videos are one of the most valuable resources that professionals or semi-professionals can utilize for this purpose. However, the current technology necessitates the use of multiple high-end cameras and expensive software for data collection and analysis, which is not a feasible solution for many players. Therefore, a more straightforward and less expensive solution is required.

A. Somani (✉) · A. Malapati
Department of Computer Science & Information Systems, BITS Pilani Hyderabad Campus, Hyderabad, Telangana 500078, India
e-mail: adityasomani2003@gmail.com

S. Bhattacharyya and D. James (eds.), *Recent Trends in Sports Engineering*, Lecture Notes in Mechanical Engineering, https://doi.org/10.1007/978-981-97-8297-0_3

The most advanced algorithm available now for tracking tiny, fast-moving objects, such as tennis balls, employs a convolution–deconvolution network. Tracknet [1, 2] has been successful in tracking the ball when its movements are clear, and each frame captures the ball at every time step. However, Tracknet is challenged when the video quality is poor, resulting in the ball being captured incompletely or not at all.

We propose a solution to this problem using BallTraj, an encoder–decoder model based on heat maps. BallTraj first learns the trajectory patterns of the ball in question and then overcomes the challenges posed by hazy and incomplete images to detect occluded balls. Our proposed approach is expected to provide a more effective and affordable solution for players who require a simpler and cheaper alternative to the current technology.

The idea is to employ a deep learning-based approach instead of conventional object detection techniques based on features such as color and shape or methods such as HOG [7] and Haar Cascade [8]. The relatively long shutter time of consumer-grade cameras renders these techniques ineffective, and learning the patterns and trajectories of ball movement is a more promising approach.

BallTraj, our proposed model, takes multiple frames consecutively as input to learn the ball's trajectory patterns. The input images are concatenated and passed through an encoder containing seven convolution layers and three pooling layers before being fed into a fully connected layer that generates a latent space of size 128. The model then extrapolates the images back to produce a single heat map, which represents the ball's position in the image.

Summarizing, our aim with this paper is as follows:

- Address the challenges in tennis post-match and practice analysis.
- Overcome challenges in current algorithms by detecting occluded balls and tracking them accurately in scenarios with poor video quality.
- Unlike traditional object detection algorithms, we aim to use deep learning techniques with multiple consecutive frames and learn the inherent patterns in the ball trajectory.

Our primary contributions include an annotated dataset curated using alpha blending techniques to generate more samples, which imitate the ball streak patterns on a tennis court and propose a heat map-based encoder–decoder-based network BallTraj, with skip connections for information propagation.

The subsequent sections are as follows: Sect. 2 gives a concise overview of prior research that addresses the target task. Section 3 outlines the specific characteristics and specifications of the dataset that is being proposed in order to successfully complete this task. The process has been elucidated in Sect. 4. Sections 5, 6, and 7 provide an experimental assessment together with the resulting outcomes. The conclusion of the work is explained in Sect. 8.

2 Literature Survey

One crucial area of computer vision is object tracking, it is an application of deep learning. The program usually starts with an initial set of object detection and then tracks the object as they move across frames of a video. Multiple Object Tracking (MOT) identifies multiple objects in a single frame and then tracks them. This sort of tracking is extremely important for applications like autonomous driving. SRK ODESA [3] comprises three primary components—an object detector, similarity estimation, and data association/similarity checking. The object detector employs the faster RCNN model, for similarity estimation, SRK ODESA utilizes three neural networks. This process enables the system to identify and track objects with high accuracy. Finally, the system uses the Hungarian algorithm and Kalman filter for data association and similarity checking.

Goturn [4] is an object detection algorithm that requires offline training for specific object classes. It tracks an object by regressing its location using the target object's image in the previous frame. The outputs of the network are the bottom-right and top-left coordinates of the object in the current frame. The model learns motion smoothness by oversampling small box movements obtained from a Laplace distribution of mean 0.

Online trackers belong to a class of object-tracking algorithms that produce immediate predictions, thereby lacking the ability to leverage future frames for refining the results. An example of such an algorithm is MDNet [5], which adopts a two-stage training process to improve its accuracy in tracking specific objects. Specifically, MDNet employs a fine-tuning technique by training a subset of its convolutional layers on specific domains while preserving the common layers to extract generic features. It is particularly useful for field like robotics with a live feed involved.

Single neural-network-based You Only Look Once (YOLO) [6] is an object detection algorithm that is used for both classification and localization. The system employs a fully convoluted neural network to divide an image into a grid and subsequently applies convolutional layers to each cell inside the grid to estimate the probabilities of the object's class and its bounding boxes. The YOLO algorithm designates a single predictor as "responsible" for forecasting an item by selecting the prediction with the greatest Intersection over Union (IOU) value compared to the ground truth. It then utilizes Non-Maximum Suppression (NMS) to eliminate redundant predictions.

Recurrent YOLO (ROLO) [9] is a modification of YOLO that uses Long Short-Term Memory (LSTM) [10] networks to track objects over time. ROLO uses LSTM networks to find the trajectory of the bounding box by learning historical patterns, making it particularly suitable for visual object tracking. It has been proven to be particularly useful in sports analysis.

Single Shot Detection (SSD) [11] is an object detection algorithm that detects multiple objects in a single input. Next, the algorithm calculates the offset of each bounding box relative to its original position. Different feature maps are created for each bounding box using the bounding box regression technique, based on VGG [12]. The algorithm uses a linear combination of two types of loss, Confidence and

Location, in its final loss term.

$$\textit{Multi Box Loss} = \textit{Confidence Loss} + \alpha * \textit{Location loss}. \tag{1}$$

SSD has proven to be highly accurate and efficient, achieving real-time performance with high accuracy on various datasets. Its implementation is straightforward and involves fewer computations compared to other object detection algorithms.

Recently, there has been notable advancement in the development of object-tracking algorithms specifically designed for tennis applications. Object Detection and Tracking Based on Trajectory in Broadcast Tennis Video [19] is a technique that depends exclusively on computer vision and involves applying a median filter to each frame's image to reduce noise. A background model is then calculated, and the foreground is obtained by subtracting the background, followed by obtaining an edge image. Fast-moving items in the foreground are detected by nalysing the difference between frames using the logical AND operator. Another algorithm for tennis is TrackNet, which employs a convolution–deconvolution pair with a convolution architecture similar to VGG-16 and combines it with a deconvolution network to obtain high-resolution output.

This study proposes BallTraj, an encoder–decoder pair convolution neural network with latent space and skip connections, which predicts the location of a tennis ball using consecutive frames as the input. Consecutive frames help the network capture the trajectory patterns in ball movements and make a prediction based on the same.

3 Datasets

In our study, we utilized an initial dataset comprising over 19,000 images captured from 9 tennis matches with different court surfaces (hard and clay courts) and different lighting conditions (daylight and floodlights). This was a modified form of the dataset used in Tracknet [18]. To ensure the effectiveness of our model in diverse court settings, the videos were divided into 100 segments, as a whole containing images from all possible tennis court scenarios. For each frame, we generated a corresponding CSV file containing four attributes:

a. **Frame Name** refers to the name of the file associated with the frame.
b. **Visibility Class** denotes the ball's current visibility state within the frame, taking one of three values: 0 for the absence of the ball, 1 for the ball's presence and easy identification, and 2 for the ball's obstruction by an opaque object.
c. **(X, Y) Coordinates**, representing the ball's location in the frame, with the end of the streak being considered as the ball's location coordinates in frames where the tennis ball is not captured as a circle.
d. **Trajectory Pattern**, indicating the ball's motion status, with values of 0, 1, and 2 representing in-play, contact with the racquet, and bouncing, respectively.

This extensive dataset enabled us to accurately train and test our deep learning model and evaluate its performance under diverse conditions. The dataset's wide coverage of different court scenarios and lighting conditions enhances the model's accuracy, making it more reliable for tracking high-speed tennis balls in real-world scenarios.

3.1 Data Generation

The key challenge encountered in existing models pertains to the identification of ball streaks during a match. As a solution, we propose the use of multi-layer alpha blending [15] to create synthetic data, which involves blending two or more images together to produce smoother transitions between them.

Alpha blending employs a weighted pixel sum value for both images, where a parameter alpha within the range of 0 and 1 determines the transparency or opacity of the mask applied to each image. Specifically, for the bottom image, the applied mask is 1-alpha, resulting in the following equation:

$$Output\ Pixel\ Value = \alpha * Top\ Pixel\ Value + (1 - \alpha) * Bottom\ Pixel\ Value \quad (2)$$

To generate synthetic data that is relevant to our research problem, we utilize sets of three consecutive images at times t, t + 1, and t + 2 for alpha blending, which produces a single image as the output. It is crucial that all three images capture the ball in play and that the ball is visible (Fig. 2).

We begin by blending the images at times t and t + 1, with the image at time t serving as the bottom image while the image at time t + 1 is the top image. Using an alpha value of 0.2, we blend the two images to generate an intermediate image. This now serves as the bottom image and is blended with the image at timestamp t + 2, again using an alpha value of 0.2 to generate an output image. We maintain all the parameters of the output image, as this represents the ball's final position in the case of a streak.

We thereby add a total of 13,073 images to our dataset, which are combined and added as a separate clip. In total, our dataset comprises 32,908 images.

4 BallTraj

This section describes the framework that we have developed to identity the trajectory patterns of a tennis ball and track its high-speed movement. Figure 4 shows the overall architecture of model.

The flow of the network includes images being passed through the encoder, then being flattened out to be converted into latent space and after that, being increased in size via the decoder network to get a heat map representation as the output. In the process, the image also passes through a couple of skip connections as explained in the following sections to increase the information flow in the network.

4.1 Preprocessing

The preprocessing step involves generating the ground truth for the algorithm. To accomplish this, we create a heat map from the original images, where the white color indicates the probability of the presence of a tennis ball and the rest of the heat map is black. We generate this heat map using a 2D Gaussian distribution, which is defined by two parameters: the mean and the variance. The center of the distribution is represented by the mean, while variance represents the distribution's width or spread (Fig. 3).

To obtain the center of the distribution, we utilize the ball's location's x- and y-coordinates in the frame, which are present in our dataset. We set the variance to 10, which is generally equivalent to the size of a tennis ball in the frame.

$$f(x, y) = \frac{1}{2\pi\sigma_x\sigma_y}\exp\left(-\frac{1}{2}\left(\frac{(x-\mu_x)^2}{\sigma_x^2} + \frac{\left(y-\mu_y\right)^2}{\sigma_y^2}\right)\right) \quad (3)$$

To set the variance as 10, we use both σx and σy as $\sqrt{10}$ They need to be equal because we wish to get a circular shape in output. Increasing σx or σy spreads the distribution out while decreasing the same, the distribution gets more concentrated and narrowed. All the images in 100 clips are converted to heat map using 2D Gaussian distribution and stored, this is used to train the model.

4.2 Network and Training

BallTraj is a type of encoder–decoder network that utilizes a latent space to connect both modules. This network is adaptable to take any number of consecutive frames as input. However, the network is currently trained to receive three consecutive frames as input, with the output being a corresponding heat map. This approach [13] is chosen because it enables the model to consider the recent trajectory of the ball when calculating the heat map.

BallTraj, by default, processes images of size 360 × 640 as input. The encoder module is comprised of seven convolutional blocks and three pooling blocks, which are followed by a transition into the latent space. The decoder module, on the other hand, consists of eight convolutional blocks and three upsampling blocks. The kernel initialization was random uniform across the network. With the exception of the decoder's last convolutional layer, the architectures of the encoder and decoder are

Table 1 Layer-wise architecture of the encoder

Layer	Number of kernels	Kernel dimension	Padding	Strides	Activation
BallTraj Convolution 1	64	(3,3)	same	1	ReLU + BN
BallTraj Convolution 2	64	(3,3)	same	1	ReLU + BN
BallTraj Pool 1	(2,2) with Strides = 2				
BallTraj Convolution 3	128	(3,3)	same	1	ReLU + BN
BallTraj Convolution 4	128	(3,3)	same	1	ReLU + BN
BallTraj Pool 2	(2,2) with Strides = 2				
BallTraj Convolution 5	256	(3,3)	same	1	ReLU + BN
BallTraj Convolution 6	256	(3,3)	same	1	ReLU + BN
BallTraj Convolution 7	256	(3,3)	same	1	ReLU + BN
BallTraj Pool 3	(2,2) with Strides = 2				

identical. This layer resizes the output to correspond with the dimensions of the input image and reduces the number of feature maps to one.

The latent space is basically a fully connected network which connects the final convoluted image to a 1D layer of 128 perceptrons. This 1D layer is further connected to a 2D array of size 45 × 80 with 256 feature maps, again through a fully connected network. The model also includes two skip connections, which will be elaborated upon later in this paper (Table 1).

4.3 *Encoder Architecture*

Latent Space: The output of the encoder module comprises 256 feature maps, each with a size of 45 × 80. In order to enter the latent space, the feature maps are first flattened. This results in a 1D layer of dimensions 45 × 80 × 256, which is equal to 9,21,600. After that, the flattened layer is passed through a second fully connected layer with a 128 as the output size, before being reconstructed to its original dimensions using another fully connected layer. As the input to the decoder module must be a 2D array, the reconstructed layer is reshaped to its original dimensions of 45 × 80 × 256. The decoder takes reshaped layer as its input (Table 2).

Table 2 Layer-wise architecture of the decoder

Layer	Number of kernels	Kernel dimension	Padding	Strides	Activation
BallTraj Upsamling 1	(2,2)				
BallTraj Convolution 1	256	(3,3)	same	1	ReLU + BN
BallTraj Convolution 2	256	(3,3)	same	1	ReLU + BN
BallTraj Convolution 3	256	(3,3)	same	1	ReLU + BN
BallTraj Upsamling 2	(2,2)				
BallTraj Convolution 4	128	(3,3)	same	1	ReLU + BN
BallTraj Convolution 5	128	(3,3)	same	1	ReLU + BN
BallTraj Upsamling 3	(2,2)				
BallTraj Convolution 6	64	(3,3)	same	1	ReLU + BN
BallTraj Convolution 7	64	(3,3)	same	1	ReLU + BN
BallTraj Convolution 8	256	(3,3)	same	1	

Decoder Architecture

Skip Connections: The compression of data into smaller dimensions followed by the expansion back to its original size can potentially result in the loss of information. To address this issue, skip connections or residual connections [16] can be implemented to allow the model to bypass one or more network layers and propagate information directly to later layers. This creates a shortcut that preserves high-level input data features that would otherwise be lost during the compression process.

The inclusion of skip connections in an encoder–decoder model enhances the ability of the model to integrate abstract high-level features learned by the encoder with specific information present in the original input, leading to improved performance. Two skip connections that link encoder and decoder maps of the same size have been added to the BallTraj model in order to improve information flow between the encoder and decoder.

Following layers are connected by the skip connections:

- Convolution Layer 4 of the encoder and Convolution Layer 5 of the decoder.
- Convolution layer 7 of the encoder and Convolution Layer 3 of the decoder.

Activation and Loss Function: The neural network outputs a heat map with continuous values ranging from 0 to 255 for each pixel. The data array is denoted by H(i,j,k). It corresponds to specific coordinates, adhering to the constraint (0,0) ≤ (i, j) ≤ (639,359), and spanning over the range of 0 to 255 for parameter k. There are 256 possible grayscale values for depth level k, and the probability distribution for each one of them is calculated by the softmax layer. It expresses the probability of depth k at the point (i, j) and is represented as P(i, j, k). The mathematical formulation of the softmax function is provided as follows:

$$P(i,j,k) = \frac{e^{H(i,j,k)}}{\sum_{k'=0}^{255} e^{H(i,j,k')}} \tag{4}$$

Categorical cross-entropy loss function is used to train the network. Specifically, let F(i, j) represent the model output at coordinates i and j, while G characterizes the distribution of labels.

The ground truth label is defined as follows:

$$G(i,j,k) = \begin{cases} 1 if F(i,j) = k \\ 0 otherwise \end{cases} \tag{5}$$

Next, the total loss is calculated as the negative product of G(i, j, k) and F(i, j, k)'s natural logarithm:

$$TotalLoss = -\sum_{i,j,k} G(i,j,k) \log((i,j,k)) \tag{6}$$

During the model's training, we employ the adadelta optimizer [14], which adaptively modifies the learning rate without requiring manual hyperparameter adjustment (Table 3).

Table 3 Hyperparameters of the model

Parameters	Value
BallTraj learning rate	1
BallTraj batch size	16
BallTraj steps in each epoch	200
BallTraj epochs	500

5 Prediction

During the testing phase, BallTraj produces a heat map in which the depth at each coordinate (x,y) corresponds to the likelihood of the ball's presence at that location in the input frame. To obtain a binary heat map from the output, a threshold is applied to the pixel values. If a pixel's value is higher than the threshold, its value is set to 255; if its value is lower or equal to the threshold, its value is set to 0. In this work, a threshold of 128 was used as it provided a reasonable balance between detection accuracy and computational efficiency. After generating the binary heat map, the HoughCircle [17] detection method is employed to identify circles present in the frame. The centroid is the output if only one circle is detected. However, in cases where multiple circles are detected, a buffer is maintained to store all centroid values. The distance between each centroid and the centroid of the previous frame is calculated, and if the minimum distance is less than 20 pixels, that centroid is selected as the output.

6 Experiments

6.1 Comparative Analysis

In this research, a test dataset was formed by randomly selecting 20% images from the previously discussed dataset. The other 80% were employed for training purposes. To assess the models' effectiveness, the positioning error was chosen as the benchmark parameter. This metric is widely used as the standard benchmark for systems that involve predicting an object's position.

As the diameter of a tennis ball is approximately five pixels, any positioning error less than five pixels is considered an accurate result. The positioning error for the Tracknet and BallTraj models was analyzed, and the results are presented in the following graph (Fig. 5) (Table 4).

Table 4 Positioning error

Count	BallTraj (%)	Retrained TrackNet (%)	TrackNet original (%)
0	8.19	7.22	7.09
1	24.25	22.09	22.78
2	28.95	29.44	24.81
3	23.9	24.71	21.32
4	8.98	10.18	12.1
5	3.8	4.38	6.68
> 5	1.93	1.99	5.22

Table 5 Comparative analysis

Network	Mean error (in pixels)	Variance	Precision	Recall
BallTraj	1.969	14.391	0.981	0.918
Retrained Tracknet	2.221	109.023	0.979	0.925
Tracknet	2.585	146.444	0.947	0.172

Examining both the graphical representation and the tabulated data, it becomes evident that BallTraj exhibits a higher percentage of instances falling within positioning errors of 1 and 2. Conversely, the relative percentage decreases as positioning errors increase. In accordance with this observation, we classify cases as either True Positive or False Positive. Specifically, those with a positioning error of 5 or less are categorized as True Positives, while instances with a positioning error exceeding 5 are labeled as False Positives.

This, however, is not the most noteworthy aspect of BallTraj. Its exceptional performance extends to the reduction of both the mean positioning error and the variance in positioning error. In essence, this implies that BallTraj not only offers superior accuracy in predicting the ball's location but also presents a narrower, more tightly distributed bell curve, facilitating more precise and streamlined predictions.

6.2 *Limitations and Challenges*

The principal constraint encountered in the proposed system pertains to computational resources, attributed to the deep learning nature of BallTraj. Its operation demands a substantial computational infrastructure, rendering real-time processing unattainable.

A pivotal aspect of the development process involved addressing challenges related to ball streaks, necessitating brainstorming sessions to devise effective strategies for synthetically replicating such phenomena within the training data. Moreover, enhancing information propagation posed a significant challenge, prompting exploration into modifications of the encoder–decoder pair architecture. While experimentation with DenseNet architectures was conducted, the integration of skip connections yielded notably superior results (Table 5).

7 Ablation Analysis

BallTraj exhibits remarkable efficacy in scenarios where the ball appears as a streak in the video. This remarkable performance can be attributed to the combined effect of sophisticated data augmentation techniques and the thoughtful architectural design.

By incorporating data augmentation, BallTraj enhances the robustness of its predictions, enabling accurate tracking of the ball even in challenging conditions where the ball appears as a streak due to motion blur or rapid movement.

The architectural design of BallTraj plays a crucial role in its success. Unlike traditional approaches that rely solely on individual frames, BallTraj leverages information from multiple consecutive frames as input. This allows the model to capture temporal dependencies and exploit the trajectory of the ball. By considering the spatio-temporal context, BallTraj not only recognizes the presence of ball streaks but also estimates the trajectory of the ball, leading to more precise and reliable predictions of its location.

Moreover, BallTraj exhibits consistent performance even when confronted with low-quality camera footage. This resilience to image degradation is of paramount importance in real-world scenarios where the availability of high-quality video data may be limited. By maintaining accurate predictions despite the challenges posed by low-quality cameras, BallTraj demonstrates its practical applicability in various settings, including sports analysis, video surveillance, and motion tracking applications.

Figures 1 and 6 showcase a visual comparison between the predictions generated by BallTraj and TrackNet specifically in situations involving ball streaks. Notably, BallTraj outperforms TrackNet by accurately predicting the ball's position along its trajectory. This capability of BallTraj holds great promise for further analysis of the tracked data. By providing accurate and reliable information about the ball's location over time, BallTraj offers valuable insights for in-depth analysis, performance evaluation, and decision-making processes in various domains.

Fig. 1 An example from US Open 2011 Final between Rafael Nadal and Novak Djokovic, where BallTraj predicts the location of the ball correctly while TrackNet gives a false positive at the centre mark on the baseline

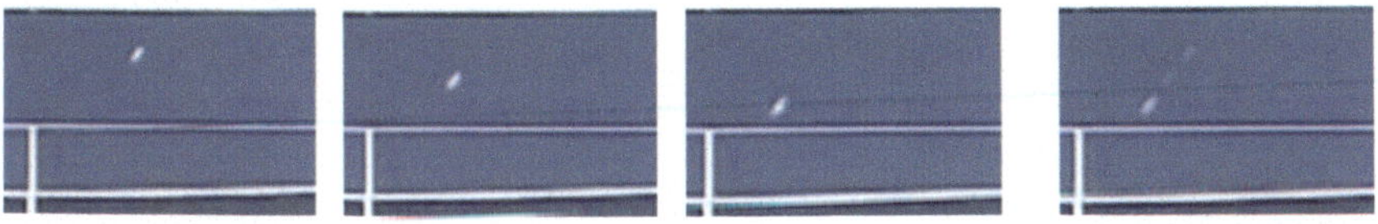

Fig. 2 Image at t, t + 1, t + 2 and the resultant image

Fig. 3 Heat map example, the circle represents the ball location

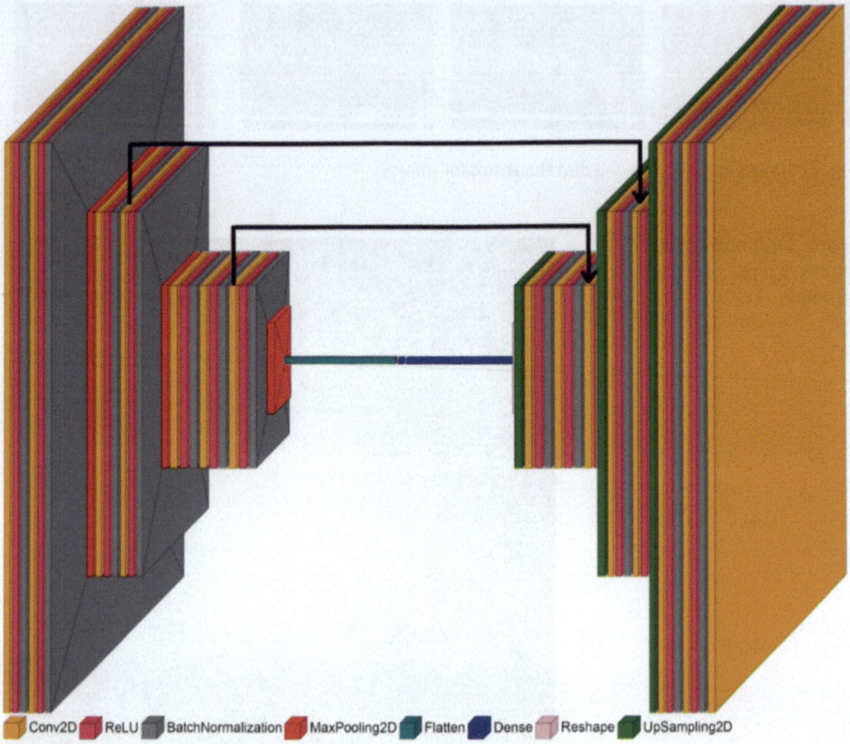

Fig. 4 BallTraj architecture

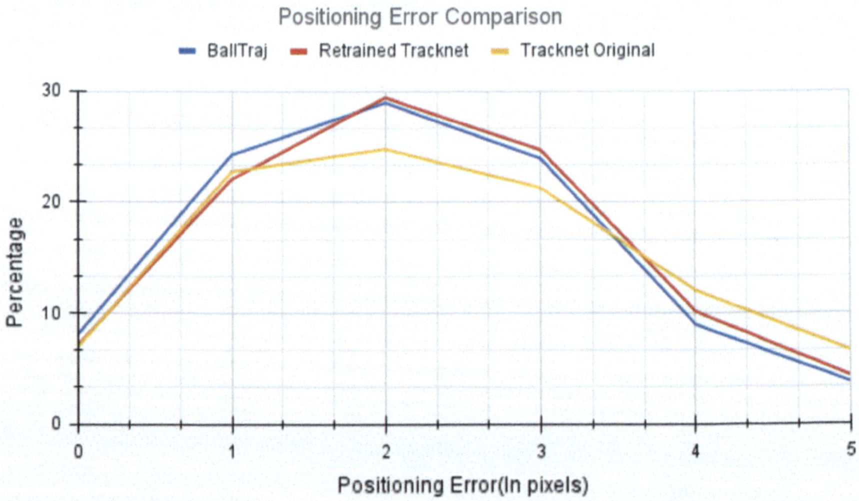

Fig. 5 Error distribution compared To current SOTA

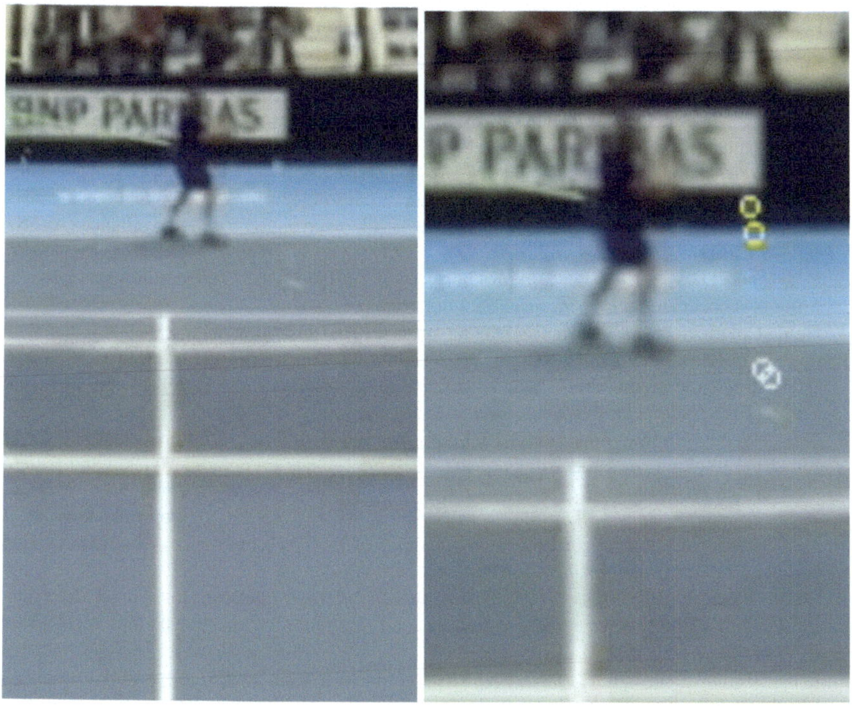

Fig. 6 Tracknet vs. BallTraj on a point between Ernests Gulbis and Mikael Ymer at a challenger event in Rennes, France in 2018, where BallTraj succeeds despite significantly worse camera quality compared to Fig. 1

8 Conclusion

The key findings of this study are as follows: firstly, the BallTraj deep learning model effectively detects and tracks small, high-speed tennis balls in videos, demonstrating remarkable accuracy and interpretability. Secondly, the heat map-based architecture, consisting of an encoder–decoder pair connected via a latent space, enables the model to accurately identify trajectory patterns of fast-moving objects. Thirdly, BallTraj can handle challenging scenarios such as ball lobbing for a smash, which poses difficulties for many existing models. Lastly, the results show that BallTraj has the potential to be a valuable tool for tennis professionals and coaches, as well as researchers in computer vision and sports analytics. The major limitation is the processing time of deep learning-based algorithms, an algorithm that can run in real time while preserving the accuracy would be extremely useful for such scenarios in the future.

References

1. Huang Y-C (April 2018) TrackNet: Tennis Ball Tracking from Broadcast Video by Deep Learning Networks," Master Thesis, advised by Tsì-Uí İk and Guan-Hua Huang. National Chiao Tung University, Taiwan
2. Huang, Yu-Chuan, I-No Liao, Ching-Hsuan Chen, Tsì-Uí İk, Wen-Chih Peng (2019) Tracknet: A deep learning network for tracking high-speed and tiny objects in sports applications. In: 2019 16th IEEE international conference on advanced video and signal based surveillance (AVSS), pp 1–8. IEEE
3. Borysenko D, Mykheievskyi D, Porokhonskyy V (2019) ODESA: object descriptor that is Smooth Appearance-wise for object tracking tasks
4. Held, David, Sebastian Thrun, Silvio Savarese (2016) Learning to track at 100 fps with deep regression networks. In: Computer Vision–ECCV 2016: 14th European Conference, Amsterdam, the Netherlands, October 11–14, 2016, Proceedings, Part I 14, pp 749–765. Springer International Publishing
5. Nam, Hyeonseob, Bohyung Han (2016) Learning multi-domain convolutional neural networks for visual tracking. In: Proceedings of the IEEE conference on computer vision and pattern recognition, pp 4293–4302
6. Redmon, Joseph, Santosh Divvala, Ross Girshick, Ali Farhadi (2016) You only look once: Unified, real-time object detection. In: Proceedings of the IEEE conference on computer vision and pattern recognition, pp 779–788
7. Dalal, Navneet, Bill Triggs (2005) Histograms of oriented gradients for human detection. In: 2005 IEEE computer society conference on computer vision and pattern recognition (CVPR'05), 1, pp 886–893. IEEE
8. Papageorgiou, Constantine P, Michael Oren, Tomaso Poggio (1998) A general framework for object detection. In: Sixth international conference on computer vision (IEEE Cat. No. 98CH36271), pp 555–562. IEEE
9. Ning, Guanghan, Zhi Zhang, Chen Huang, Xiaobo Ren, Haohong Wang, Canhui Cai, Zhihai He (2017) Spatially supervised recurrent convolutional neural networks for visual object tracking. In: 2017 IEEE international symposium on circuits and systems (ISCAS), pp 1–4. IEEE
10. Hochreiter S, Schmidhuber J (1997) Long short-term memory. Neural Comput 9(8):1735–1780
11. Liu, Wei, Dragomir Anguelov, Dumitru Erhan, Christian Szegedy, Scott Reed, Cheng-Yang Fu, Alexander C Berg (2016) Ssd: Single shot multibox detector. In Computer Vision–ECCV 2016: 14th European Conference, Amsterdam, The Netherlands, October 11–14, 2016, Proceedings, Part I 14, pp 21–37. Springer International Publishing
12. Simonyan, Karen, Andrew Zisserman (2014) Very deep convolutional networks for large-scale image recognition. *arXiv preprint* arXiv:1409.1556
13. Belagiannis, Vasileios, Andrew Zisserman (2017) Recurrent human pose estimation. In: 2017 12th IEEE international conference on automatic face & gesture recognition (FG 2017), pp 468–475. IEEE
14. Zeiler, Matthew D (2012) Adadelta: an adaptive learning rate method. *arXiv preprint* arXiv: 1212.5701
15. Salvi, Marco, Karthik Vaidyanathan (2014) Multi-layer alpha blending. In: Proceedings of the 18th Meeting of the ACM SIGGRAPH Symposium on Interactive 3D Graphics and Games, pp 151–158
16. He, Kaiming, Xiangyu Zhang, Shaoqing Ren, Jian Sun (2016) Deep residual learning for image recognition. In: Proceedings of the IEEE conference on computer vision and pattern recognition, pp 770–778
17. Hough Circle Method. https://docs.opencv.org/3.4/d4/d70/tutorial_hough_circle.html
18. TrackNet Dataset. https://nycu1-my.sharepoint.com/:u:/g/personal/tik_m365_nycu_edu_tw/ETCr6-M0e1VDhGCdMbvljcsBu31AJTO5xa_1cW8pHa7niA?e=55tLJ9
19. Archana M, Kalaisevi Geetha M (2015) Object detection and tracking based on trajectory in broadcast tennis video. Procedia Comput Sci 58: 225–232

Effect of Wind Direction with Surrounding Environment on Shuttlecock Wind Drift and Ventilation in the Multipurpose Indoor Stadium

S. M. Vignesh, G. Nallavan, and R. Ramakrishnan

Abstract Shuttlecock easily deviates from the actual path due to indoor stadium ventilation known as wind drift, which leads to enormous setbacks for elite players in the tournament. To understand the impact of environment and wind direction on wind drift and cross-ventilation, Computational Fluid Dynamics (CFD) analysis is performed in the urban environment. Tamil Nadu Physical Education and Sports University (TNPESU) Multipurpose stadium and campus are subjected to the CFD study and validated with field measurement data. The study is carried out for six wind resource configurations based on the wind rose and campus surrounding terrain. The velocity contour and area-average velocity are studied for the wind drift at the plane height of 1.5 m from the ground. The major stadium area has less wind drift, area-average velocity is between 0.7 and 1.7 m/s. The air change rate per hour (ACH) is measured to quantify the stadium ventilation, ranging from 1.6 to 2.6 h^{-1}, which is lower than the recommended rate. The ventilation is insufficient even though the TNPESU stadium is facilitated with lower wind drift.

Keywords Shuttlecock · Wind drift · Stadium · Cross-ventilation · Computational Fluid Dynamics (CFD) · Field measurement

1 Introduction

In stadium construction, Engineering plays a huge role in the art of design and comfort for the spectators and players. Indoor arenas or stadiums are widely preferred for sports sensitive to windy environments. In the case of Badminton, a shuttlecock is used instead of a ball, which is lightweight (4.5 g–6 g) and has a bluff body [1–3]. The shuttlecock tends to deviate from the actual path [4, 5] easily by the small gust of wind

S. M. Vignesh (✉) · G. Nallavan · R. Ramakrishnan
Department of Sports Technology, Tamil Nadu Physical Education and Sports University, Chennai, India
e-mail: smvignesh@gmail.com

S. Bhattacharyya and D. James (eds.), *Recent Trends in Sports Engineering*, Lecture Notes in Mechanical Engineering, https://doi.org/10.1007/978-981-97-8297-0_4

flow in the stadium, known as wind drifts, due to the natural ventilation system or air-conditioning (HVAC) [6–11]. Even though shutting down the ventilation system to avoid wind drift is not a solution, it leads to a lack of ventilation for the players and spectators. Since badminton is an endurance and stamina-demanding sport, a lack of ventilation will cause dehydration and muscle cramp injuries for the player [7]. Constructing the stadium with minimum wind drift in the court area and efficient ventilation is challenging. In this research paper, the indoor multipurpose stadium from Tamil Nadu Physical Education and Sports University (TNPESU) (see Figs. 1 and 2) is used for the case study to understand the impact direction of wind and surrounding environment efforts on wind drift and ventilation.

Fig. 1 The exterior view of the TNPESU multipurpose indoor stadium

Fig. 2 The interior view of the TNPESU multipurpose indoor stadium

In general, abundant research has been carried out in isolated a model with different configurations by varying roof type, opening location and size, opening shape, and louver using an experiment (wind tunnel or field study) and Computational Fluid Dynamics (CFD) [11–20]. Amsterdam stadium [21–23] is studied with a CFD and validated in field data for ventilation, pollution dissipation and pedestrian comfort. However, no potential study has been found for indoor stadium wind drift issues and ventilation [21–27]. This study uses the CFD to understand the indoor stadium's cross (natural)-ventilation, a coupled simulation between indoor and outdoor airflow simulation. For CFD simulation, the standard guidelines are followed, which is well discussed in one of our literature papers, Hoof et al. [21]. The significant uniqueness of this research is that CFD results are used to understand the influence of wind drift inside the badminton stadium concerning environmental conditions like wind speed and direction. Section 2 explains a CAD modeling of the TNPESU campus and multipurpose stadium interior with grid independence study and simulation setup descriptions. Section 3 compares the obtained simulation results of the qualified grid model case (second grid model) with a field study for validation. Further, the CFD simulation of different ventilation and wind drift cases is discussed. Section 4 discusses the conclusion from the simulation and validated results of the TNPESU multipurpose stadium.

2 CAD Modeling and Simulation

2.1 Stadium and Surrounding Environmental Modeling

The indoor multipurpose stadium is located inside the Tamil Nadu Physical Education and Sports University campus area of 125 acres at Melakottaiyur village, 50 km from Chennai city. The stadium is surrounded by the campus buildings (administrator building, academic building, hostel, staff quarters, canteen, and other sports infrastructures including another stadium, velodrome, and synthetic 400 m running track), Police Housing board, water bodies, and uneven rough landscape with colossal vegetation. For this study, the buildings are modeled within a radius of 1500 m inside the campus, where 4the landscape is planned while modeling. The building windows, balcony, interiors, and surrounding vegetation are not modeled because they contribute less impact on the airflow inside the stadium and will benefit during grid generation [21]. The indoor multipurpose stadium is modeled with a ventilation opening with doors and windows with interior envelopes. The gallery seating and other indoor infrastructures should be included, which have less interaction and are suitable for smooth grid generation. For CAD modeling, Google Maps (see Fig. 3) information, authorized campus model sketches, and field survey data are used to generate the cleaned and watertight 3D model (see Figs. 4 and 5) for the CFD analysis. The indoor multipurpose stadium is built in barrel roof configuration with an area of 1,862m^2, which holds more than 300 + spectators. The ventilation opening

Fig. 3 The Google Map view of the TNPESU Campus

Fig. 4 The CAD model of TNPESU Campus

is on the lateral side of the stadium, where the air exchange is from the northeast to the southwest direction.

The simulation is carried out in the hexagonal domain (see Fig. 6) with a side dimension of 1500 m and a height of 45 m from the ground with a blockage ratio of 3%. The hexagonal domain [28] is selected over the cuboidal domain, where the hexagonal domain gains the advantage of the change in wind direction without redevelopment in grid generation and demands less computational resources. In the hexagonal domain, boundary conditions are three inlet-velocity with Atmospheric Boundary Layer (ABL) conditions based on the roughness length of terrain in all

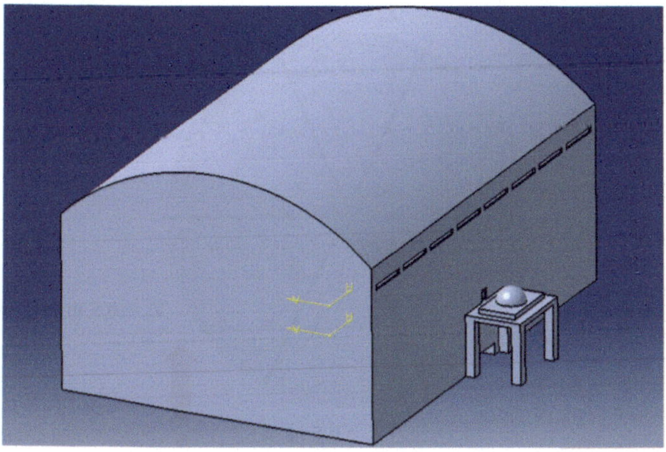

Fig. 5 The CAD model of TNPESU indoor stadium

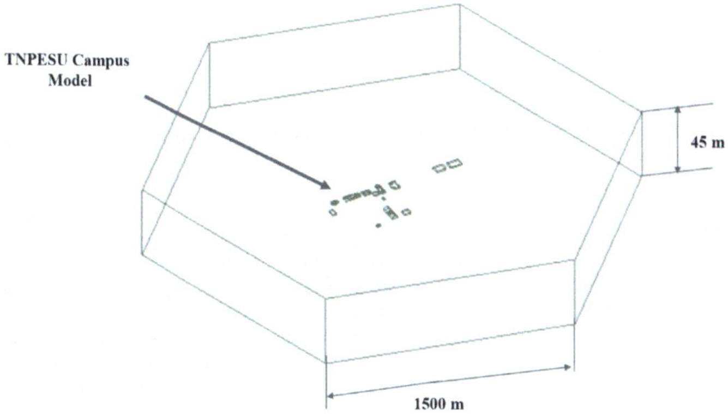

Fig. 6 Computational domain

directions (see Fig. 7); three pressure-outlet with zero atmospheric pressure conditions; ground and building as a wall and top (sky) of the domain is symmetry where the no-slip condition is applied [28].

The roughness length parameter for the Inlet-velocity boundary condition is used from the roughness length mapping carried out from all the directions of the computational domain at a 10 km radius. The roughness length will change the pattern of the ABL and the wind velocity for each direction is obtained from the wind rose data [22 and 29]. The annual wind resource data (see Fig. 8) and wind rose (see Fig. 9) of the Chengalpattu district are referred from Meteoblue, a meteorological service provided by the University of Basel, Switzerland. Based on that data, eight wind resource configurations are selected: 8 m/s in the south direction, 5.5 m/s in the south direction, 5.5 m/s in the north direction, 5.5 m/s in the east direction, 5.5 m/s

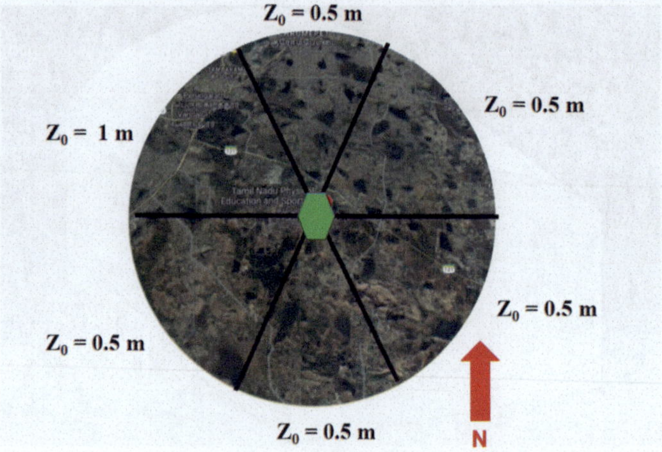

Fig. 7 Roughness length mapping

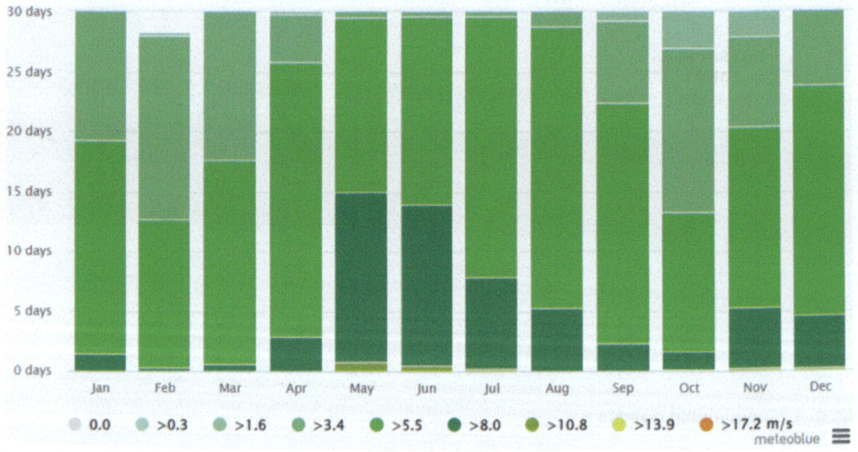

Fig. 8 Annual wind resource for Chengalpattu District, Tamil Nadu

in the east direction and 5.5 m/s in the southwest direction. The other configuration is omitted due to wind resources having low wind speeds, which will not provide a potential contribution to indoor stadium ventilation and wind drift.

The Fluent meshing and solver are used for CFD analysis; the poly-hex-core (see Figs. 10 and 11) is used for the grid generation for accurate results with less demanding computational power and time. The grid independence study carried out with three different numbers of cells generation model. In three cases, the number of cells in the first case is 5,65,852 cells, the second case is 7,04,408 cells, and the last case is 9,00,564 cells. The results and details of these three cases will be discussed later in Sect. 3. The building models and stadium interior maintain high

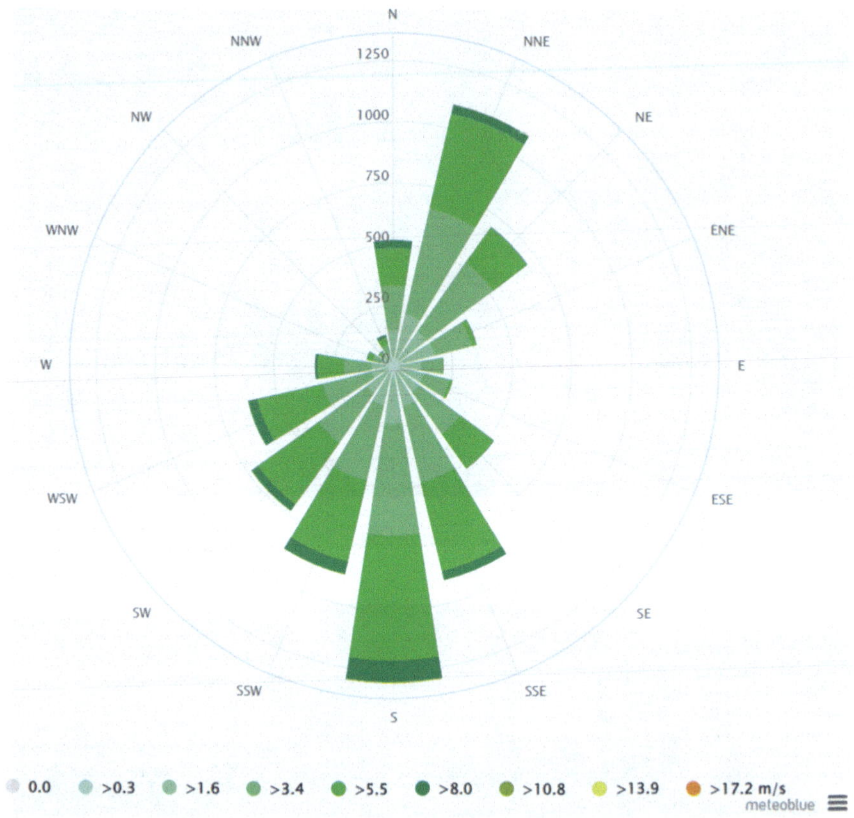

Fig. 9 Wind Rose for Chengalpattu District

grid quality to capture the fluid variable accurately. The 3D steady RANS—Reynolds Average Navier–Strokes is used with the Realizable k-epsilon model; the solution setup and turbulence model are chosen primarily based on Hoof et al. [21]. The SIMPLE algorithm activated with residuals of X, Y, and Z directional momentum is 10^{-7} and continuity, turbulence, kinetic energy, and epsilon is 10^{-5} [22 and 29]. This setup is used for all the simulation configurations and monitors until 8000 iterations to reach the convergence. The efficient grid model is selected for other configurations based on the grid independence study and field study validation results. The simulation is performed in the steady and isothermal assumption, where the temperature of the atmosphere and other metrological parameters are not considered to reduce the computational demand for the analysis. The study is performed only with the influence of the wind, with a factor of roughness with respect to direction. It agrees with the guidelines [21] of computational wind engineering, as our prime focus is on the influence of wind drift on ventilation through the natural environment (cross-ventilation).

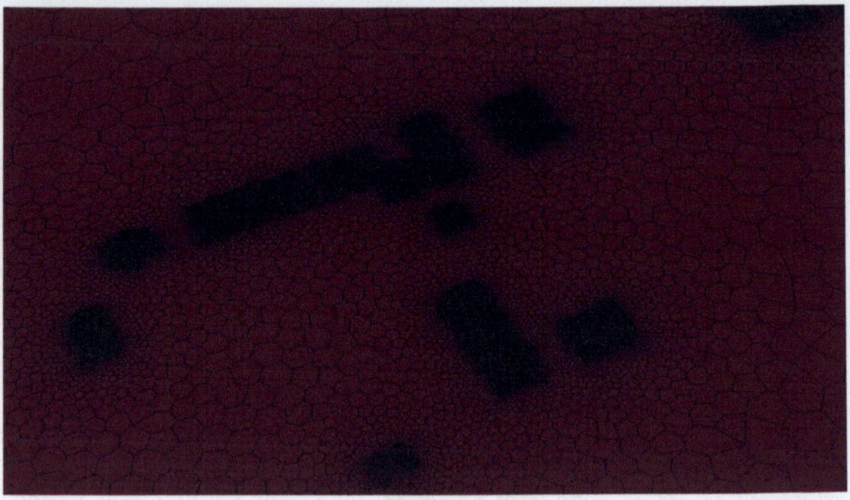

Fig. 10 Localized grid generation on urban environment

Fig. 11 Close view of grid generated stadium

3 CFD Validation and Results

3.1 Qualified Grid Model Case Validation with Field Data

The second case (7,04,408 cells) is a qualified grid model compared to the first and third cases; the selection is made by comparing the mass flow rate. The second case shows only a 0.8% deviation from the third grid model and 7.8% from the first grid model. The 8 m/s south direction wind resource is selected for the validation study. Selecting high wind velocity will be convenient to perform a field experiment for validation [21]. The experiment conducted with 8 m/s south direction wind between 23 and 25 June 2023. This period is selected because it is a dominant wind speed from the annual wind data. The digital anemometer with a vane that can turn to the direction of wind which is placed in the central entrance gate in the middle of the path at a height of 0.7 m. The average wind velocity (see Fig. 12) is measured in different time intervals (8.00–10.00, 11.00–13.00, 14.00–16.00) for 3 days between 23 and 25 June 2023. The maximum wind velocity per minute is recorded at every minute from the anemometer. The mean wind velocity is calculated from the data collection using Auto-regressive Integrated Moving Average (ARIMA) [29] time series analysis models. The obtained wind velocity from the statistical model is 4.3 m/s at the main gate during 8 m/s wind in the southern direction for 3 days. The experiment was conducted to compare (see Fig. 13) with CFD results with a second grid model of the

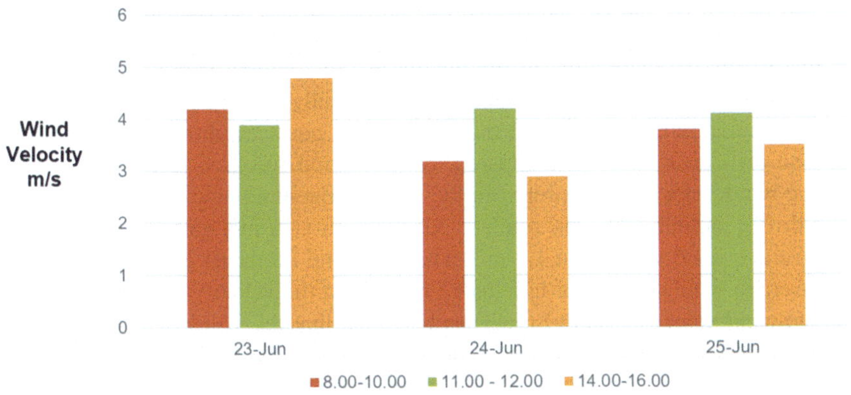

Fig. 12 Average wind velocity in main gate is recorded in anemometer for different time interval between 23 and 25 June

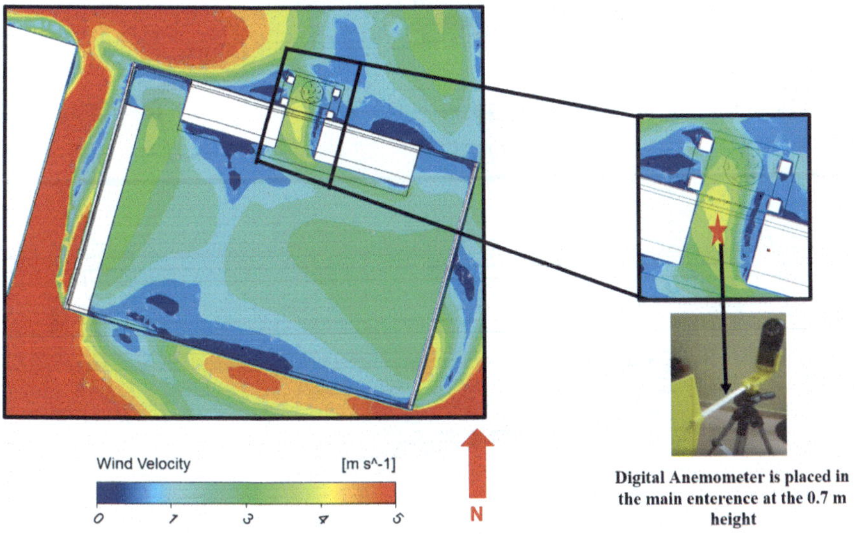

Fig. 13 The location comparison of CFD result and field measurement result in the main gate entry

same wind resource; the CFD result at the experiment location is 3.8 m/s, which is acceptable accuracy (12% error) with an experiment in urban environment analysis in CFD. After validation, the same gird model and solution setup are used to perform a numerical analysis of the other seven wind resource cases.

4 Results for Different Wind Direction

The velocity field (see Figs. 14, 15, 16, 17, 18, and 19) is measured in a horizontal plane 1.5 m above the ground. In all wind resources, the velocity is between 0 and 2.5 m/s for most of the area. In the middle of the southern side of the stadium, wind velocity is much higher in some cases, especially in southeast wind direction due to the main gate opening. In 5 m/s in north and northeast directions, gain more no wind drift area (V = 0) compared to other wind resource results. The area-average wind velocity is measured in the same horizontal plane to understand the wind drift quantitatively. The south wind direction at 8 m/s gains the maximum area-average wind velocity at 1.7 m/s. The north and northeast gain the minimum area-average wind velocity of less than 1 m/s. From the velocity contour and area-average wind velocity (Table 1), we can conclude that the TNPESU multipurpose stadium possesses less wind drift in the court area, where the minimum is 0.7 m/s and the maximum is 1.7 m/s. The wind drift is predicted to be lesser, and the influence

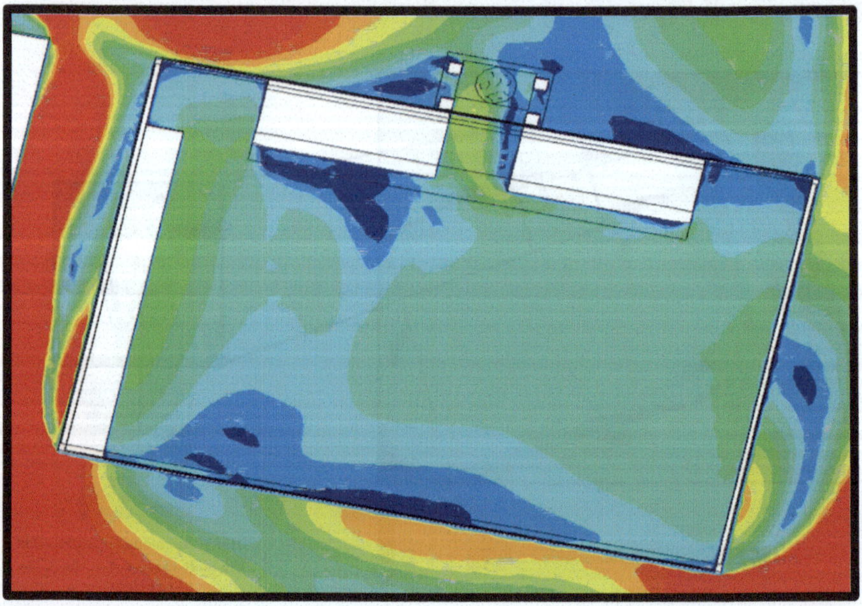

Fig. 14 The velocity contour is visualized at 1.5 m from the ground in the south wind direction at 8 m/s

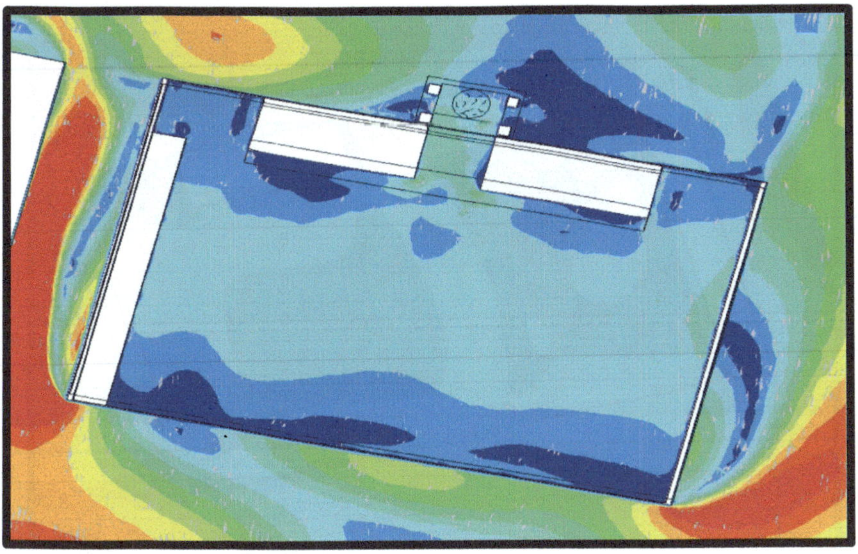

Fig. 15 The velocity contour is visualized at 1.5 m from the ground in the south wind direction at 5.5 m/s

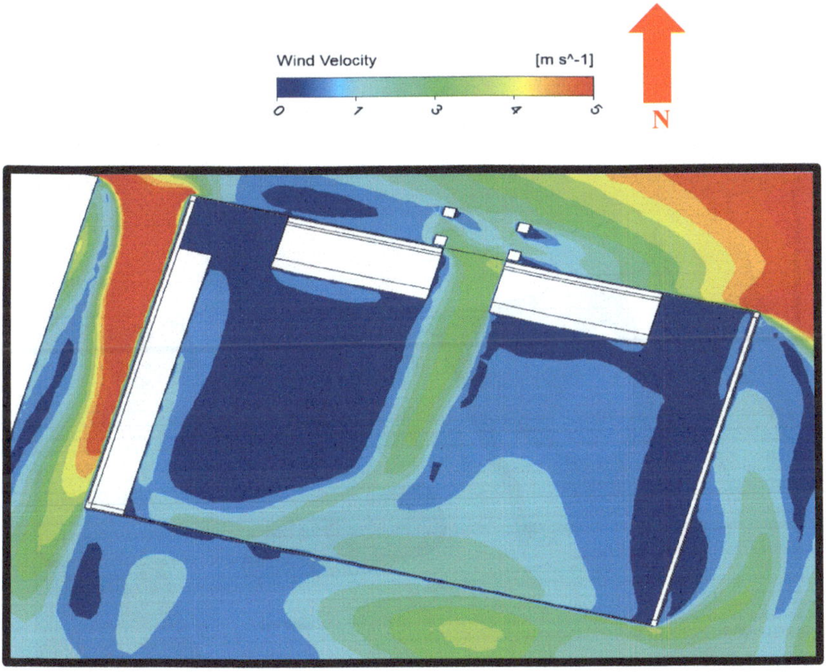

Fig. 16 The velocity contour is visualized at 1.5 m from the ground in the north wind direction at 5.5 m/s

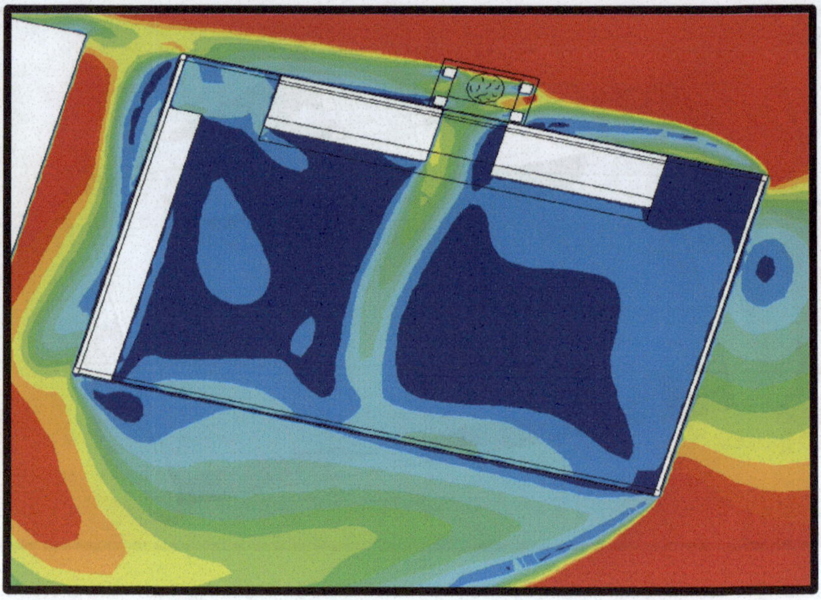

Fig. 17 The velocity contour is visualized at 1.5 m from the ground in the northeast wind direction at 5.5 m/s

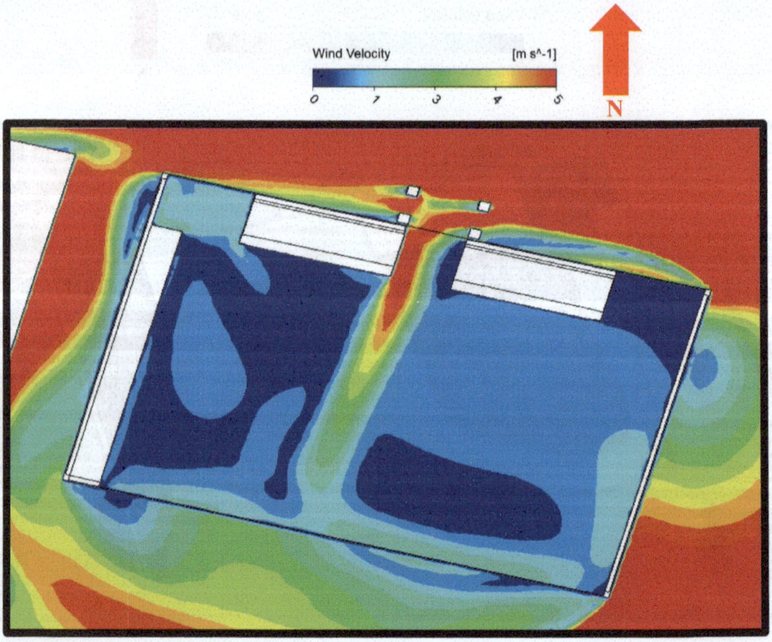

Fig. 18 The velocity contour is visualized at 1.5 m from the ground in the southeast wind direction at 5.5 m/s

over the shuttlecock during the tournament will be expected to be less with this wind velocity in the stadium.

The study's primary goal is to understand the ventilation of the stadium, not only the wind drift. Thc Air Change Rate per Hour (ACH) h^{-1} (see Fig. 20) is measured in all the selected wind resources, where north and northeast directions of 5.5 m/s gain the lowest among other wind resources equal to and less than 1.5 ACH. The other wind resource is near and more than a 2 h^{-1}; the ACH varies from 1.6 to 2.6 h^{-1}. The recommended ACH is between 3 and 5 h^{-1} for any commercial or

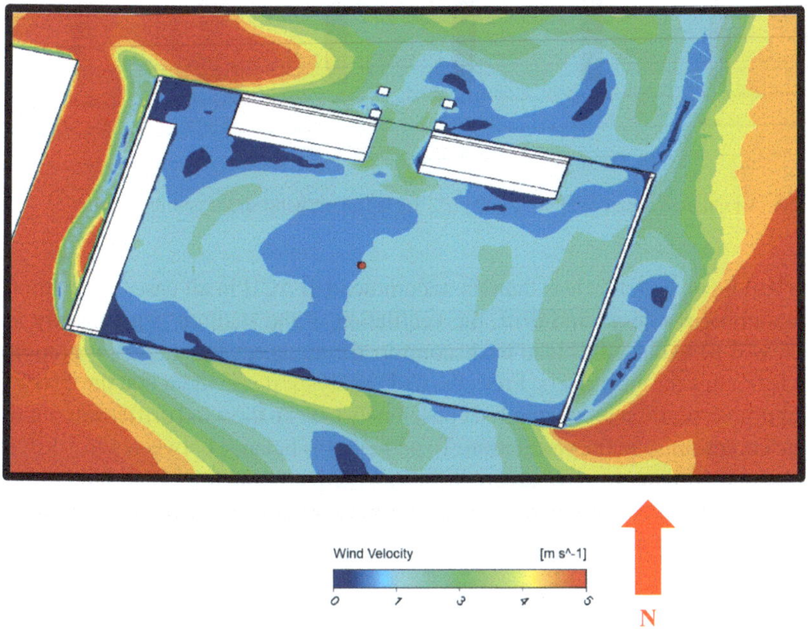

Fig. 19 The velocity contour is visualized at 1.5 m from the ground in the southwest wind direction at 5.5 m/s

Table 1 Area-average velocity is measured in the horizontal plane of velocity contour located at height 1.5 m from the ground

S.no	Wind velocity (m/s)	Wind direction (-)	Area-average velocity (m/s)
1	8	South	1.7
2	5.5	South	1.1
3	5.5	North	0.8
4	5.5	Northeast	0.7
5	5.5	Southeast	1.5
6	5.5	Southwest	1.2

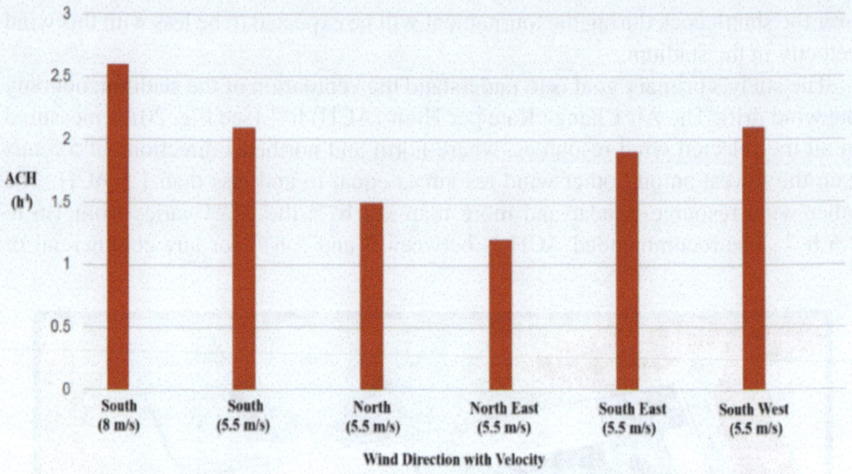

Fig. 20 Air Change Rate per Hour (ACH) h^{-1} for different wind speed and direction

large buildings, which is less than the recommended ACH in all cases. Even though the stadium gains less wind drift, the ventilation of the stadium is potentially less, which will be expected to lead to discomfort for the spectators and dehydration or muscle cramps for the players. From the CFD analysis, it is evident that the TNPESU multipurpose stadium is constructed consciously to keep the wind drift lower but with a huge compromise in the ventilation. To address this balance between wind drift and ventilation, it's vital to adopt an engineering practice through CFD and field testing for the design and construction of the stadium. Especially in the interest of reducing wind drift without compromising the ventilation for the badminton tournament.

5 Conclusion

The CFD analysis is conducted on the TNPESU indoor multipurpose stadium and surrounding environment. The main focus is understanding the stadium cross-ventilation and wind drift dominant in the court area. In CFD, both ventilation (ACH) and wind drift (velocity field) at the horizontal plane are 1.5 m horizontally from the ground is analyzed. The grid independen study is carried out in three different grid sizes; the second grid model (7,04,408 cells) is selected because of an optimized computational power and time demand. The selected grid size is used for all other wind configurations in the hexagonal domain. This domain is selected due to the advantage of changing boundary conditions without meshing again for every wind configuration. The field study is conducted with a digital anemometer at the main gate entrance for 3 days (23–25 June in three intervals) and the average velocity is obtained for the validation with CFD results. The results validate the CFD case (8 m/s wind velocity in the south direction) and obtain an accuracy of 88% with an experiment. The CFD process, like grid generation, turbulence model, and solution setup, are obtained from the literature and best practice guidelines for urban simulation. The inlet velocity boundary condition is Atmospheric Boundary Layer (ABL), where the roughness parameter and wind speed change based on the terrain and wind rose.

The following conclusion has been obtained from the velocity field, area-average velocity, and ACH results for the TNPESU multipurpose stadium ventilation and wind drift.

- The CFD tool could obtain realistic results for wind drift and ventilation. Compared with the experiment, the result obtained justifiable accuracy (88%) with the subjected assumption in CFD simulation.
- The wind drift is estimated to be between 0 and 2.5 m/s at the plane 1.5 m from the ground in most of the area inside the stadium. However, the high wind velocity is estimated mainly from the main gate opening to the inside of the stadium in the south wind configuration. Southern wind accelerates into the stadium through the main gate opening.
- The area-average velocity estimated a measurable wind drift in the stadium at the 1.5 m plane from the ground. The value is between 0.7 m/s and 1.7 m/s, implying that wind drifts in the stadium are lower and influence over the shuttlecock during the tournament will be negligible in most wind configurations.
- Wind velocity plays a role in ventilation and wind drift, but less effort in a slight change in direction. The 5 m/s for south, southeast, and southwest are almost similar area-average velocity and ACH, and the same is true for north and northeast.
- The north and northeast receive less ventilation (ACH) and have a significantly lower wind drift area (area-average velocity). This is due to the laboratory building and other high infrastructure on the north side of the campus blocking the airflow into the stadium.
- The south, southwest, and southeast directions are well-dominated in ventilation (ACH) due to less campus infrastructure in the southern direction.
- The experiment is performed with one anemometer in a single location. To understand more about CFD accuracy, experiments need to be carried out in multiple locations.
- Even though the CFD results are validated with field experiments, the analysis is carried out in isothermal and steady-state conditions. In the future, the above studies need to be carried out in transient and with energy model to overcome these limitations.

References

1. Badminton World Federation, Kuala Lumpur, Malaysia, https://bwfbadminton.com/, last accessed 2023/11/21

2. World Badminton Community: Badminton updates and information collected by the worldwide Badminton Community, http://www.worldbadminton.com/, last accessed 2023/11/21
3. International Badminton Federation Rules (IBF), Complete version of the IBF rules, https://badmintonottawa.com/basicRules.htm, last accessed 2023/11/21
4. Lin CSH, Chua CK, Yel JH (2015) Analysis and simulation of badminton shuttlecock flight through parameter identification of a slow-speed serve shot. Proc Inst Mech Engin, Part P: J Sports Engin Technol 229(4),. Sage Journals
5. Badmintons Best, Can Badminton be Played Outdoors?, https://www.badmintonsbest.com/blog/can-badminton-be-played-outdoors, last accessed 2023/11/21.
6. Bangkok Post, Air-con controversy not cool for Badminton, 15 Aug 2013, https://www.bangkokpost.com/world/364758/air-con-controversy-not-cool-for-badminton, last accessed 2023/11/21
7. South China Post, Air-conditioning row an uncool distraction at badminton world championships, 16 Aug 2013, https://www.scmp.com/sport/china/article/1296952/air-conditioning-row-uncool-distraction-badminton-world-championships, last accessed 2023/11/21
8. Business Standard, No evidence of badminton aircon cheating: official, 26 Sept 2014, https://www.business-standard.com/article/pti-stories/no-evidence-of-badminton-aircon-cheating-official-114092601361_1.html, last accessed 2023/11/21
9. USA Today, Get the drift? Badminton players face indoor winds in Rio, 13 Aug 2016, https://www.usatoday.com/story/sports/olympics/2016/08/13/get-the-drift-badminton-players-face-indoor-winds-in-rio/88672886/, last accessed 2023/11/21
10. The Star, Air conditioning 'drifts' pose problems for Olympic badminton players, 13 Aug 2016, https://www.thestar.com/sports/olympics/2016/08/13/air-conditioning-drifts-pose-problems-for-olympic-badminton-players.html?rf, last accessed 2023/11/21
11. Sport Star, Wind a worry for P.V. Sindhu, 18 Aug 2018, https://sportstar.thehindu.com/asian-games-2018/interviews/wind-a-worry-for-pv-sindhu-badminton-asian-games-2018/article24724992.ece, last accessed 2023/11/21
12. Ramponi R, Blocken B (2012) CFD Simulation of cross-ventilation for a generic isolated building: impact of computational parameters. In: Building and environment, vol 53. Science Direct, pp 34–48
13. Ramponi R, Blocken B (2012) CFD simulation of cross-ventilation flow for different isolated building configurations: Validation with wind tunnel measurements and analysis of physical and numerical diffusion effects. J Wind Engin Indust Aerodyn 104–106:408–418. Science Direct
14. Tominagaa Y, Blocken B (2015) Wind tunnel experiments on cross-ventilation flow of a generic building with contaminant dispersion in unsheltered and sheltered conditions. In: Building and environment, vol 92. Science Direct, pp 452–46
15. Peren JI, Hooff TV, Leite BCC, Blocken B (2015) CFD analysis of cross-ventilation of a generic isolated building with asymmetric opening positions: Impact of roof angle and opening location. In: Building and environment, vol 85. Science Direct, pp 263–276
16. Perén JI, Hooff TV, Leite BCC, Blocken B (2015) Impact of eaves on cross-ventilation of a generic isolated leeward sawtooth roof building: windward eaves, leeward eaves and eaves inclination. In: Building and environment, vol 92. Science Direct, pp 578–590
17. Tominaga Y, Blocken B (2016) Wind tunnel analysis of flow and dispersion in cross-ventilated isolated buildings: Impact of opening positions. J Wind Engin Indust Aerodyn 155:74–88. Science Direct
18. Kosutova K, Hooff TV, Vanderwel C, Blocken B, Hensen J (2016) Cross-ventilation in a generic isolated building equipped with louvers: Wind-tunnel experiments and CFD simulations. In: Building and environment, vol 154. Science Direct, pp 263–280
19. Pabiou H, Salort J, Ménézo C, Chillà F (2015) Natural cross-ventilation of buildings, an experimental study. In: 6th international building physics conference, IBPC on energy procedia, vol 78. Science Direct, pp 2911–2916
20. Hooff TV, Blocken B, Aanen L, Bronsema B (2011) A venturi-shaped roof for wind-induced natural ventilation of buildings: Wind tunnel and CFD evaluation of different design configurations. In: Building and environment, vol 46. Science Direct, pp 1797–807

21. Blocken B, Hooff TV, Aanen L, Bronsema B (2011) Computational analysis of the performance of a venture-shaped roof for natural ventilation: venturi-effect versus wind-blocking effect. Comput Fluids 48(1):202–213. Science Direct
22. Hooff TV, Blocken B (2010) Coupled urban wind flow and indoor natural ventilation modeling on a high-resolution grid: a case study for the Amsterdam Arena stadium. Environ Model Softw 25(1):51–65. Science Direct
23. Hooff TV, Blocken B (2010) On the effect of wind direction and urban surroundings on natural ventilation of a large semi-enclosed stadium. Comput Fluids 39(7):1146–1155. Science Direct
24. Hooff TV, Blocken B (2012) Full-scale measurements of indoor environmental conditions and natural ventilation in a large semi-enclosed stadium: possibilities and limitations for CFD validation. J Wind Engin Indust Aerodyn 104–106:330–34. Science Direct
25. Culley MSP, Pascoe J (2015) Computerised design of stadiums. In: Stadium and arena design, 2nd institution of civil engineers (ICE). Emerald Publishing, London, pp 119–132
26. Ghani S, ElBialy EA, Bakochristou F, Gamaledin SMA, Rashwan MM, Hughes B (2017) Thermal performance of stadium's Field of Play in hot climates. Energy Build 139:702–718. Science Direct
27. Khalil E, El-Bialy EM, Abdel-Maksoud W, Me A (2017) Flow patterns in an air-conditioned playground zone of football stadium. J Appl Mech Engin 6(4), Walsh Medical Media
28. Sofotasiou P (2017) Aerodynamic optimization of sports stadiums towards wind comfort. Ph.D. Thesis, University of Sheffield, Sheffield, UK, (2017). http://etheses.whiterose.ac.uk/id/eprint/17820 , last accessed 2023/11/21
29. Toparlar Y, Blocken B, Vos P, Heijst GJFV, Janssen WD, Hooff TV, Montazeri H, Timmermans HJP (2015) CFD simulation and validation of urban microclimate: A case study for Bergpolder Zuid, Rotterdam. In: Building and environment, vol 83. Science Direct, pp 79–90
30. Rizkya I, Syahputri K, Sari RM, Siregar I, Utaminingrum J (2019) Autoregressive integrated moving average (ARIMA) model of forecast demand in distribution centre. In: Annual conference on industrial and system engineering (ACISE) on IOP conference series: materials science and engineering, vol 598. IOP Publishing, pp 012071

Comparison of Body Composition and Cardiorespiratory Functional Differences Among Track Athlete

Chennakeshavulu Vemula, Aishwarya Andugula, Mekam Maheshwar, and Challa Suresh

Abstract

Aim and Objectives The aim of the study was to describe anthropometric and body composition characteristics, body composition in track athletes (short-, middle- and long-distance athletes), and compare these variables with cardiorespiratory functional differences among short, middle, and endurance runners which will enable them in finding out the body composition and also the cardiorespiratory functional differences (https://pubmed.ncbi.nlm.nih.gov/31968680/).

Methodology A total of 50 state and national athletes aged 18–28 years grouped as sprinters, middle- and long-distance runners. The athletes who were healthy, having at least 3 years of practice and willing to participate were included and only male athletes were included in the study. Total of 20 variables were analyzed which includes age, height, weight, body composition, physiological parameters, field tests, and recovery parameters. Data procurement was done with the help of Bioelectrical Impedance Analysis (BIA), Cardiopulmonary Exercise Testing (CPET), 12 min Cooper test, medicine ball throw, and hop test (right and left legs) (https://pubmed.ncbi.nlm.nih.gov/31968680/).

Results One-way ANOVA was applied to for the following variables and the data was statistically analyzed using SPSS 20.0 software version between the groups and the results were obtained.

Discussion There was a significant difference observed in SMM %, VO2 max, RF max, HRR, Hop test (right & left legs), and RHR at the end of 3 min, whereas all other variables were insignificant (https://pubmed.ncbi.nlm.nih.gov/30677531/).

C. Vemula (✉) · A. Andugula · M. Maheshwar
Department of Sports Sciences, Extension and Training Division, ICMR—National Institute of Nutrition, MYAS, Tarnaka, Hyderabad 500007, India
e-mail: chennalaxmi@gmail.com

C. Suresh
Department of Cell Biology, ICMR—National Institute of Nutrition, Tarnaka, Hyderabad 500007, India

S. Bhattacharyya and D. James (eds.), *Recent Trends in Sports Engineering*, Lecture Notes in Mechanical Engineering, https://doi.org/10.1007/978-981-97-8297-0_5

Conclusion The study showed that few parameters of body composition, cardiorespiratory parameters are significant and the higher the fat percentage the lesser is the VO2 max and higher the muscle mass higher is the VO2 max (Greene et al. in J Sci Med Sport 7:373–383, 2004).

Keywords Dehydration effects · Physiological performance · Cognitive changes · Exercise intensity

1 Introduction

The history of exercise and sports dates back to 100 years and the oldest definition of exercise was found from Charaka Samhita which has been percolated from the world's oldest record of medicine practice [4]. Physio kinetics is the study of how exercise alters the structure and function of the body. It is basically the study of how the body adapts physiologically to the acute or short-term stress of exercise, and the chronic or long-term stress of physical training. Harre defined sports training, based on scientific knowledge, is a pedagogical process of sports perfection through which systematic effect of psycho-physical performance ability and performance readiness aims at leading the sportsman to high and the highest performance. An athlete's success depends on the combination of ability, tactical, technical, physical fitness, and psychological individualities [5]. It is a well-known fact that athletes practicing in different sports have different body dimensions reflecting both body proportionality and body composition track athletes are unique population which requires assessment within themselves and to correlate this assessment to their performance [6]. Track athletes include short-, middle- and long-distance runners and huddlers. Some events may require higher body mass compared to other events and also the body composition varies in these athletes. [7]. The higher the muscle mass in the lower body, the higher the benefit, especially for track athletes [8]. In endurance and ultra-endurance athletes, a variety of different anthropometric characteristics (body mass, body height, and skeletal muscle mass) and physiological variables (maximum oxygen uptake, anaerobic threshold, ventilatory threshold, and lactate threshold) have been identified as important predictor variables for the performance of an athlete [9]. Anthropometric measurements and bioelectrical impedance analysis play a vital role in the estimation of body composition. The easy access to the subcutaneous fat layer and the non-invasive nature of the skin-fold measurement have led to application of these in various equations. On the other hand, assessment of body composition using bioelectrical impedance analysis is based on the electrical properties of the tissues. Here fat acts as an insulator that stops the passage of the current through the tissues. [10] BIA technique is used to predict body composition based, on the electrically conductive properties of the body and involves measuring of the impedance (Z) to the flow of a low electrical current at a fixed frequency. The BIA device can be of single frequency, when it operates at a frequency of 50 kHz or multi-frequency, when a wide range of frequencies are used. It measures body composition swiftly and conveniently in a non-invasive manner.

BIA technique has been recommended as a useful tool to analyze body composition based upon the close, direct relationship between impedance and total body water. BIA is a total body water estimation tool that is based upon the role of electrolytes as conductors. The impedance of adipose tissue id relatively high due to the limitation of body fluid in that type of tissue. In contrast, it is reported that fat-free mass, which includes everything that is not body fat, has a relatively lower impedance. Therefore, BIA is used as a method to assess both the fat mass and the fat-free mass. The principle of BIA is that lean tissue (LT) consisting of water and electrolytes, is a good conductor, while fat, which does not have water, is a poor conductor of electricity. H. C. Lukaski et al. Cardiorespiratory parameters were obtained using spirometer equipment which includes forced vital capacity, maximum volume of ventilation were obtained.

The aim of the study was to check the anthropometric characters like size, shape, proportion, body composition, and cardiorespiratory characters such as heart rate, stroke volume, breath frequency, vital capacity in short-, middle-, and long-distance runners and standardize all these parameters for talent identification and nurture.

2 Research Methodology

Journals, articles, books,, and various other international scholar social networking sites served as the literature survey repository of the study. In order to include all the prior studies and to analyze the work carried out in this particular field the literatures were taken as early as 1995 and were reviewed for the study. The key terms were identified prior to searching to cover all the articles relevant to the aim of the study The following keywords identified for the literature search include anthropometry, body composition, cardiorespiratory parameters, VO_2 max, and track athletes being keenly interested in sports and sports performance and after a thorough review of literature the topic was selected where body composition and cardiorespiratory fitness plays a vital role in performance of an athlete, especially in track athletes and very few studies were available on Indian athletes considering one or two parameters of both body composition and cardiorespiratory fitness. Ethical clearance was obtained from the ICMR-NIN Scientific Research Committee. A total of 50 male state and national athletes were included in the study. Depending on the statement of the problem variables were identified which are anthropometric measurements (height, weight, heart rate, blood pressure, and SpO2), body composition parameters (minerals, SMM%, percentage of body fat, SMM in trunk and lower limbs, and visceral fat area), cardiorespiratory fitness parameters (MVV, VO2 max, respiratory frequency at rest and maximum exercise conditions, and heart rate at rest and maximum exercise condition), field test parameters (medicine ball throw, jump test (right & left legs)), recover parameters RHR (1& 3 min), and recovery SpO2 (1 & 3 min). Then the samples were chosen from various sports academies in and around

Hyderabad. The inclusion criteria for the participants to be involved in this study were the athletes practicing at least for 3 years, only male athletes were included, and healthy and non-injured athletes were included in the study. The test process was explained to the athletes and the written consent form was obtained from the athletes. The samples were chosen randomly from the sports academies. Laboratory and field tests were performed on athletes as base line assessment. Consent was taken before laboratory and field based assessments. After the written consent submission, anthropometric measurements such as height, weight were obtained by using stadiometer and weighing scale (Seca) respectively, earth rate, SPO_2 were measured with portable pulse oximetry, blood pressure was measured with automated sphygmomanometer, and body composition parameters were obtained using Bioelectrical Impedance Analysis (InBody770) with all necessary conditions to follow before performing BIA.. The athletes were instructed about the test and the procedure of the BIA. It was performed by making the athlete stand on the equipment holding electrodes in hands and foot placed on markings present on the equipment which passes electric current through the body. There are different range of voltage of current passing through the body which is very minimal. Here the cells act as absorbent of water and fat acts as an insulator where the absorption of water is less by fat. This test takes around 30 s of time and the results are obtained immediately and can be checked in the computer attached to the InBody 770 equipment. Cardiorespiratory parameters like FVC, MVV, heartrate, recovery SpO2, and VO_2 max, were obtained using spirometer and CPET (cardiopulmonary exercise testing) before the estimation of these parameters the athletes were instructed about the pre-test conditions required for the test and were detailed about the procedure of performing the test. The athletes were being fitted by the mask, nose piece, and also the heart rate monitors and were made comfortable and adjustable to the laboratory environment. Then the athletes were asked to breathe in and out normally for few respirations and then they were instructed to take a deep breath as much as possible and were instructed to exhale immediately as much as they can exhale and the results were obtained. This procedure is repeated thrice for obtaining accurate results. For estimating VO_2 max in field, 12-min Cooper test was performed by the athletes and the results were noted down and for estimating VO2 max in laboratory conditions cardiopulmonary exercise testing was done on athletes using the NIN Protocol (standardized with 20 m shuttle run test) by using a treadmill which is connected to a computer where the results are documented by breath by breath indirect calorimetry. This data was arranged according to the format required for analyzing statistically the data is divided into different groups like short-, middle-, and long-distance runners and also according to the parameters like anthropometric, body composition, respiratory parameters, and field tests. This data is statistically analyzed using SPSS software 20.0 and analysis of variance was applied to the data and the results were obtained. VO_2 max of 12 min of Cooper test was obtained using VO2 max = (22.351X distance covered in kilometer)—11.288.

The temperature during the tests in an external environment ranged from 20 to 28 °C and the humidity was between 55 and 65%.

In laboratory, the temperature was maintained at about 18 °C.

3 Result

From the Tables 1 and 2, it was observed that there is a significant difference observed in SMM %, VO2 max, RF max, HRR, Hop test (right & left legs), and RHR at the end of 3 min, whereas all other variables were insignificant.

4 Discussion

The study was aimed to evaluate the body composition variables, to compare these variables with the cardiorespiratory fitness among track athletes, to understand the variance of body composition, and to obtain pulmonary lung function parameters, VO_2 max of track athletes. Initially, it was hypothesized that there would be a significant difference in body composition in track athletes. It was assumed that there would be a significant difference in cardiorespiratory parameters in track athletes. It was correlated that there will be a significant difference in performance abilities in track athletes later the finding observed from the study were there was a significant difference observed in SMM %, VO_2 max, RF max, HRR, and RHR at the end of 3 min, whereas all other variables were insignificant. As it was hypothesized that there will be a significant difference in body composition the significant difference was observed in SMM % as a whole muscle mass of the whole body where was no difference observed in SMM Trunk, lower limbs. In cardiorespiratory parameters, VO_2 max (both laboratory and field tests) was significant and rest of the parameters such as respiratory frequency, maximum volume of ventilation, forced vital capacity, and heartrate at rest and maximum exercise conditions did not show any significant differences, whereas, in field tests such as medicine ball throw, hop test (right & left legs), and 12 min Cooper running/jogging/walking tests, there was a significant difference observed. When recovery status was checked in athletes in laboratory after cardiopulmonary exercise testing, there was a significant difference observed in recovery SpO2 checked at the end of first minute and recovery heart rate at the end of third minute significant difference was observed, whereas recovery SpO2 checked at the end of third minute and recovery heart rate observed at the end of first minute did not show any significant difference. These differences were observed because of the wide range of differences observed in body composition observed in these athletes as these sports demand different body types such as endurance sports such as long-distance runners (5 and 10 kms) need an essential amount of fat stores, whereas short-distance running such as 100, 200, and 400 m running and hurdles require a body type where there are adequate stores of glucose in the body. Middle-distance runners which comprises running between 800 and 3000 m where there is a use of mixed energy systems the body composition is different. The high percentage of body fat in long-distance runners is one of the reasons for significant differences observed in cardio respiration.

Table 1 Depicts the values of mean and standard deviation of short-distance, middle-distance, and long-distance runners and also the total of these athletes

Age	Short distance	26	20.77	4.033	0.791	19.14	22.40	15	29
Height	Middle distance	12	21.83	3.927	1.134	19.34	24.33	17	28
	Long distance	12	24.67	1.826	0.527	23.51	25.83	22	28
	Total	50	21.96	3.881	0.549	20.86	23.06	15	29
	Short distance	26	172.2000	5.29898	1.03922	170.0597	174.3403	161.70	182.00
	Age	Short distance	26	20.77	4.033	0.791	19.14	22.40	15
	Long distance	Middle distance	12	21.83	3.927	1.134	19.34	24.33	17
	Total	Long distance	12	24.67	1.826	0.527	23.51	25.83	22
Weight	Short distance	26	63.5462	8.07486	1.58361	60.2846	66.8077	50.00	80.00
	Middle distance	12	60.9500	8.21434	2.37128	55.7309	66.1691	51.60	82.50
	Long distance	12	60.4167	6.63117	1.91425	56.2034	64.6299	54.20	79.70
	Total	50	62.1720	7.77227	1.09916	59.9631	64.3809	50.00	82.50
Minerals	Short distance	26	3.7458	0.50192	0.09843	3.5430	3.9485	2.92	4.79
	Middle distance	12	3.4950	0.43829	0.12652	3.2165	3.7735	2.79	4.56

(continued)

Table 1 (continued)

	Long distance	12	3.5075	0.48625	0.14037	3.1986	3.8164	2.92	4.93
	Total	50	3.6284	0.48988	0.06928	3.4892	3.7676	2.79	4.93
SMMPercentage	Short distance	26	47.8200	2.17484	0.42652	46.9416	48.6984	42.70	52.40
	Middle distance	12	46.4100	1.78104	0.51414	45.2784	47.5416	43.10	48.60
	Long distance	12	46.1967	2.56780	0.74126	44.5652	47.8282	42.20	50.20
	Total	50	47.0920	2.27968	0.32240	46.4441	47.7399	42.20	52.40
BFP	Short distance	26	15.2500	3.86732	0.75844	13.6880	16.8120	8.50	23.90
	Middle distance	12	17.2000	2.90047	0.83729	15.3571	19.0429	13.40	21.90
	Long distance	12	17.3250	3.95408	1.14145	14.8127	19.8373	10.80	23.60
	Total	50	16.2160	3.75001	0.53033	15.1503	17.2817	8.50	23.90
SMMTrunk	Short distance	26	23.2385	2.63182	0.51614	22.1754	24.3015	19.00	28.30
	Middle distance	12	21.8333	2.25160	0.64998	20.4027	23.2639	17.90	26.50
	Long distance	12	21.4750	2.61816	0.75580	19.8115	23.1385	18.80	28.70
	Total	50	22.4780	2.62027	0.37056	21.7333	23.2227	17.90	28.70

(continued)

Table 1 (continued)

SMMLowerlimbs	Short distance	26	17.0108	1.98727	0.38974	16.2081	17.8134	13.33	21.24
	Middle distance	12	16.2192	2.30343	0.66494	14.7556	17.6827	11.65	20.72
	Long distance	12	16.2350	2.59679	0.74963	14.5851	17.8849	13.83	23.01
	Total	50	16.6346	2.20823	0.31229	16.0070	17.2622	11.65	23.01
VisceralFatArea	Short distance	23	36.2565	16.15782	3.36914	29.2694	43.2437	12.40	65.30
	Middle distance	12	38.5500	14.73983	4.25502	29.1848	47.9152	16.70	72.50
	Long distance	12	38.2833	11.88160	3.42992	30.7341	45.8325	19.80	59.70
	Total	47	37.3596	14.55251	2.12270	33.0868	41.6324	12.40	72.50
FVC	Short distance	26	3.9835	0.73486	0.14412	3.6866	4.2803	2.07	5.38
	Middle distance	12	3.9225	0.52206	0.15071	3.5908	4.2542	3.32	5.24
	Long distance	12	4.3525	0.70354	0.20309	3.9055	4.7995	3.43	5.61
	Total	50	4.0574	0.69029	0.09762	3.8612	4.2536	2.07	5.61
MVV	Short distance	26	132.1231	20.24801	3.97096	123.9447	140.3014	79.80	173.10
	Middle distance	12	134.8417	14.12596	4.07781	125.8665	143.8169	119.80	169.90

(continued)

Table 1 (continued)

	Long distance	12	133.0750	28.84000	8.32539	114.7509	151.3991	66.80	174.00
	Total	50	133.0040	21.02209	2.97297	127.0296	138.9784	66.80	174.00
VO2Max	Short distance	26	48.6346	4.66793	0.91546	46.7492	50.5200	41.80	59.90
	Middle distance	12	49.7417	6.30994	1.82152	45.7325	53.7508	42.20	59.90
	Long distance	12	60.2750	3.80553	1.09856	57.8571	62.6929	55.30	66.00
	Total	50	51.6940	6.87306	0.97200	49.7407	53.6473	41.80	66.00
RFRest	Short distance	26	24.7654	4.72321	0.92630	22.8576	26.6731	16.40	34.90
	Middle distance	12	25.4500	3.76937	1.08812	23.0551	27.8449	18.30	31.30
	Long distance	12	24.1417	4.70386	1.35789	21.1530	27.1304	16.30	31.90
	Total	50	24.7800	4.44394	0.62847	23.5170	26.0430	16.30	34.90
RFMax	Short distance	26	53.6346	15.23861	2.98854	47.4796	59.7896	27.10	88.20
	Middle distance	12	62.3750	11.28483	3.25765	55.2050	69.5450	40.30	75.00
	Long distance	12	71.7250	13.32913	3.84779	63.2561	80.1939	52.20	92.30
	Total	50	60.0740	15.60424	2.20677	55.6393	64.5087	27.10	92.30

(continued)

Table 1 (continued)

HeartRateRest	Short distance	26	65.3462	4.59515	0.90118	63.4901	67.2022	57.00	74.00
	Middle distance	12	60.6667	5.22813	1.50923	57.3449	63.9885	55.00	72.00
	Long distance	12	67.5000	5.66488	1.63531	63.9007	71.0993	54.00	76.00
	Total	50	64.7400	5.49883	0.77765	63.1772	66.3028	54.00	76.00
HeartRateMax	Short distance	26	190.9615	13.79415	2.70525	185.3900	196.5331	160.00	214.00
	Middle distance	12	192.2500	14.85154	4.28727	182.8138	201.6862	162.00	214.00
	Long distance	12	185.2500	18.93110	5.46494	173.2218	197.2782	143.00	208.00
	Total	50	189.9000	15.30673	2.16470	185.5499	194.2501	143.00	214.00
MedicineBallThrow	Short distance	26	10.2508	2.30147	0.45136	9.3212	11.1804	7.10	14.40
	Middle distance	12	9.1317	0.90770	0.26203	8.5549	9.7084	7.54	10.91
	Long distance	12	8.3167	1.04018	0.30027	7.6558	8.9776	6.70	9.90
	Total	50	9.5180	1.95069	0.27587	8.9636	10.0724	6.70	14.40
JumpTestRightLeg	Short distance	26	12.8912	1.31618	0.25812	12.3595	13.4228	10.95	16.10
	Middle distance	12	11.2383	1.62712	0.46971	10.2045	12.2722	8.19	13.93

(continued)

Table 1 (continued)

	Long distance	12	10.9458	1.38227	0.39903	10.0676	11.8241	9.00	13.25
	Total	50	12.0276	1.65588	0.23418	11.5570	12.4982	8.19	16.10
JumpTestLeftLeg	Short distance	26	12.7415	1.31223	0.25735	12.2115	13.2716	10.20	15.00
	Middle distance	12	11.5850	1.48703	0.42927	10.6402	12.5298	9.10	14.10
	Long distance	12	10.7292	1.48116	0.42758	9.7881	11.6703	8.50	13.25
	Total	50	11.9810	1.61138	0.22788	11.5231	12.4389	8.50	15.00
CooperTest	Short distance	26	2628.6538	283.36040	55.57155	2514.2021	2743.1056	2000.00	3100.00
	Middle distance	12	2903.3333	375.31401	108.34382	2664.8702	3141.7965	2400.00	3790.00
	Long distance	12	3439.1667	362.62198	104.67995	3208.7677	3669.5657	2980.00	4050.00
	Total	50	2889.1000	460.70453	65.15346	2758.1692	3020.0308	2000.00	4050.00
VO2Maxx	Short distance	26	46.9423	6.44411	1.26379	44.3395	49.5451	33.40	58.00
	Middle distance	12	53.3000	8.61035	2.48559	47.8292	58.7708	42.30	73.60
	Long distance	12	65.4000	7.88958	2.27753	60.3872	70.4128	55.70	78.10
	Total	50	52.8980	10.43752	1.47609	49.9317	55.8643	33.40	78.10

(continued)

Table 1 (continued)

RecoveryHeartRate1min	Short distance	26	179.1154	18.35500	3.59971	171.7016	186.5291	141.00	211.00
	Middle distance	12	177.2500	26.82985	7.74511	160.2031	194.2969	113.00	211.00
	Long distance	12	177.8333	17.43994	5.03448	166.7525	188.9141	132.00	194.00
	Total	50	178.3600	20.06089	2.83704	172.6588	184.0612	113.00	211.00
RecoverySpo21min	Short distance	26	93.8077	2.05950	0.40390	92.9758	94.6395	90.00	98.00
	Middle distance	12	91.9167	2.67848	0.77321	90.2148	93.6185	87.00	97.00
	Long distance	12	89.7500	7.67671	2.21607	84.8725	94.6275	67.00	96.00
	Total	50	92.3800	4.45334	0.62980	91.1144	93.6456	67.00	98.00
RecoveryHeartRate3Min	Short distance	26	133.1923	13.93419	2.73272	127.5642	138.8204	112.00	166.00
	Middle distance	12	133.3333	10.06645	2.90593	126.9374	139.7292	120.00	157.00
	Long distance	12	120.2500	9.20598	2.65754	114.4008	126.0992	108.00	139.00
	Total	50	130.1200	13.12365	1.85596	126.3903	133.8497	108.00	166.00
RecoverySpo23Min	Short distance	26	96.4615	1.63048	0.31976	95.8030	97.1201	91.00	98.00
	Middle distance	12	96.5833	0.79296	0.22891	96.0795	97.0872	95.00	98.00

(continued)

Table 1 (continued)

	Long distance	12	97.0833	1.08362	0.31282	96.3948	97.7718	95.00	98.00
	Total	50	96.6400	1.35164	0.19115	96.2559	97.0241	91.00	98.00

Table 2 Depicts the significance and F values of short-, middle-, and long-distance athletes

		Sum of squares	Df	Mean square	F	Sig.
Age	Between groups	124.971	2	62.486	4.791	0.013
	Within groups	612.949	47	13.041		
	Total	737.920	49			
Height	Between groups	19.997	2	9.998	0.298	0.744
	Within groups	1577.183	47	33.557		
	Total	1597.180	49			
Weight	Between groups	103.990	2	51.995	0.856	0.432
	Within groups	2856.011	47	60.766		
	Total	2960.001	49			
Minerals	Between groups	0.747	2	0.374	1.594	0.214
	Within groups	11.012	47	0.234		
	Total	11.759	49			
SMMPercentAge	Between groups	28.981	2	14.490	3.018	0.058
	Within groups	225.671	47	4.802		
	Total	254.651	49			
BFP	Between groups	50.640	2	25.320	1.864	0.166
	Within groups	638.428	47	13.584		
	Total	689.067	49			
SMMTrunk	Between groups	32.095	2	16.048	2.478	0.095
	Within groups	304.331	47	6.475		
	Total	336.426	49			
SMMLowerlimbs	Between groups	7.666	2	3.833	0.779	0.465
	Within groups	231.271	47	4.921		
	Total	238.938	49			

(continued)

Table 2 (continued)

		Sum of squares	Df	Mean square	F	Sig.
VisceralFatArea	Between groups	55.230	2	27.615	0.125	0.882
	Within groups	9686.443	44	220.146		
	Total	9741.673	46			
FVC	Between groups	1.406	2	0.703	1.505	0.232
	Within groups	21.943	47	0.467		
	Total	23.349	49			
MVV	Between groups	60.761	2	30.381	0.066	0.936
	Within groups	21,593.718	47	459.441		
	Total	21,654.479	49			
VO2M	Between groups	1172.698	2	586.349	24.131	0.000
	Within groups	1142.011	47	24.298		
	Total	2314.708	49			
RFRest	Between groups	10.282	2	5.141	0.252	0.778
	Within groups	957.398	47	20.370		
	Total	967.680	49			
RFMax	Between groups	2770.592	2	1385.296	7.108	0.002
	Within groups	9160.524	47	194.905		
	Total	11,931.116	49			
HeartRateRest	Between groups	300.069	2	150.034	5.968	0.005
	Within groups	1181.551	47	25.139		
	Total	1481.620	49			
HeartRateMax	Between groups	355.038	2	177.519	0.750	0.478
	Within groups	11,125.462	47	236.712		
	Total	11,480.500	49			

(continued)

Table 2 (continued)

		Sum of squares	Df	Mean square	F	Sig.
MedicineBallThrow	Between groups	33.070	2	16.535	5.067	0.010
	Within groups	153.384	47	3.263		
	Total	186.455	49			
JumpTestRightLeg	Between groups	40.907	2	20.453	10.287	0.000
	Within groups	93.449	47	1.988		
	Total	134.355	49			
JumpTestLeftLeg	Between groups	35.726	2	17.863	9.175	0.000
	Within groups	91.505	47	1.947		
	Total	127.230	49			
CooperTest	Between groups	5,396,948.282	2	2,698,474.141	25.349	0.000
	Within groups	5,003,236.218	47	106,451.834		
	Total	10,400,184.500	49			
VO2Maxx	Between groups	2799.766	2	1399.883	25.920	0.000
	Within groups	2538.383	47	54.008		
	Total	5338.150	49			
RecoveryHeartRate1min	Between groups	32.949	2	16.475	0.039	0.961
	Within groups	19,686.571	47	418.863		
	Total	19,719.520	49			
RecoverySpo21min	Between groups	138.575	2	69.287	3.908	0.027
	Within groups	833.205	47	17.728		
	Total	971.780	49			
RecoveryHeartRate3Min	Between groups	1538.325	2	769.162	5.238	0.009
	Within Groups	6900.955	47	146.829		
	Total	8439.280	49			

(continued)

Table 2 (continued)

		Sum of squares	Df	Mean square	F	Sig.
RecoverySpo23Min	Between groups	3.225	2	1.613	0.878	0.422
	Within groups	86.295	47	1.836		
	Total	89.520	49			

Track event is a sport that includes athletic contests based on running. Track and field are categorized under the umbrella sport of athletics, which also includes road running, cross-country running, and race walking. Scientific evaluation in terms of health and sports performance helps in aiding the overall excellence of the athlete by providing insights on various body-related parameters such as physique, body composition, biomechanics, physiology, etc. This study totally revolves around where track and field athletes stand in terms of event-specific health as well as performance scientific parameters and how the body composition is affecting the cardiorespiratory parameters of the athletes. This work involved a sample collection of 50 ($n = 50$) active track athletes like sprinters, middle- and long-distance runners. Anthropometric measurements such as height and weight were measured using Seca GmbH & Co manufactured equipment, body composition was measured using InBody 770 equipment which is the most advanced equipment available which uses impedance for the measurement and has the ability to measure body mass, fat and fat-free mass, protein content, mineral content, segmental lean analysis, and visceral fat area. Lung function capacity was measured using spirometer where the athlete was given a filter and their nose was closed using a nose clip and they were asked to perform the test with the help of the mouth. VO2 max was measured using Cosmed CPET equipment which uses the modified NIN protocol on treadmill. Later, recovery status was checked using pulse oximeter. It was concluded that in body composition mineral content, skeletal muscle mass (trunk and lower limbs) percentage of body fat and visceral fat were significant, whereas skeletal muscle mass as a whole was not significant. In cardiorespiratory parameters, VO2 max, respiratory frequency (maximum exhaustion condition), and heart rate (rest) were significant, whereas forced vital capacity, maximum volume of ventilation, respiratory frequency (rest), and heart rate (maximum exhaustion condition) were not significant. In field tests, all the parameters were significant. In recovery status, recovery SpO2 checked at the end of first minute and recovery pulse checked at the end of third minute were significant, whereas recovery HR checked at the end of first minute and recovery SpO2 checked at the end of third minute were not significant. The final conclusion of the work is to perform regular sports-specific scientific evaluation for individualized training plan and nutritional plan to achieve optimum performance.

5 Limitations

(1) The study includes only male athletes and female athletes were excluded from the study.
(2) Due to time limitation, the study was conducted in one particular city and the sample size was not according.
(3) There was no uniformity in athletic count from the track events.
(4) As the study includes only male athletes, we could not compare or see the difference in fat percentage and cardiorespiratory functional differences among these groups.
(5) The study was conducted on different days, different times, and in different environmental conditions (both laboratory- and field-based tests) where a slight difference in temperature and humidity was observed.
(6) These might lead to slight variations in the results of the athletes.

6 Conclusion

Track event is a sport that includes athletic contests based on running. Track and field are categorized under the umbrella sport of athletics, which also includes road running, cross-country running, and race walking. Scientific evaluation in terms of health and sports performance helps in aiding the overall excellence of the athlete by providing insights on various body-related parameters such as physique, body composition, biomechanics, physiology, etc. This study totally revolves around where track and field athletes stand in terms of event specific health as well as performance scientific parameters and how the body composition is affecting the cardiorespiratory parameters of the athletes. This work involved a sample collection of 50 (n = 50) active track athletes like sprinters, middle- and long-distance runners. Anthropometric measurements such as height and weight were measured using Seca GmbH & Co manufactured equipment, body composition was measured using InBody 770 equipment which is the most advanced equipment available which uses impedance for the measurement and it has the ability to measure body mass, fat and fat-free mass, protein content, mineral content, segmental lean analysis, and visceral fat area. Lung function capacity was measured using spirometer where the athlete was given a filter and their nose was closed using a nose clip and they were asked to perform the test with the help of the mouth. VO_2 max was measured using Cosmed CPET equipment which uses the modified NIN protocol on treadmill. Later, recovery status was checked using pulse oximeter. It was concluded that in body composition mineral content, skeletal muscle mass (trunk and lower limbs), percentage of body fat and visceral fat were significant, whereas skeletal muscle mass as a whole was not significant. In cardiorespiratory parameters VO_2 max, respiratory frequency (maximum exhaustion condition), and heart rate (rest) were significant, whereas forced vital capacity, maximum volume of ventilation, respiratory frequency (rest), and heart rate (maximum exhaustion condition) were not significant. In field tests, all the parameters were significant. In recovery status, recovery SpO2 checked at the end of first minute and recovery pulse checked at the end of third minute were significant, whereas recovery HR checked at the end of first minute and recovery SpO_2 checked at the end of third minute were not significant. The final conclusion of the work is to perform regular sports-specific scientific evaluation for individualized training plan and nutritional plan to achieve optimum performance.

7 Future Research

(1) The study should include both male and female athletes.
(2) The study should be conducted on a large sample.
(3) The study should include athletes from all over the nation so that a comparison can be done between these athletes and if there is any difference in the results due to the change in climatic conditions.
(4) A comparison should be between male and female athletes regarding fat percentage and cardiorespiratory functional differences.
(5) The above-mentioned study was a hybrid model (where both laboratory- and field-based tests were performed) only anyone of these tests can be performed and can be compared to check the variations in the results of the athletes.

8 A***bbreviations

BFP	Body Fat Percentage
BIA	Bioelectrical Impedance Analysis
CPET	Cardiopulmonary Exercise Testing
FVC	Forced Vital Capacity
MVV	Maximum Volume of Ventilation
RF REST	Respiratory Frequency in Rest Conditions.
RF MAX	Respiratory Frequency at Maximum Intensity of Exercise.
SMM	Skeletal Muscle Mass
SMM trunk	Skeletal Muscle Mass in trunk region
SMM lower limbs	Skeletal Muscle Mass in lower limbs.
VO2 max	Maximum Consumption of oxygen at highest intensity exercise.

References

1. https://pubmed.ncbi.nlm.nih.gov/31968680/ (abstract)
2. https://pubmed.ncbi.nlm.nih.gov/30677531/

3. Greene DA, Naughton GA, Briody JN, Kemp A, Woodhead H, Farpour-Lambert N (2004) Musculoskeletal health in elite male adolescent middle-distance runners. J Sci Med Sport 7(3):373–383. https://doi.org/10.1016/s1440-2440(04)80032-x
4. Dengel DR, Keller KA, Stanforth PR, Oliver JM, Carbuhn A, Bosch TA (2020) Body composition and bone mineral density of Division 1 Collegiate Track and field athletes, a consortium of college athlete research (C-CAR) study. J Clin Densitometry: Off J Int Soc Clin Densitometry 23(2):303–313. https://doi.org/10.1016/j.jocd.2019.07.008
5. Akbar S, Soh KG, Jazaily Mohd Nasiruddin N, Bashir M, Cao S, Soh KL (2022) Effects of neuromuscular training on athletes physical fitness in sports: a systematic review. Front Physiol 13:939042. https://doi.org/10.3389/fphys.2022.939042
6. Dengel DR, Keller KA, Stanforth PR, Oliver JM, Carbuhn A, Bosch TA (2020) Body composition and bone mineral density of Division 1 Collegiate Track and field athletes, a consortium of college athlete research (C-CAR) study. J Clin Densitometry: Off J Int Soc Clin Densitometry 23(2):303–313. https://doi.org/10.1016/j.jocd.2019.07.008
7. Nevill A, Brown D, Godfrey R, Johnson P, Romer L, Stewart AD, Winter EM (2003) Modeling maximum oxygen uptake of elite endurance athletes. Med Sci Sports Exerc 35(3):488–494. https://doi.org/10.1249/01.MSS.0000053728.12929.5D
8. Tesch PA, Karlsson J (1985) Muscle fiber types and size in trained and untrained muscles of elite athletes. J Appl Physiol (Bethesda, Md: 1985) 59(6):1716–1720. https://doi.org/10.1152/jappl.1985.59.6.1716
9. Knechtle B (2014) Relationship of anthropometric and training characteristics with race performance in endurance and ultra-endurance athletes. Asian J Sports Med 5(2):73–90
10. Bernal-Orozco MF, Posada-Falomir M, Quiñónez-Gastélum CM, Plascencia-Aguilera LP, Arana-Nuño JR, Badillo-Camacho N, Márquez-Sandoval F, Holway FE, Vizmanos-Lamotte B (2020) Anthropometric and body composition profile of young professional soccer players. J Strength Cond Res 34(7):1911–1923. https://doi.org/10.1519/JSC.0000000000003416
11. Hirsch KR, Smith-Ryan AE, Trexler ET, Roelofs EJ (2016) Body composition and muscle characteristics of Division I track and field athletes. J Strength Cond Res 30(5):1231–1238. https://doi.org/10.1519/JSC.0000000000001203[11]
12. Khalil SF, Mohktar MS, Ibrahim F (2014) The theory and fundamentals of bioimpedance analysis in clinical status monitoring and diagnosis of diseases. Sensors (Basel) 14(6):10895–10928. https://doi.org/10.3390/s140610895. PMID: 24949644; PMCID: PMC4118362
13. Wells JC, Fewtrell MS (2006) Measuring body composition. Arch Dis Child 91(7):612–617. https://doi.org/10.1136/adc.2005.085522. PMID: 16790722; PMCID: PMC2082845
14. Achamrah N, Colange G, Delay J, Rimbert A, Folope V, Petit A, Grigioni S, Déchelotte P, Coëffier M (2018) Comparison of body composition assessment by DXA and BIA according to the body mass index: a retrospective study on 3655 measures. PLoS One 13(7):e0200465. https://doi.org/10.1371/journal.pone.0200465
15. Bartels EM, Sørensen ER, Harrison AP (2015) Multi-frequency bioimpedance in human muscle assessment. Physiol Rep 3(4):e12354. https://doi.org/10.14814/phy2.12354.PMID:25896978; PMCID:PMC4425960

Effect of Training on Selected Anthropometric and Physiological Parameters Among Swimming Athletes of Junior Age Category (<16 Yrs)

Chennakeshavulu Vemula, Advika Manya Mandala, M. Maheshwar, and C. Suresh

Abstract Swimming is an internationally recognized and popular sport, swimming has evolved over the years with the inclusion of real-time analytics, body analysis, and a lot of cell- and body-level research around the world, but particularly in India, the amount of research on it is very minimal and with the popularity for the sport raising, it was apt for me to start research on it. To evaluate and compare physiological and body composition parameters among different ages and genders during one mesocycle in swimmers. The study involved a sample collection from 60 (n = 60) active national swimmers in the under 16 age categories of which 33 were male and 27 were female. Anthropometric measurements, body composition, and VO2 max parameters were analyzed by using universally accepted and used equipment for the purpose of tests such as anthropometry, body composition, lung capacity, etc., on athletes. Anthropometrical measurements are the basic measurements such as height, weight, BP, SPO2, etc. Body composition tests are conducted using the InBody 770. Spirometry test is used to conduct forced vital capacity. Post spirometry test a breath-by-breath analysis test (CPET) is conducted for VO2 max. The sample was further classified on the basis of their age and gender and a t-test was used for two independent samples in order to compare the classified groups. The statistical analysis was done with the help of software named IBM SPSS Statistics 20. From the study, we were able to conclude that parameters such as anthropometry, body composition, and lung functions play a vital role in the performance of an athlete depending on their age, gender, and training protocol, also working on these parameters would surely help them improve their performance.

Keywords CPET · Swimming · SPO_2 · Performance

C. Vemula · A. M. Mandala (✉) · M. Maheshwar · C. Suresh
MYAS-NIN Department of Sports Science, ICMR-NIN, ET Division Tarnaka, Hyderabad 500007, India
e-mail: manyaadvikamandala14@gmail.com

S. Bhattacharyya and D. James (eds.), *Recent Trends in Sports Engineering*, Lecture Notes in Mechanical Engineering, https://doi.org/10.1007/978-981-97-8297-0_6

1 Introduction

Swimming is an internationally recognized and popular sport, swimming has evolved over the years with the inclusion of real-time analytics, body analysis, and a lot of cell- and body-level research around the world, but particularly in India, the amount of research on it is very minimal and with the popularity for the sport raising, it was apt for investigators to start research on it. Scientific evaluation in terms of health and sports performance helps in aiding the overall excellence of the athlete by providing insights into various body-related parameters such as physique, body composition, biomechanics, physiology, etc. The major study involved in scientific assessment is the study of adaptations to exercise, training, lifestyle, etc. The study totally revolves around how the body reacts to sports and physical activities from anthropometry to whole-body functional levels in Indian national swimmers [11].

1.1 Aim

To evaluate and compare physiological and body composition parameters among different ages and genders during one mesocycle of training in national-level swimmers.

1.2 Objectives

- To determine the anthropometry of swimming athletes.
- The following anthropometric parameters were considered for the study as they were observed to parameters that could result in large variance between samples.

Body weight.

Standing height.

Sitting height.

Arm span.

Right-hand grip.

Left-hand grip.

Skeletal muscle mass %

Body fat %

Fvc

VO2 Max

- To understand the variance of body composition.
- To know physiological and pulmonary lung function (FVC) levels.

2 Methadology

See Fig. 1.

2.1 Selection Criteria

Swimming requires a unique combination of strength, flexibility, and body proportions, which can provide valuable insights into how these physical characteristics influence performance in the Water Selecting swimming athletes under the age of 16 for a study of anthropometry can provide insights into how growth and development during adolescence impact physical characteristics relevant to swimming performance. This age group allows us to study the effects of puberty on body composition, limb length, and other anthropometric factors that may influence swimming ability [9].

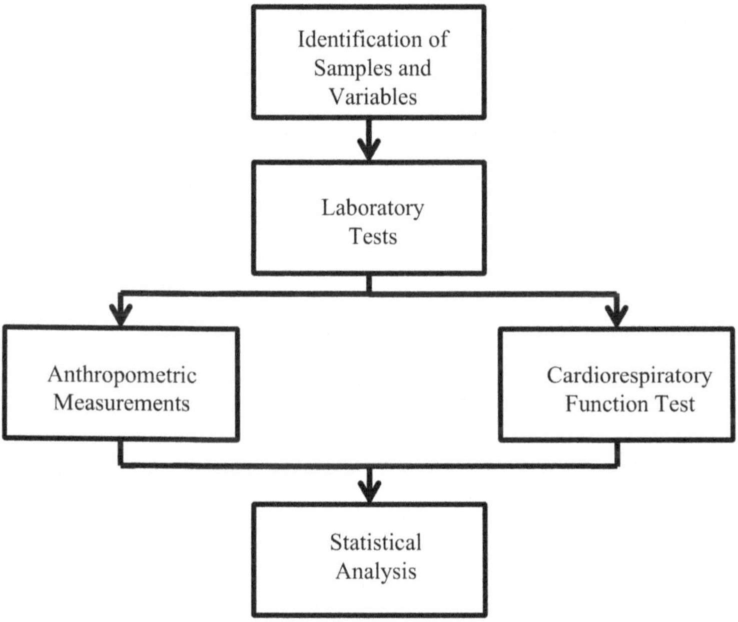

Fig. 1 Representation of flow of study

2.2 *Identification of Samples and Variables*

The athletes of the age group less than 16 years were selected from across different clubs. The basis of selection included that they have been into the sport for a minimum of 2 years and have represented at various national or international level competitions.

The study involved a sample collection from 60 (n = 60) active national-level swimming athletes of which 33 were male and 27 were female.

2.3 *Variables*

Dependent Variable:

- Anthropometric measurements
- Body composition
- VO_2 max, FVC.

Independent Variable

- General and swim-specific training and lifestyle.

3 Laboratory Tests

- Laboratory tests such as BCA, InBody, and CPET were conducted twice on the complete sample and the average of both the tests was considered for further statistical analysis.
- Universally accepted and used equipment for the purpose of tests such as anthropometry, body composition, lung capacity, etc., on athletes.

3.1 *Anthropometrical Measurements*

- Anthropometrical measurements are the basic measurements such as height, weight, etc., are conducted and these tests are conducted using SECA GmbH & Co manufactured equipment.

3.2 *Body Composition*

- Body composition tests are conducted using the InBody 770, the equipment has the ability to measure fat mass, muscle mass, and body water levels within a minute. The equipment mainly uses impedance for measurement purpose.

3.3 Cardiorespiratory Function Test

- Spirometry test is used to conduct the lung capacity test. In this test, the spirometer measures the amount of air that an athlete can execute forced breath in and exhale with force after 2–3 normal breath. Post spirometry test a breath-by-breath VO2 max analysis test (CPET) was conducted based on the modified Bruce treadmill graded exercise test protocol. BP, SPO2, and grip strength were also performed on swimmers [1].

4 Statistics Analysis

- Descriptive statistics
- Sample size: 60
- Sample participants: Athletes involved in the same sport(swimming)
- classified on the basis of their age and gender
- A t-test was conducted to determine if there is a significant difference between the means of two groups (male and female).
- Statistical Analysis: T-test was used for two independent samples in order to compare the classified groups.
- Mean, standard deviation, and p-value were calculated.
- **Statistical Software**: IBM SPSS Statistics 20.0

IBM SPSS statistics 20.0 is a widely used software package for studies involving Health Sciences, using this particular software package enables us to clearly understand the difference in value (p-value) among the groups, also since it is enhanced for health sciences the software provides more accurate values.

5 Result

The category involves athletes below 16 yrs of age. This category is again split into male and female sub-classes. In this group, it was noted that anthropometrical measurements were not significant, since this is a growth phase for both male and female. Girls undergo puberty earlier than boys by about 1–2 years, and generally finish the stages of puberty quicker than males due to their differences in biology. Body composition such as body fat and skeletal muscle girls tend to show higher body fat than males and females, the reason for the higher VO2 max in males is due to the higher muscle mass as visible in the above graph as the p-value is closer to 1 the deviation is higher thereby suggesting that the parameter is not significant on the other hand parameters having p values closer to 0 are significant and are to be considered for effective performance (Fig. 2 and Table 1).

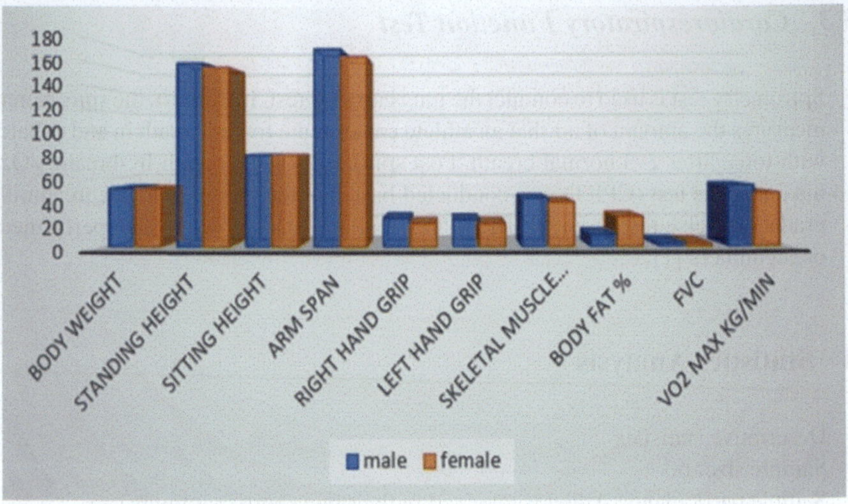

Fig. 2 Swimmers < 16 years gender category

Table 1 Swimmers < 16 years gender category

Male (n = 33)			Female (n = 27)		
	Mean	Std. deviation	Mean	Std. deviation	P Value
Body weight	51.55	10.01	51.15	9.04	0.874
Standing height	162.12	10.54	158.15	7.51	0.105
Sitting height	81.18	5.28	81.11	3.70	0.953
Arm span	174.58	12.76	167.78	8.25	0.020
Right-hand grip	25.39	8.27	20.89	5.57	0.018
Left-hand grip	24.27	8.42	20.63	5.77	0.061
Skeletal muscle mass %	43.16	4.87	39.63	3.16	0.002
Body fat %	10.67	4.59	26.41	5.67	0.000
FVC	3.42	0.83	2.89	0.69	0.010
VO2 Max kg/min	53.91	6.92	46.96	4.95	0.000

6 Conclusion

From the study, we were able to conclude that parameters such as anthropometry, body composition, and lung functions play a vital role in the performance of an athlete depending on their age, gender, and training protocol. Through the statistical analysis, it is very evident that there are a few anthropometric parameters such as skeletal muscle mass percentage, body fat %, FVC, VO2 max, etc., which are vital and need to be focused on to improve the overall performance of the athletes.

References

1. Caspersen CJ, Powell KE, Christenson GM (1985) Physical activity, exercise, and physical fitness: definitions and distinctions for health-related research. Public Health Rep 100(2):126–131. PMID: 3920711; PMCID: PMC1424733
2. Kasper Korey MD (2019) Sports training principles. Current Sports Med Rep 18(4):95–96. https://doi.org/10.1249/JSR.0000000000000576
3. Britannica T (2022) Editors of Encyclopaedia. Swimming. Encyclopedia Britannica. https://www.britannica.com/sports/swimming-sport
4. https://gaisf.sport/members/international-swimming-federation/. https://olympics.com/en/news/the-history-of-olympic-swimming
5. Rozi G, Thanopoulos V, Geladas N, Soultanaki E, Dopsaj M, Anthropometric characteristics and physiological responses of high level swimmers and performance in 100 m freestyle swimming. https://www.researchgate.net/publication/325338467_Anthropometric_characteristics_and_physiological_responses_of_high_level_swimmers_and_performance_in_100_m_freestyle_swimming
6. Lazovic-Popovic B, Zlatkovic-Svenda M, Durmic T, Djelic M, Djordjevic Saranovic S, Zugic V (2016) Superior lung capacity in swimmers. https://doi.org/10.1016/j.rppnen.2015.11.003
7. (https://www.sciencedirect.com/science/article/pii/S2173511515001979)
8. Energy systems in swimming Ferran A. Rodríguez1 and Alois Mader2 1 Institut Nacional d'EducacióFísica de Catalunya, Universitat de Barcelona, Spain 2 Deutsche Sporthochschule Köln, Cologne, Germany https://www.researchgate.net/publication/256696190_Energy_systems_in_swimming
9. Dos Santos MA, Henrique RS, Salvina M, Silva AH, Junior MA, Queiroz DR, Duncan MJ, Maia JA, Nevill AM (2021) The influence of anthropometric variables, body composition, propulsive force and maturation on 50m freestyle swimming performance in junior swimmers: an allometric approach. J Sports Sci. https://doi.org/10.1080/02640414.2021.1891685
10. Dopsaj M, Zuoziene IJ, Milić R, Cherepov E, Erlikh V, Masiulis N, di Nino A, Vodičar J (2020) Body composition in international sprint swimmers: are there any relations with performance? Int J Environ Res Public Health 17(24):9464. https://doi.org/10.3390/ijerph17249464
11. Moura T, Costa M, Oliveira S, Júnior MB, Ritti-Dias R, Santos M (2014) Height and body composition determine arm propulsive force in youth swimmers independent of a maturation stage. J Hum Kinet 10(42):277–284. https://doi.org/10.2478/hukin-2014-0081.PMID:25414760;PMCID:PMC4234767
12. Stager J, Cordain L, Becker TJ (1984) Relationship of body composition to swimming performance in female swimmers. J Swim Res 1:21–26
13. Cortesi M, Gatta G, Michielon G, Di Michele R, Bartolomei S, Scurati R (2020) Passive drag in young swimmers: effects of body composition, morphology and gliding position. Int J Environ Res Public Health 17(6):2002. https://doi.org/10.3390/ijerph17062002
14. Rossi FE, Ricci-Vitor AL, Sabino JP, Vanderlei LC, Freitas IF Jr (2014) Autonomic modulation and its relation with body composition in swimmers. J Strength Condition Res 28(7):2047–2053
15. Rozi G, Thanopoulos V, Geladas N, Soultanaki E, Dopsaj M (2018) Anthropometric characteristics and physiological responses of high level swimmers and performance in 100 m freestyle swimming. Movement Sport Sci - Sci &Motricité 2018(3):3–7. https://doi.org/10.1051/sm/2018007

References

Analyses of Respiratory Performance of Sportspersons

Subhajit Banerjee, Sudhir Chandra Murmu, Himadri Chattopadhyay, and Abhijit Dutta

Abstract This paper examines the respiratory performances of sportspersons from the perspective of thermodynamics. Performing in a comfortable environment not only widens the respiratory smoothness of athletes but also boosts their performance. By psychometric analysis of the air-conditioned stadium environment, the type and quality of air conditioning can be forecasted. A new term, metabolic respiratory efficiency (MRE) is introduced in this paper which formulates the ratio of respiratory workdone to the energy supplied through glucose by metabolism. The effect of limited and excess metabolic oxygen supply on the MRE has been observed and in context of this, the percentage of glucose required for smooth breathing process with beforesaid air condition is also assessed. To enhance convenience, the ambient is dehumidified without overcooling the temperature in most indoor stadiums. A comparison of all these parameters is made for a number of sports to analyse the dependence of performance on variations of respiratory conditions and the type of air conditioning system required for indoor air-conditioned stadiums for respective sports.

Keywords Respiratory efficiency · Metabolism · Regulated air supply · Psychrometry · Cooling load

---

S. Banerjee (✉) · S. C. Murmu · H. Chattopadhyay
Department of Mechanical Engineering, Jadavpur University, Kolkata, West Bengal 700032, India
e-mail: bsubhajit15@gmail.com

S. C. Murmu
e-mail: murmusudhir@gmail.com

H. Chattopadhyay
e-mail: himadri.chattopadhyay@jadavpuruniversity.in

S. Banerjee
Department of Mechanical Engineering, Greater Kolkata College of Engineering & Management, Baruipur, Kolkata, West Bengal 743387, India

A. Dutta
Department of Mechanical Engineering, MCKV Institute of Engineering, Liluah, Howrah, West Bengal 711204, India
e-mail: abhijitdutta@mckvie.edu.in

S. Bhattacharyya and D. James (eds.), *Recent Trends in Sports Engineering*, Lecture Notes in Mechanical Engineering, https://doi.org/10.1007/978-981-97-8297-0_7

1 Introduction

Breathing and exercise has always been closely linked in athletic training to ensure efficient pulmonary ventilation. Every developed biological system executes an energy management system called biogenetics, i.e. extracting useful energy from the outside, converting, storing it and using it for muscle contraction, substrate transport, protein synthesis and other utilising processes. The breathing process can be divided into three different components, inhalation of ambient air, conditioning of the inhaled air and exhalation of the conditioned air to the environment. When the lung expands, ambient air enters the body through the nostril and passes through the tracheal tree to the lungs. When the lung expands, the breathing air comes out of the body through the same tract. This total process is performed by the respiratory muscle activity which is carried out by the energy from glucose metabolism in the body. During athletic activities, the total capacity of lung (TLC) is subjected under a significant volume increment for holding the increased air intake. It results in more frictional workloss thus more muscular work for smoothness of breathing [8].

Many studies have been carried out in the field of performance analysis of biological systems to understand the natural phenomena in living systems. Loer [21], Mady et al. [20, 22, 23] studied the exergetic efficiency of the human metabolism and reported that for performing muscular work, 60% of the exergy coming from nutrient is used through ATP production efficiency is not affected by the degree of physical activity. Metabolic efficiency in glucose consumption of diaphragm muscle for healthy and COPD patients been observed by Catak et al. [6]. Loer et al. [20] reported that the human lung consumes about 5% of the total body oxygen intake. The conversion of chemical energy in thermomechanical energy during muscle contraction has been reported by Hemcker et al. [16]. Bit et al. [5] investigated the air flow in the tracheal flow for a bronco-pulmonary patient and pointed out the variation of pressure, velocity and kinetic energy at different levels of bifurcation. The breathing rate increases with inspiratory air intake Corne et al. [7]. The gradual reduction of tracheal resistance to air flow in the respiratory system up to the 23rd level [27] which has been quantified by Banerjee et al. [3] for different physiological states of human body. The effect of age and gender on the thermodynamic performance of lungs has been reported by Dutta et al. [10–12] also investigated the role of psychrometric conditions and frictional effects on entropy formation in the human lung using two different models.

The state of survey highlights that there is a void in the in the field of respiratory efficiency analysis for different sports. In the present work, the authors are interested in quantifying the performance of the human respiratory system in light of the first law of thermodynamics. The quality of indoor air conditioning needed for flawless performance of the player during sport has also been assessed in this study. This work also reports the impact of seasonal variation of ambient air on the conditioning factors for different sports.

2 Model Description

Breathing efficiency analysis is a complicated phenomenon. The subject is extremely important for both medical and athletic wellness. The basic parameters for understanding the mechanics are airflow rate, breathing rate (cycles per minute), lung volume and physiological status. This work includes respiratory performance analysis of players from four different sport segments, i.e. marathon running, cycling, hockey and football. The flow rate of air corresponding to the mentioned sports are 115 lpm [29], 118 lpm [5], 200 lpm [19] and 205 lpm [2], respectively. The corresponding breathing rates are 50/min, 30/min, 35/min [31] and 53/min [9], respectively, along with the vital capacity of 5.11 lit, 5.26 lit [25], 5.27 lit [30] and 5.78 lit [25]. Vital capacity of lungs is almost 80% of the total lung capacity (TDC) [1]. The effect of limited and excess metabolic oxygen supply on MRE is also assessed in the range of ± 20% at 5% interval. The body temperature is considered to be 36.6 °C [26]. The inter-diaphragmatic pressure is taken as 8 mm of Hg during sports [28]. In the spectators of a match, the frequencies of breathing per minute are taken as 15 [33], 40 and 11 [32] for calm people, excited spectators and disappointed fans, respectively. In this study, three ambient conditions are considered, i.e. average (2 5°C, 40% RH), summer (42 °C, 75% RH) and winter (16°C, 30% RH) for assessing the seasonal variation of enthalpy difference of breathing air, cooling load and dehumidification load for players.

3 Mathematical Formulation

Body muscles perform specific work by consuming energy from glucose metabolism. In the presence of O_2, glucose breaks down and provides energy in the form of ATP and CO_2 and H_2O are produced as by-products. Equation (1) represents the stoichiometric equation of aerobic cellular respiration [17].

$$C_6H_{12}O_6 + 6O_2 = 6CO_2 + 6H_2O \tag{1}$$

In the present study, human respiratory muscles also get energy from said process to perform the operation of breathing. From the literature survey, it is observed that 1.83×10^{-5} mol of glucose are required to carry out one complete breathing cycle [6].

A new term, Metabolic Respiratory Efficiency (MRE) is introduced in this context which formulates the ratio of respiratory workdone to the energy supplied through glucose by metabolism (Fig. 1).

$$MRE = \frac{Workdone\ by\ lungs/breathing\ cycle}{Energy\ supplied\ by\ glucose/breathing\ cycle} \tag{2}$$

Each breathing cycle comprises of two phases, i.e. inspiration and expiration. The workdone required for a complete breathing cycle is similar to workdone of lung. When the lung expands, the ambient air enters the respiratory tract through nasal cavity and reaches the alveoli passing through the 23rd bifurcation level. The intercostal muscle has to do muscular workdone to expand the lung. Then again lung collapse causes the air to get back to the environment through the respiratory tract as expiration. The lung collapse does not require external workdone being similar to elastic recoil (Dutta et al, 2022).

The workdone of a complete breathing cycle can be written as in Eq. (3) [14].

$$\textbf{Workdone}_{\textbf{Lung}} = \textbf{Expansive work} + \textbf{Loss of equivalent work}$$
$$WD_{Lung} = (p \times dv) \times n \times t + \left[(h_l \times \rho \times g) \times n \times dv \times t\right] \quad (3)$$

In Eq. (3), p is the inter-diaphragmatic pressure in N/m^2, dv is the total lung volume in m3, and n is the number of breaths per minute. h_l represents the entire pressure head loss caused by tracheal resistance during a single breathing cycle. The air density (ρ) and gravitational acceleration (g) are measured in kg/m^3 and m/s^2, respectively. The loss equivalent work is the amount of additional work performed by the respiratory muscle to compensate for the work loss caused by fictitious pressure head loss in the tracheal tree during inspiration and expiration.

The calculation of frictional head loss in respiratory tract is done through the procedure followed by Banerjee et al. [3].

In this context, the percentage of glucose required for respiration process is depicted in Eq. (4)

$$\%\ of\ glucose\ required\ in\ respiration\ process = \frac{Glucose\ used\ in\ respiration\ process/min}{Glucose\ produced\ in\ metabolic\ reaction/min} \times 100\% \quad (4)$$

All the above observations are made taking the ambient air condition at 1 bar and 25 °C.

Performing in a comfortable environment not only widens the respiratory smoothness of athletes but also boosts their performance. By psychometric analysis of the air-conditioned stadium environment, the type and quality of air conditioning can be forecasted.

Generally, air exhaled is hotter than inhaled air. During the expiration, air leaving the body is saturated with water vapour, so the enthalpy of exhaled air is greater than that of inhaled air.

$$\begin{aligned} &\textit{Enthalpy of moist air} \\ &\qquad = \textit{Enthalpy of 1 kg of dry air} \\ &\qquad + \textit{Enthalpy of water vapour associated with 1 kg of dry air} \\ &h = 1.005 \times T_{db} + \omega \times [2500 + 1.88 \times T_{db}] \end{aligned} \tag{5}$$

In Eq. (5), T_{db} and ω are the dry bulb temperature and specific humidity of the air in °C and kg/kg of dry air, respectively.

The enthalpy difference between inhaled and exhaled air (Δh) at average ambient condition, i.e. at 25 °C and 40% RH is also calculated by Eq. (6)

$$\Delta h = h_2 - h_1 \tag{6}$$

In Eq. (6), h_1 and h_2 are representing the enthalpies of inhaled and exhaled air in KJ/kg, respectively.

In a challenging environment, either scorching heat or chilling cold, the ambient is dehumidified without overcooling the temperature in most indoor stadiums to enhance convenience. In this study, three ambient conditions are considered, i.e. average (25 °C, 40% RH), summer (42 °C, 75% RH) and winter (16 °C, 30% RH).

Dehumidification load can be defined as the amount of water to be removed from the closed stadium to maintain the ambient air in the average condition. Dehumidification load per hour for each person is shown in Eq. (7)

Dehumidification Load

= Specific humidity difference between inhaled and exhaled air

± Specific humidity difference between ambient condition and required condition

$$\textit{Dehumidification Load} = [(\omega_2 - \omega_1) \pm (\omega_4 - \omega_3)] \times n \times 60 \tag{7}$$

ω_1, ω_2, ω_3 and ω_4 represent the specific humidity of inhaled air, exhaled air, external environment and conditioned indoor environment in kg/kg of dry air, respectively.

Specific humidity of moist air (ω) can be found out by Eq. (8)

$$\omega = \frac{\varphi \times p_{vs}}{1.6 \times p_a} \tag{8}$$

In Eq. (8), ϕ is the relative humidity of the air. p_{vs} and p_a are the partial pressures of saturated water vapour and dry air in bar, respectively.

Cooling load is defined as the amount of heat to be removed from the environment to keep ambient condition pleasant. Cooling load per hour for a single player is framed in Eq. (9). The desired air condition to be maintained inside the stadium is considered to be 25 °C and 40% relative humidity for each sport during summer and winter.

$$\begin{aligned} \textit{Cooling Load} &= \textit{Enthalpy difference between inhaled and exhaled air} \\ &\pm \textit{Enthalpy difference between ambient condition and required} \\ &\textit{condition} \\ \textit{Cooling Load} &= [(h_2 - h_1) \pm (h_4 - h_3)] \times n \times 60 \end{aligned} \tag{9}$$

In Eq. (9), h_4 and h_3 are the enthalpies of air in KJ/kg in specific temperature at ambient air condition and desired air condition, respectively.

Equation (9) can also be used for assessing seasonal variation of the cooling load for the spectators separately. Irrespective of the sports, the stadium is considered to hold 15,000 spectators during a 1 h match. The total match is divided into three specific time zones, i.e. starting, climax and end. It is considered that, during the starting of the match, 50% of the spectators are excited and the rest are enjoying the match, whereas, at climax, 80% spectators are excited and 20% are in rest, i.e. they are just enjoying the match and during the ending, among the spectators 50% are excited, 30% are in rest and 20% are disappointed.

4 Results

The present work includes the respiratory performance analysis of sportspersons for different sports, i.e. marathon running, cycling, hockey and football. The variation of metabolic respiratory efficiency (MRE) and percentage of glucose consumed for respiration work at limited and excess air supply is carried out. The enthalpy difference between inhaled and exhaled air, dehumidification load for a single player and cooling load for stadium have also been assessed.

4.1 *Variation of Metabolic Respiratory Efficiency (MRE) at Limited and Excess Air Supply*

The variation of metabolic respiratory efficiency (MRE) at limited and excess air supply for four different sports is depicted in Fig. 4. During analysis, the density of the breathing air is kept at 1.19 kg/m^3 at 25 °C and the dynamic viscosity of the air is taken as 0.000018 kg /m-s.

Figure 2 depicts that at a particular ambient condition, the metabolic respiratory efficiency increases with increase in air supply for all the considered sports. In human body, the respiration system is working just like heat engine which takes up energy from glucose ($C_6H_{12}O_6$) consumption and does the muscular work for respiration. The efficiency is also in the range of efficiency of a heat engine, i.e. 30%–50%. The breathing workdone increases as the frictional work loss is more for enhanced air volume inhaled. But at the same time, glucose energy supplied for the respiratory workdone is fixed. So with increased air inhalation, the breathing muscle does more workdone to complete a breathing cycle at same glucose energy supply which causes MRE to increase with increased air supply.

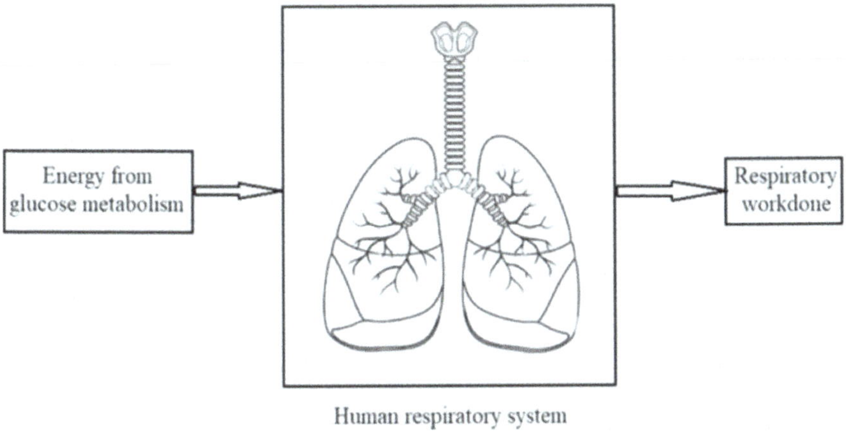

Fig. 1 Schematic diagram of human respiratory system

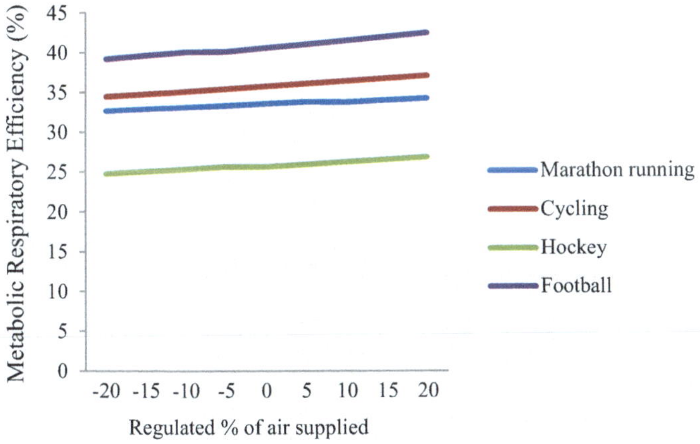

Fig. 2 Variation of metabolic respiratory efficiency (MRE) at limited and excess air supply

4.2 Variation of the Percentage of Glucose Required for Respiration Process at Regulated Air Supply

Figure 3 represents the percentage of glucose consumption for respiratory work at regulated air supply for different sports.

From Fig. 3, it is observed that glucose consumption for respiration is decreased with increased air supply. As the MRE increases with increased air supply, the glucose ($C_6H_{12}O_6$) consumption for respiratory work also decreases as more air consumption leads to enhanced metabolism of glucose.

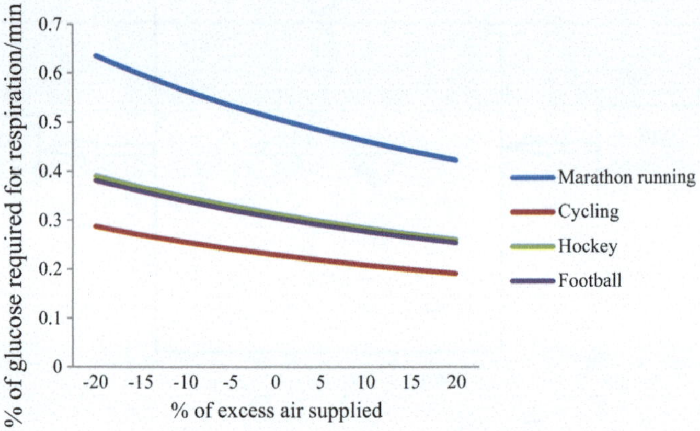

Fig. 3 Variation of the percentage of glucose consumption for respiratory work at regulated air supply

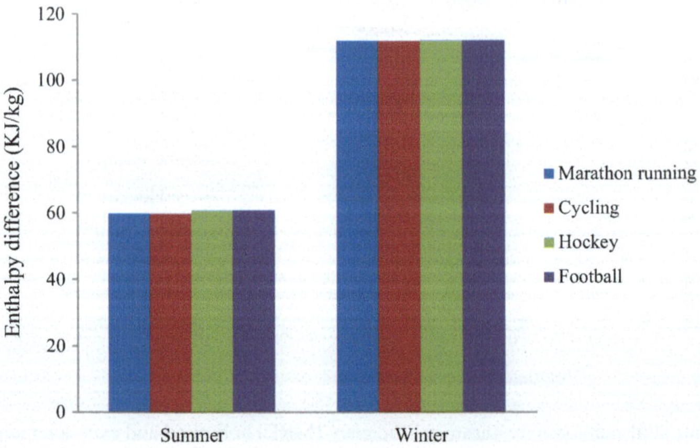

Fig. 4 Seasonal variation of enthalpy difference between exhaled and inhaled air

In marathon running, the amount of air inhaled per breath by the athlete is less than the other sports. The metabolic respiratory efficiency (MRE) is also the least among the other considered sports which results in more breathing work. Hence, the percentage of glucose consumption is higher for a marathon runner in comparison with other athletes.

4.3 Variation of Enthalpy Difference Between Exhaled and Inhaled Air During Summer and Winter

Since the air coming out of body is fully saturated with water vapour at body temperature 36.6 °C [26], exhaled air definitely possesses more heat energy than inhaled air irrespective of inhaled air condition. Figure 4 shows the variation of enthalpy difference between exhaled and inhaled air during summer (42 °C, 75% RH) and winter (16 °C, 30% RH).

From Fig. 4, it can be commented that irrespective of the nature of sports the said enthalpy difference is quiet smaller in summer than that of in winter. At an average environment, i.e. at 25 °C and 40% relative humidity, for all the sports considered the enthalpy difference between inhaled and exhaled air is about 98 kJ/kg. In winter, this difference experiences a hike of 15% and, in summer, this difference is reduced by 39% for all the sports. During summer, the ambient temperature as well as relative humidity is very high. The air leaving the body is also saturated at body temperature 36.6 °C. So, the enthalpy difference between the inhaled and exhaled air is less. But in winter, temperature and relative humidity of ambient air are very less which further results in greater enthalpy difference between inhaled and exhaled air.

4.4 Variation of Dehumidification Load of Each Player at Summer and Winter

The seasonal variations in dehumidification load in kg/hr for a single player for different sports are represented by Fig. 5.

In Fig. 5, it can be noticed that dehumidification load is more in summer for all the sports. The dehumidification load in winter is decreased by 24% than that of in summer. During dehumidification process, an attempt is made to reduce humidity keeping the temperature constant. In summer, more moisture including excess humidity of the exhaled air needs to be removed from the ambient to attain the desired condition. But in winter, only exhaled air needs to be dehumidified and the desired ambient can be achieved by adding some extra moisture to the conditioned space. Humid air enters the body during inspiration and after getting fully saturated with the water vapour in the lungs, comes out during expiration. So, the increase of the moisture in external ambient completely depends on the number of

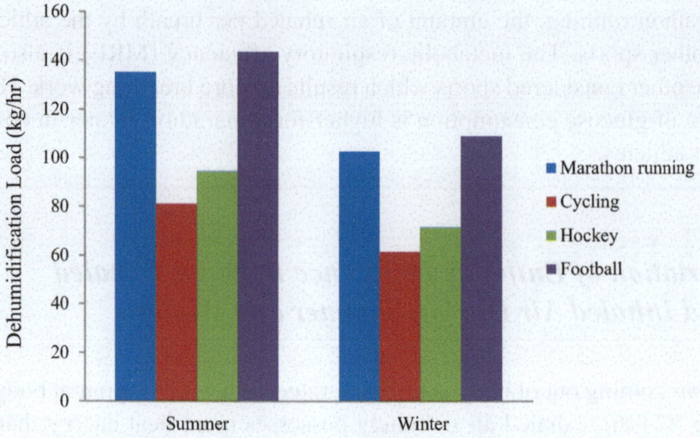

Fig. 5 Seasonal variations in dehumidification load for a single player for different sports

breathing cycles executed by the person in a specific time period. Among the sports, dehumidification load for football irrespective of the season is highest, whereas the cycling, hockey and marathon running are having reduced dehumidification load by 43%, 33% and 5%, respectively.

4.5 Variation of Cooling Load of a Single Player at Summer and Winter

Extreme environment causes inconvenience in the smooth performance of players. The indoor stadium is aimed to be maintained at an ambient condition of 25 °C and 40% relative humidity. To achieve the said ambient condition, sensible heat as well as latent heat is to be adjusted. The seasonal variations in cooling load in KJ/kg-hr for a single player for different sports are represented by Fig. 6.

From Fig. 6, it is observed that for all the sports, the cooling load per hour is greater in winter as compared to summer. During winter, the temperature difference between inhaled and exhaled breathing air is 46% higher as compared to summer. This is the main reason for increasing cooling load in winter. It is noticed that, to maintain the ambient of the stadium in winter (16 °C, 30% RH) at 25 °C and 40% RH, the amount of heat removal from the stadium rises by near about 44% than that of in summer for all the sports.

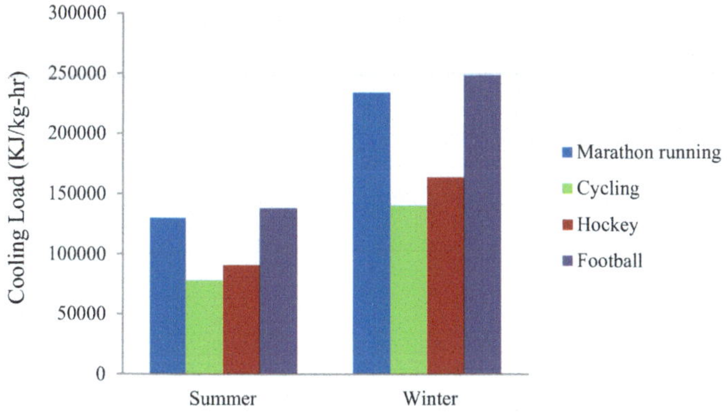

Fig. 6 Variation of cooling load for single player in summer and winter

4.6 Variation of Cooling Load for Spectators in Summer and Winter

Since each stadium is having a huge number of spectators for each match, the cooling load for spectators is more significant than that of the players. The indoor stadium is aimed to be maintained at an ambient condition of 25 °C and 40% relative humidity. To achieve the said ambient condition, sensible heat as well as latent heat is to be adjusted. The seasonal variations in cooling load in KJ/kg-hr for spectators of a stadium with 15,000 seat capacity are depicted by Fig. 7.

Figure 7 clearly shows that the cooling load for the spectators during summer is considerably higher than in winter. Summer experiences an enhanced cooling load by 93% as compared with winter due to significant psychrometric (temperature and humidity) difference between ambient air and conditioned air.

At the climax, the cooling load is highest, i.e. 38% of the total cooling load, whereas, at the starting and ending of the match, only 32% and 30% of the total

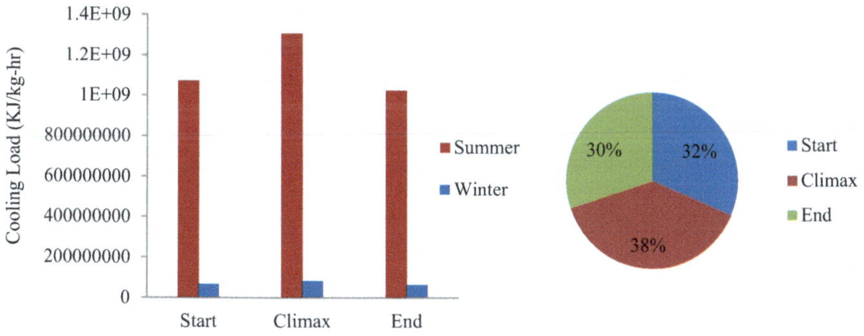

Fig. 7 Seasonal variations in cooling load for spectators in stadium

cooling load for spectators is contributed, respectively. This can be attributed to the degree of excitement of the spectators during the match. During climax, 80% spectators are assumed to be excited for the match which raises their exhaled air temperature, whereas during staring and ending, only about 50% spectators are excited for the match. The degree of excitement controls their breathing frequency which further causes noticeable differences in cooling load.

5 Conclusion

This paper presents the metabolic efficiency of human respiratory system and the impact of seasonal climate variation on the indoor air conditioning for different sports. The following conclusions are drawn from the entire analyses.

The metabolic respiratory efficiency (MRE) increases with increase in air supply for all the considered sports. The efficiency is in the range of efficiency of 30%–50% similar to a heat engine which does the mechanical work by consuming chemical energy from glucose metabolism.

As the MRE increases with increased air supply, the glucose ($C_6H_{12}O_6$) consumption for respiratory work also decreases since metabolism of total glucose intake increases with increased breathing air consumption.

Irrespective of the nature of sports the enthalpy difference between exhaled and inhaled air is quite smaller in summer than that in winter. In winter, this difference experiences a hike of 15% and in summer this difference is reduced by 39% for all the sports.

For all the sports, the cooling load per hour is greater in winter as compared to summer. It is noticed that to maintain the ambient of the stadium in winter (16 °C, 30% RH) at 25 °C and 40% RH, the amount of heat removal from the stadium rises by 70% than that in summer due to large temperature and humidity difference between the external environment and the regulated environment in the stadium.

Dehumidification load is greater in summer than that in winter irrespective of sports because during winter only exhaled air needs to be dehumidified and the desired ambient can be achieved by adding some moisture to the conditioned space.

Summer experiences an enhanced cooling load by 93% as compared with winter due to a significant psychrometric (temperature and humidity) difference between ambient air and conditioned air.

6 Limitations and Future Scopes

The respiratory efficiency is a very important physiological parameter that is currently not well monitored in a range of sports healthcare settings. In the drive to attain convenient performance monitoring at sports ground, the ambient of the stadium needs to be governed in terms of temperature and humidity.

The study focuses on the first law of efficiency of the respiratory system. The research can further be updated for second law efficiency or exergy analysis for the respiratory efficiency analysis of sportsmen. The cooling load for the stadium can further be assessed by combining the influence of players and spectators for different sports. The results presented here are not limited to the atmospheric environment at sports grounds, however, can also be extrapolated to any large-scale gatherings of people.

References

1. American Lung Association (2017) Lung capacity and aging, Lung health and diseases. https://www.lung.org/lung-health-diseases/how-lungs-work/lung-capacity-and-aging. Last accessed on 23/11/2023
2. Atan, T, Akyol, P, CEBI, M.: Comparison of respiratory functions of athletes engaged in different sports branches. Turkish J Sports Exercise 14(3):76–81
3. Banerjee S, Dutta A, Chattopadhyaya H (2023) Airflow analyses of human respiratory tract at different physiological conditions. J Mech Med Biol 23(3):2350044
4. Bekar A (2020) Focus and breathing on the bike. www.roadbikerider.com, Newsletter issue No. 932 (2020).
5. Bit A, Chattyopadhay H, Nag D (2009) Study of airflow in the trachea of a broncho-pulmonary patient using CT data. Int J Biomech 31–36
6. Catak J, Ozilgen M, Yilmaz B (2018) Thermodynamic analysis of human respiratory (diaphragm) skeletal muscles. In: Conference: European Respiratory Society (ERS) international congress 2018. France, Paris, pp PA2447
7. Corne S, Gillespie D, Roberts D, Younes M (1997) Effect of inspiratory flow rate on respiratory rate in intubated ventilated patients. Am J Respir Crit Care Med 155:304–308
8. Troosters T, Dupont L, Bott J, Hansen K (2016) Your lungs and exercise. Breathe 12(1):97–100
9. Di Paco A, Catapano GA, Vagheggini G, Mazzoleni S, Micheli M, Ambrosino N (2014) Ventilatory response to exercise of elite soccer players. Multidisciplin Respirat Med 9(1):20
10. Domkundwar A (2004) A course in refrigeration and air conditioning, 7th edn, Dhanpat Rai and Co. (P) Ltd, Delhi
11. Dutta A, Chattopadhyay H, Biswas A (2019) A comparative study on the entropy generation in the human respiratory tract based on Hess–Murray law and Weibel experimented result. J Mech Med Biol 19(6):1950046(1–12)
12. Dutta A, Chattopadhyay H, Yasmin H, Gorji MR (2019b) Entropy generation in the human lung due to effect of psychrometric condition and friction in the respiratory tract. Comput Methods Progr Biomed 180(1–8):105010
13. Dutta A, Chattopadhyay H (2020) Exergetic analysis of human respiratory system including effect of age and gender. Int J Exergy 31(4):370–385
14. Dutta A, Chattopadhyay H (2021) Performance analysis of human cardiorespiratory system based on the second law of thermodynamics. Int J Exergy 34(1):29–49
15. Dutta A, Chattopadhyay H, Catak J (2023) Thermodynamic performance analysis of human cardiovascular system for normal and cardiomegaly conditions. J Mech Med Biol 23(1):2350002
16. Hemker HC, Hermens W (1971) Theoretical considerations on a thermodynamic model of muscle action. Cardiovasc Res 1:84–86
17. Hopkins E, Sanvictores T, Sharm S (2022) Physiology, acid base balance. National library of medicine. StatPearls Publishing LLC, PMID: 29939584, Florida, US
18. Lam CJ, Chan AL (2001) CFD analysis and energy simulation of a gymnasium. Build Environ 36:351–358

19. Lumme A, Haahtela T, Ounap J, Rytila P, Obase Y, Helenius M, Remes V, Helenius I (2003) Airway inflammation, bronchial hyper responsiveness and asthma in elite ice hockey players. Eur Respir J 22(1):113–117
20. Loer SA, Scheeren TWL, Tarnow J (1997) How much oxygen does the human lung consumes? Anesthesiology 86(3):532–537
21. Mady CEK, Ferreira MS, Yanigihara JI, Saldiva PHN, Oliveira Junior S (2011) Second law of thermodynamics and human body. Therm Eng 10(1–2):88–95
22. Mady CEK, Henriques IB, Oliveira Junior S (2013) Exergy analysis of human body and lifespan: a first approach. In: 22nd international congress of mechanical engineering (COBEM 2013), November 3–7. Ribeirao Preto, SP, Brazil
23. Mady CEK, Oliveira Junior S (2013) Human body exergy metabolism. Int J Therm 16(2):73–80
24. Mady CEK, Henriques IB, Oliveira Junior S (2014) Method to perform exergy analysis to the human body and daily exergy balance. In: 15th Brazilian congress of thermal science and engineering, November 10–13. Belem, PA, Brazil
25. Mazic S, Lazovic B, Djelic M, Suzic-Lazic J, Djordjevic-Saranovic S, Durmic T, Soldatovic I, Zikic D, Gluvic Z, Zugic V (2015) Respiratory parameters in elite athletes – does sport have an influence? Revista Portuguesa De Pneumologia (English Edition) 21(4):192–197
26. Obermeyer Z, Samra JK, Mullainathan S (2017) Individual differences in normal body temperature: longitudinal big data analysis of patient records. BMJ 359:j5468
27. Reis AH, Miguel AF, Ayedin M (2004) Constructal theory of flow architecture of the lungs. Med Phys 31(5):1135–1140
28. Schurch S, Goerke J, Clements JA (1976) Direct determination of surface tension in the lungs. Proc Natl Acad Sci USA 73(12):4698–4702
29. Zoladz JA, Nieckarz Z (2021) Marathon race performance increases the amount of particulate matter deposited in the respiratory system of runners: an incentive for "clean air marathon runs". PeerJ 9:e11562
30. Ramos I, Barreira S, Viana R (2020) Inspiratory muscle training on lung function of male roller hockey players: a randomized controlled trial pilot study. Porto Biomed J 5(5):e075
31. Sarkar S, Chatterjee S, Dey SK (2019) Effect of 8 weeks high intensity interval training on maximum oxygen uptake capacity and related cardio-respiratory parameters at anaerobic threshold level of indian male field hockey players. European J Phys Educat Sport Sci 5(5)
32. Celso MM, Patrick, K, Jonathan, G.: Real-time expression of affect through respiration. 21(3–4) (2010).
33. Veres PR, Faber P, Drewnick F, Lelieveld J, Williams J (2013) Anthropogenic sources of VOC in a football stadium: assessing human emissions in the atmosphere. Atmos Environ 77:1052–1059
34. Trianti-Stourna E, Spiropoulou K, Theofilaktos C, Droutsa K, Balaras CA, Santamouris M (1998) Energy conservation strategies for sports centers. Part A sports halls. Energy Build 27:108–121
35. Stathopoulou OI, Assimakopoulos VD, Flocas HA, Helmis CG (2008) An experimental study of air quality inside large athletic halls. Build Environ 43:834–848
36. Junker MT, Koller MC (2000) An assessment of indoor air contaminants in buildings with recreational activity. Sci Total Environ 246:139–152
37. Chow WK, Fung WY, Wonga LT (2002) Preliminary studies on a new method for assessing ventilation in large spaces. Build Environ 37:145–152
38. Priyadarsini R, Luther MB (2013) Thermal and ventilation performance of a naturally ventilated sports hall within an aquatic centre. Energy Build 58:111–122

Determination of Energy Expenditure and Body Composition Among Kabaddi and Volleyball Players

Chennakeshavulu Vemula, Pranathi Kalvakolu, Maheshwar Mekam, and Suresh Challa

Abstract

Background The examination of an Athlete's health status at any age is extremely important, and in recent years, body composition analysis has emerged as a key component, especially when it comes to enhancing the performance. Determination of energy expenditure is one of the important components of sports training in order to estimate the energy intake and prevent injury. Objective: To determine the energy expenditure and body composition among the kabaddi and volleyball players.

Materials and Methods Fourteen team players (7 kabaddi, 7 volleyball players) were assessed for basic anthropometric measurements, body composition, and energy expenditure by using SECA stadiometer, InBody Bioimpedance Analysis 770, and Quark COSMED K5 were used, respectively.

Results The mean energy expenditure of the volleyball players on game-specific training day was reported to be 4800 ± 415.56 kcals, whereas on the gym session day was found to be 4531 ± 382.25 kcals. The mean energy expenditure of the kabaddi players on moderate-intensity training was found to be 4946 ± 306.19 kcals, whereas high-intensity training was found to be 5184 ± 383.445 kcals. Though the total energy expenditure of the kabaddi players was found to be more, training energy expenditure of the volleyball players was found to be more.

Conclusion The total energy expenditure depends on various factors like physical and physiological demands of the particular sport and RMR. As the RMR of kabaddi players was more, the total energy expenditure of the kabaddi players was found to be more.

C. Vemula (✉) · P. Kalvakolu · M. Mekam
Department of Sports Sciences, ICMR-NIN, MYAS-NIN, Tarnaka, Hyderabad 500007, India
e-mail: chennalaxmi@gmail.com

S. Challa
Department of Cell Biology, ICMR-National Institute of Nutrition, Tarnaka, Hyderabad 500007, India

S. Bhattacharyya and D. James (eds.), *Recent Trends in Sports Engineering*, Lecture Notes in Mechanical Engineering, https://doi.org/10.1007/978-981-97-8297-0_8

Keywords Energy expenditure · Indirect calorimetry · Body composition · Kabaddi · Volleyball · COSMED K5

1 Introduction

A crucial aspect of an athlete's well-being, physical fitness, and performance optimization is a meticulously crafted diet tailored to meet their specific requirements, encompassing resting energy expenditure, thermic effect of food, non-activity energy expenditure, and activity energy expenditure [1]. The athlete's body mass, composition, and function are susceptible to negative consequences when daily dietary energy intake falls short or exceeds energy expenditure, emphasizing the importance of striking a delicate balance [1]. Elite athletes experience substantial changes in TEE, primarily driven by the energy expended during athletic endeavors, with TEE reaching up to five times BMR during intense continuous activities over several weeks [2]. In contrast, recovery days, pre-competition taper days, or off-season activities entail significantly lower energy expenditure, possibly even comparable to sedentary behavior [2]. Body composition analysis holds paramount importance in sporting contexts, offering insights into an athlete's physiological makeup and providing a basis for strategic modifications aligned with sports-specific goals [3]. The body composition, encompassing bone, muscle, tissue, and water, is assessed through various compartment models, with the two-compartment model being the most fundamental, evaluating fat mass and fat-free mass [3]. For practicing athletes, energy intake emerges as a critical factor, with insufficient calorie consumption linked to diminished physiological functioning, micronutrient deficiencies, altered body composition, and bone mineral density changes [4]. Therefore, calculating Total Energy Expenditure (TEE) becomes imperative for determining energy requirements and achieving energy balance [4]. RMR exhibits variability based on the sport practiced and years of training, with active individuals displaying greater efficiency compared to age-matched sedentary counterparts [5]. In elite endurance athletes, RMR accounts for 38–40% of TEE, while female endurance athletes demonstrate RMR representing 54 and 42% of TEE [5–7]. The thermic effect of food (TEF), constituting 6–10% of total energy consumption, encompasses energy expended above RMR for food digestion, assimilation, and absorption [8]. The most variable component, the thermic effect of physical exertion, can range from 10 to 15% for sedentary individuals to as much as 50% for those engaged in active pursuits [8]. Kabaddi is an intermittent sport that demands both anaerobic and aerobic endurance in addition to a well-developed body. There is no information on the determination of energy requirements of national and international kabaddi players except for physical characteristics, somatotype and cardiorespiratory variables, and physiological responses during match play [9, 10] The current study of Indian national volleyball and kabaddi players was conducted in order to: (1) examine anthropometric measures; (2) examine body composition analysis; and (3) calculate the energy expenditure of different training days.

2 Research Methodology

Study Design

In the investigation, a cross-sectional study was conducted, comprising seven kabaddi players and seven volleyball players aged between 14 and 25 years. The selection criteria ensured that all athletes had a minimum of 2 years of training in their respective sports. Participants were screened for health conditions and injuries, confirming their overall well-being. Ethical approval was obtained from an institutional review board, aligning with the principles of the Declaration of Helsinki and STROBE guidelines. Athletes meeting the criteria were informed about the testing procedures, potential discomforts, and provided written consent. The training protocols of both kabaddi and volleyball players were meticulously observed over a period of 6 days.

Participants

Participants in this research numbered 14 ($n = 14$). The inclusion criteria for participation are detailed as follows:

1. Sex: Male
2. Age: 16–25 years
3. Bmi:
   - Kabaddi players: 22.943 ± 2.374
   - Volleyball players: 20.429 ± 4.209; within the normal range for adult males
4. Health Status:
   - No acute/chronic diseases.
   - Athletes were free from injuries at the time of the procedures.

Subject involvement encompassed individuals sourced from state government organizations, all of whom had attained a national level of play.

Procedures

Observation of training protocol: For 6 days, the volleyball and kabaddi players followed the same training regimen which they used to follow before, under the same living and working conditions.

Physical measurements: A standard technique was followed in order to perform the fundamental anthropometric measures. [11] Using a SECA213 stadiometer, the athlete's height was measured to the nearest 0.1 cm while preserving the Frankfurt plane. After emptying the bladder, body mass was measured with the least amount of clothing on, using a SECA digital scale, to the nearest 0.1 kg. The weight of the athletes was taken in a post-absorptive state.

Body composition: Body composition of the athletes was taken by using Bioelectrical Impedance InBody 770 (Biospace, Seoul, South Korea). Participants were advised on how to stay well hydrated before the body composition test to enable accurate measurement of the body fat percentage and FFM, which were used in additional analysis.

Resting energy expenditure: Resting energy expenditure was quantified using indirect calorimetry, employing the VO2 mask (Hans Rudolph, USA) and COSMED K5 apparatus (Cosmed, Albano Laziale, Rome, Italy) with Quark 8.1 PC software support, based on the Weir equation [12, 13]. The measurement was conducted between 7:00 and 9:00 am under post-absorptive conditions, in a thermoneutral setting, with participants in a relaxed and comfortable position [14]. The assessment lasted for 15 min, during which the respiratory quotient was monitored. Values below 0.70 or exceeding 1 were considered indicative of protocol deviations or imprecise gas measurement [15].

Energy expenditure of training activities: The time allocation pattern was taken from the participants to measure the time in minutes of each and every activity (training and non-training). For the determination of energy expenditure of training activities, K5 equipment was set up, instructions were given to the athletes by the investigator throughout the study, and consumption of oxygen consumed during the activity was observed in laptop by connecting the equipment with the laptop through Bluetooth, energy expenditure was calculated. The respiratory quotient was monitored throughout the study to understand the source of energy utilized for different activities. Total energy expenditure of the day was calculated by combining the results of the resting metabolic rate (RMR), NEAT (non-exercise activity thermogenesis), and Training activities energy expenditure, which resulted in the total energy expenditure of the athlete.

Data analysis: The data analysis was conducted using IBM SPSS Software for Windows (version 23, Inc., Armonk, New York, USA). Numeric variables are reported as means and standard deviations.

3 Results

Table 1 explains the results of the study and reveals key parameters related to the physical characteristics and energy expenditure of the kabaddi players. The mean and standard deviation of age were found to be 20.371 $\pm$ 3.09 years, indicating a relatively homogeneous age group. The participants exhibited an average height of 172.843 $\pm$ 6.55 cm and a weight of 68.586 $\pm$ 8.791 kg, with a corresponding Body Mass Index (BMI) of 22.943 $\pm$ 2.374. In terms of body composition, the study participants demonstrated an average skeletal muscle mass of 45.455 $\pm$ 3.655 kg and a body fat percentage of 18.843 $\pm$ 7.032, providing insights into their overall body composition. Additionally, the investigation delved into the participant's daily energy expenditure, distinguishing between moderate- and high-intensity activities. The moderate-intensity day training energy expenditure was measured at 5184.571

± 383.445 kcal, while the high-intensity day energy expenditure was 4946.429 ± 306.19 kcal.

Table 2 explains the results of anthropometric, body composition, and energy expenditure of volleyball players. The mean and standard deviation of age were determined to be 19.529 ± 1.849 years, reflecting a relatively narrow age range among the participants. Notably, the athletes exhibited an average height of 185.343 ± 6.913 cm and a weight of 74.129 ± 12.96 kg, resulting in a Body Mass Index (BMI) of 20.429 ± 4.209. Regarding body composition, the volleyball players demonstrated an average skeletal muscle mass of 47.78 ± 3.172 kg and a body fat percentage of 16.329 ± 5.399, underscoring the importance of lean muscle mass in their physical makeup. The study also investigated the energy expenditure associated with game-specific training and gym sessions. On game-specific days, players expended an average of 4800 ± 415.564 kcal, while gym sessions accounted for an energy expenditure of 4531 ± 382.25 kcal. These findings provide valuable insights into the physical characteristics and energy demands specific to volleyball players, contributing to a deeper understanding of the unique physiological requirements of the sport.

Pearson correlation analysis did show some significant positive and negative correlations between anthropometric measurements, body composition measurements, and energy expenditure measurements. The research findings reveal several significant correlations among various anthropometric and physiological parameters in kabaddi players. A significant positive correlation ($p < 0.05$) was observed between age and both weight ($r = 0.760$) and body fat percentage ($r = 0.872$). Additionally, a significant negative correlation ($p < 0.01$) was identified between age and skeletal muscle mass ($r = -0.908$). Further analyses demonstrated a significant positive correlation ($p < 0.05$) between weight and Body Mass Index (BMI) ($r = 0.805$), while a significant negative correlation ($p < 0.05$) was found between weight and skeletal muscle mass ($r = -0.808$). Moreover, BMI exhibited a significant negative

Table 1 Basic anthropometric measurements, body composition, and energy expenditure of the kabaddi players

	Parameters	Mean	Std. dev
Anthropometric measurements	Age (years)	20.371 ± 3.09	
	Height (cms)	172.843 ± 6.55	
	Weight (kgs)	68.586 ± 8.791	
	BMI (kg/m^2)	22.943 ± 2.374	
	SMM (kgs)	45.455 ± 3.655	
	BF%	18.843 ± 7.032	
Body composition parameters	Moderate-intensity day training EE (Kcals)	5184.571 ± 383.445	
Energy expenditure	High-intensity day training EE (Kcals)	4946.429 ± 306.195	

BMI—body mass index, SMM—skeletal muscle mass, BF%—body fat percentage, EE—energy expenditure

Table 2 Basic anthropometric measurements, body composition, and energy expenditure of the volleyball players

	Parameters	Mean ± Std. dev
Anthropometric measurements	Age (years)	19.529 ± 1.849
	Height (cms)	185.343 ± 6.913
	Weight (kgs)	74.129 ± 12.96
	BMI (kg/m^2)	20.429 ± 4.209
	SMM (kgs)	47.78 ± 3.172
	BF%	16.329 ± 5.399
Body composition parameters	Game-specific day training EE (Kcals)	5184.571 ± 383.445
Energy Expenditure	Gym session day training EE (Kcals)	4946.429 ± 306.195

BMI—body mass index, SMM—skeletal muscle mass, BF%—body fat percentage, EE—energy expenditure

correlation ($p < 0.05$) with skeletal muscle mass ($r = -0.819$). Notably, a robust negative correlation ($p < 0.01$) was detected between skeletal muscle mass and body fat percentage ($r = -0.977$), emphasizing the inverse relationship between muscle mass and body fat percentage in kabaddi players. In the realm of training dynamics, a significant positive correlation ($p < 0.01$) was established between moderate-intensity training day energy expenditure and high-intensity training day energy expenditure ($r = 0.973$). This underscores the coherence between energy expenditures on different training intensities, providing valuable insights into the training regimen of kabaddi players.

The research outcomes elucidate notable correlations pertaining to body composition and energy expenditure in volleyball players. A significant positive correlation ($p < 0.05$) was discerned between height and weight ($r = 0.795$), underscoring the association between these anthropometric parameters in the studied cohort of volleyball players. Furthermore, an exceptionally strong positive correlation ($p < 0.01$) was identified when game-specific day energy expenditure was correlated with gym sessions ($r = 0.998$). This robust correlation highlights the coherence between the energy expended during volleyball match days and the corresponding gym sessions, shedding light on the interconnectedness of specific training regimens and the demands of competitive gameplay for volleyball athletes.

4 Discussion

The sport of kabaddi blends elements of rugby, judo, wrestling, and gymnastics. Because catching, holding, locking, and jumping are key body motions in this game, having a desirable anthropometric profile, body composition, and energy intake recommendations will give you a bigger advantage in performing better and improving physiological parameters. It was discovered that the kabaddi subjects' mean weight and height were greater than those of the typical Indian population. (18)There were significant positive and negative correlations found between anthropometric measurements, body composition, and energy expenditure among kabaddi players. The age, height weight, and body fat percentage of kabaddi players were found to be 27.91 ± 3.42 years, 174.34 ± 4.33cms, 75.49 ± 5.13 kgs, and 17.56 ± 3.48 percentage, respectively, body fat percentage of kabaddi players was determined by skinfold measurements (Harpenden skinfold calipers) [9]. Sports like kabaddi are best suited for somatotype of mesomorphic endomorph physique, according to the literature [9]. Age, height, weight, and body fat percentage in kabaddi players of our study were found to be 20.37 ± 3.09 years, 172.84 ± 6.55 cms, 68.58 ± 8.79 kgs, and 18.84 ± 7.03 percentage, respectively. Body fat percentage was determined by using Bioelectrical Impedance InBody 770.There was a significant difference found between age and weight of kabaddi players in the previous study and the current study. The results of volleyball players show that there were significant positive correlations found between anthropometric measurements and energy expenditure. Age, height, weight, and body fat percentage of volleyball players were found to be 22.30 ± 1.36 years, 173.10 ± 4.19 cms, 58.87 ± 6.94 kgs, and 10.04 ± 2.98, respectively. (19) Body fat percentage of kabaddi players was determined by skinfold measurements (Holtain skinfold calipers) (19). Age, height, weight, and body fat percentage of volleyball players in our study were found to be 19.52 ± 1.849 years, 185.34 ± 6.913 cms, 74.12 ± 12.96 kgs, and 16.32 ± 5.39 percentage, respectively. A significant difference was found between age, height, weight, and body fat percentage of the previous studies and the current studies, respectively.

5 Conclusion

Kabaddi is an indigenous sport and popular in South Asia, and has evolved over the years. Though it has evolved as a sport, research is too little in this sport as it is not internationally well recognized. Research on this particular sport has not been done yet except on general physical and physiological characteristics and somatotype of kabaddi players. Volleyball is an internationally recognized sport and research was done only on the determination of body composition in volleyball players. It is important to determine body composition and energy expenditure in sports setting for maintaining suitable body composition for the relevant sport and also maintaining energy balance. As there was no research done on the determination of energy expenditure on both sports. Investigators found that it would be helpful to provide energy intake guidelines for the sport and an experimental study was conducted. The final conclusion of the work is determination of energy expenditure is important to maintain energy balance which directly influences the body composition of the players and also it is important to give sport-specific energy intake guidelines for the players.

References

1. Frączek B, Grzelak A, Klimek AT (2019) Analysis of daily energy expenditure of elite athletes in relation to their sport, the measurement method and energy requirement norms. J Hum Kinet 70:81
2. Heydenreich J, Kayser B, Schutz Y, Melzer K (2017) Total energy expenditure, energy intake, and body composition in endurance athletes across the training season: a systematic review. Sports Med-Open 3:1–24
3. Mountjoy M, Sundgot-Borgen J, Burke L, Carter S, Constantini N, Lebrun C, … Ljungqvist A (2014) The IOC consensus statement: beyond the female athlete triad-relative energy deficiency in sport (RED-S). Br J Sports Med 491–497
4. Beidleman BA, Puhl JL, De Souza MJ (1995) Energy balance in female distance runners. Am J Clin Nutr 61(2):303–311
5. Guebels CP, Kam LC, Maddalozzo GF, Manore MM (2014) Active women before/after an intervention designed to restore menstrual function: resting metabolic rate and comparison of four methods to quantify energy expenditure and energy availability. Int J Sport Nutr Exerc Metab 24(1):37–46
6. Burke L, Deakin V, Deakin V (2015) Clinical sports nutrition. Mc Graw-Hill Education/Australia
7. Dey SK, Khanna GL, Batra M (1993) Morphological and physiological studies on Indian National Kabaddi players. Br J Sports Med 27(4):237
8. Khanna GL, Majumdar P, Malik V, Vrinda T, Mandal M (1996) A study of physiological responses during match play in Indian National Kabaddi players. Br J Sports Med 30(3):232
9. Marfell-Jones MJ, Stewart AD, De Ridder JH (2012) International standards for anthropometric assessment
10. Weir JDV (1949) New methods for calculating metabolic rate with special reference to protein metabolism. J Physiol 109(1–2):1
11. Torstveit MK, Fahrenholtz IL, Lichtenstein MB, Stenqvist TB, Melin AK (2019) Exercise dependence, eating disorder symptoms and biomarkers of Relative Energy Deficiency in Sports (RED-S) among male endurance athletes. BMJ Open Sport—Exercise Med 5(1)
12. Westerterp KR (2015) Measurement of energy expenditure. In: Translational research methods for diabetes, obesity and cardiometabolic drug development: a focus on early phase clinical studies, pp 169–187
13. Compher C, Frankenfield D, Keim N, Roth-Yousey L, Evidence Analysis Working Group (2006) Best practice methods to apply to measurement of resting metabolic rate in adults: a systematic review. J Am Dietetic Assoc 106(6):881–903
14. Sodhi HS, Sidhu LS (1984) Physique and selection of sportsmen: a kinanthropometric study. Punjab Publishing House
15. Bandyopadhyay A (2007) Anthropometry and body composition in soccer and volleyball players in West Bengal, India. J Physiol Anthropol 26(4):501–505

Dehydration Impact on Physiological and Cognitive Changes in Team Athletes During Diverse Training Intensities

Chennakeshavulu Vemula, Anjali Pawar, Maheshwar Mekam, and Suresh Challa

Abstract This study scrutinizes dehydration's impact on the physiological and cognitive performance of team athletes across diverse exercise intensities. Fourteen male team players (7 volleyball and 7 kabaddi) underwent assessments including anthropometrics, cognitive tests, and urine analysis before and after high- and moderate-intensity exercises. Significant findings revealed observable effects of dehydration on Maximum Heart Rate (Max HR), Rating of Perceived Exertion (RPE), Urine Specific Gravity (USG), and Reaction Time among volleyball athletes. Kabaddi players exhibited significance in Max HR and RPE across both intensities. These outcomes underscore the influence of dehydration on physiological and cognitive aspects in team sports, emphasizing the need for tailored hydration strategies to optimize performance. This condensed abstract captures the study's focus on dehydration's impact on physiological markers and cognitive functions in volleyball and kabaddi players during varied exercise intensities, emphasizing the importance of customized hydration plans for performance optimization.

Keywords Dehydration effects · Physiological performance · Cognitive changes · Exercise intensity

1 Introduction

Water, an elemental cornerstone of life, serves as the conduit for the orchestration of multifarious physiological functions within the human body [1, 2]. Its role extends far beyond mere hydration, encapsulating a diverse array of crucial bodily processes. Central to thermoregulation, biochemical reactions, nutrient transportation, waste removal, and vascular volume maintenance, water stands as an indispensable element

C. Vemula (✉) · A. Pawar · M. Mekam
Department of Sports Sciences, ICMR-NIN, MYAS-NIN, Tarnaka, Hyderabad 500007, India
e-mail: chennalaxmi@gmail.com

S. Challa
Department of Cell Biology, ICMR-National Institute of Nutrition, Tarnaka, Hyderabad 500007, India

S. Bhattacharyya and D. James (eds.), *Recent Trends in Sports Engineering*, Lecture Notes in Mechanical Engineering, https://doi.org/10.1007/978-981-97-8297-0_9

for sustaining optimal human functionality [3]. The balance of body water, often underappreciated in its complexity, holds sway over the delicate equilibrium between health and compromise [1]. Termed "hydration," this equilibrium delineates the delicate interplay between water gain and loss, underpinning fundamental physiological states. "EU hydration" characterizes the normative body water content, while deviations—"hypohydration" and "hyperhydration"—signify deficits or excesses beyond the standard range. "Dehydration," the process of water loss, stands in stark contrast to "rehydration," the replenishment of body water [2]. This investigation embarks upon an intricate exploration of the ramifications of hydration on human physiology, delving into its profound impact on performance and cognition across diverse environmental contexts and physical exertions. Embracing a multidimensional approach, this study elucidates the distinct effects of water balance on endurance, strength, and cognitive functions. Focusing on the nuanced facets of hydration, the article unfolds to delineate the differential impact of hypo- and hyperhydration on thermoregulation, exercise performance, and cognitive acuity [1] (Kalman, 2010). Additionally, it ventures into the realm of environmental influence, dissecting the effects of cold exposure on athletic performance and the concomitant role of hydration in mitigating potential adversities [4] (Rivera et al. 1999). Moreover, this study underscores the imperative of tailored hydration strategies, underscoring the role of electrolytes, sweat rates, and fluid composition in optimal rehydration protocols. It illuminates the significance of monitoring hydration status through biomarkers, leveraging body weight changes, urine output, and plasma osmolality measurements as pragmatic indicators.

2 Research Methodology

2.1 Participants

This research adopted a targeted approach, centering on a distinct segment of the athletic population consisting of male athletes aged 17–24 actively engaged in demanding training routines, specifically comprising 7 volleyball players and 7 kabaddi players. Despite an initial goal of 30 participants, logistical constraints led to a final sample size of 14 individuals who met the stringent inclusion criteria. The inclusion criteria emphasized the active involvement of male athletes aged 18–25 in regular training sessions, specifically within team sports like volleyball and kabaddi. Exclusion criteria encompassed female athletes, individuals with injuries, and those not participating in team sports underlining the study's focus on a homogeneous subset of male athletes engaged in similar rigorous training sessions, ensuring a cohesive examination of the impact of training intensity on physiological and cognitive parameters.

2.2 Data Collection

The data collection process was conducted meticulously to ensure accuracy and relevance for subsequent analysis. The tools used for data collection included measurements of height, weight, resting heart rate, maximum heart rate, urine specific gravity test, ruler drop test, and wall-toss test. These parameters were chosen to assess dehydration and its effects on both physiological (resting heart rate, maximum rate recovery, urine specific gravity, and body mass) and cognitive variables (coordination, reaction time, RPE—Rating of Perceived Exertion, and concentration).

2.3 Parameters of the Study

The study primarily focused on evaluating physiological and cognitive variables. Physiological variables included resting heart rate, maximum rate recovery, urine specific gravity, and body mass, while cognitive variables encompassed coordination, reaction time, RPE, and concentration. These variables were chosen as key indicators to assess the impact of dehydration on both physical and cognitive performance in the selected athletes.

2.4 Statistical Analysis

Descriptive statistics were employed to analyze the data collected from the 14 athletes participating in team sports. The athletes were categorized based on their training protocols and respective games. To compare these classified groups, the T-test for two independent samples was utilized. Statistical parameters such as mean, standard deviation, and p-values were computed using IBM SPSS Statistics 20.0 to ascertain any significant differences between the groups.

3 Result

Table 1 examines the physiological parameters of volleyball athletes undertaking high-intensity (HI) and moderate-intensity (MI) training. Analysis revealed non-significant differences in Pre- and Post-training Weight and Resting Heart Rate between the HI and MI groups ($p > 0.05$). However, a significant disparity emerged in maximum heart rates post-training (HI: 184.00 vs. MI: 168.86 bpm, $p = 0.100$), indicating distinct cardiovascular responses to training intensities. Additionally, significant variations were observed in post-training urine specific gravity (USG) and dehydration percentages ($p < 0.05$) between HI (USG: 1.05, dehydration: 3.41%) and

MI (USG: 1.04, dehydration: 3.27%) groups. These findings highlight the impact of training intensity on hydration markers and cardiovascular responses among volleyball players. Displayed equations are centered and set on a separate line.

Table 2 explores cognitive and perceived exertion parameters in high-intensity (HI) versus moderate-intensity (MI) volleyball and kabaddi training. Volleyball athletes displayed significant discrepancies in reaction time post-training (HI: 0.11 vs. MI: 0.17, $p = 0.000$) and perceived exertion levels ($p < 0.05$) between intensity groups. Conversely, kabaddi athletes showcased no significant variance in cognitive metrics (concentration, coordination, and reaction time, $p > 0.05$) but exhibited a notable difference in perceived exertion (HI: 7.86 vs. MI: 5.43, $p = 0.000$). This suggests dissociation between perceived physical effort and cognitive performance among kabaddi players while highlighting varied responsiveness to stimuli and effort perception among volleyball athletes at different training intensities.

Table 1 Physiological variances and significance in volleyball and kabaddi across intensity levels pre-training and post-training

Physiological variables	Volleyball high intensity	Volleyball moderate intensity	Kabaddi high intensity	Kabaddi moderate intensity
Pre-training				
Weight (Mean ± SD)	74.11 ± 12.91	74.11 ± 12.19	68.85 ± 8.87	68.08 ± 9.12
Resting HR (Mean ± SD)	67.29 ± 9.58	67.29 ± 10.63	62.71 ± 9.92	67.93 ± 11.01
Max HR (Mean ± SD)	184.00 ± 6.65	168.86 ± 10.63	183.00 ± 5.26	161.29 ± 5.93
Post-training				
Weight (Mean ± SD)	73.68 ± 12.89	73.85 ± 12.19	68.85 ± 87.77	68.08 ± 9.12
USG (Mean ± SD)	1.05 ± 0.02	1.04 ± 0.19	1.05 ± 0.02	1.04 ± 0.19
Dehydration % (Mean ± SD)	–	–	3.41 ± 3.05	3.27 ± 2.26
P-value				
Weight	0.981	0.981	0.875	0.875
Resting HR	1.000	1.000	0.417	0.417
Max HR	0.100	0.100	0.000	0.000
USG	–	0.043	0.618	0.923
Dehydration %	–	0.048	–	–

Table 2 Cognitive variances and significance in volleyball and kabaddi across intensity levels pre-training and post-training

Physiological variables	Volleyball high intensity	Volleyball moderate intensity	Kabaddi high intensity	Kabaddi moderate intensity
Pre-training				
Concentration	11.66	11.66	14.90	16.07
Coordination	14.65	14.65	15.23	19.50
Reaction time	0.15	0.15	0.16	0.15
Post-training				
Concentration	10.04	12.57	12.66	16.07
Coordination	13.10	14.28	13.57	19.50
Reaction time	0.11	0.17	0.14	0.15
RPE	7.86	5.43	–	–
P-value				
Reaction time	0.000	–	0.955	–
RPE	0.000	–	–	–

4 Discussion

The comparison of physiological parameters among volleyball athletes engaged in high-intensity (HI) and moderate-intensity (MI) training unveiled crucial distinctions in cardiovascular responses and hydration markers. Notably, despite non-significant differences in weight and resting heart rate pre- and post-training, a significant disparity in post-training maximum heart rates between HI and MI groups pointed to distinct cardiovascular adaptations induced by varying training intensities. These findings align with studies emphasizing the impact of exercise intensity on cardiovascular responses and hydration status among athletes [5, 6]. Hydration and electrolyte balance play pivotal roles in athletic performance, especially during high-intensity activities and in warm climates. The observed variations in urine-specific gravity and dehydration percentages post-training highlight the influence of training intensity on hydration markers among volleyball players. Optimal fluid replacement strategies during and after exercise are essential to prevent hypohydration, which can impair performance and strain the cardiovascular system. Thermoregulation remains critical for athletes to prevent adverse conditions, emphasizing the need to balance body temperature while ensuring adequate hydration levels through sweat. Adaptations in elite athletes, such as increased plasma volume, aid in better thermoregulation during exercise. However, dehydration-induced increases in blood viscosity may pose risks to cardiovascular health [7]. The impact of dehydration extends beyond cardiovascular effects. Gastrointestinal distress during intense exercise is a common occurrence, although moderate physical activity may mitigate such issues [8]. Additionally, dehydration's influence on kidney function could contribute to kidney stone formation and potentially accelerate chronic kidney disease progression [9, 10]. Volleyball and kabaddi present varying physiological demands due to the diverse nature of these sports [11, 12]. Volleyball's requirements for a blend of strength, endurance, speed, and agility differ from kabaddi's reliance on anaerobic energy systems and neuromuscular coordination. Athletes in team sports like volleyball and kabaddi encounter substantial sweat losses, impacting cognitive, technical, and physical performance, especially in high heat conditions [13]. The prevalence of hypohydration among elite athletes, despite access to fluids, suggests the need for vigilant hydration strategies, considering its impact on cognitive function and mood [14]. Assessment methods like reaction time tests and perceived exertion rate scales provide non-invasive means to effectively measure hydration levels in athletes [15] (Matthew S., Aric J. Warren, et al., 2018).

5 Conclusion

The study elucidates distinct responses to varying training intensities among volleyball and kabaddi athletes, showcasing significant differences in Maximum Heart Rate (HR), Rate of Perceived Exertion (RPE), post-training Urine-Specific Gravity (USG), and reaction time. While volleyball athletes exhibited significant disparities in these parameters across intensity levels, other measured factors such as concentration, weight, and USG displayed no significance. Conversely, kabaddi athletes showcased pronounced significance in maximum HR and RPE at both high and moderate intensities, while none of the other parameters exhibited significance. These findings collectively underscore the profound impact of dehydration on the physiological and cognitive performance of athletes in both sports, emphasizing the need for tailored hydration strategies to optimize performance and well-being across varied athletic demands and intensities.

References

1. Sawka MN, Pandolf KB (1990) Effects of body water loss on physiological function and exercise performance. In: Gisolfi CV, Lamb DR (eds) Perspectives in exercise science and sports medicine, vol 3. Fluid Homeostasis During Exercise, pp 1–38
2. Lieberman MD (2007) Social cognitive neuroscience: a review of core processes. Annu Rev Psychol 58:259–289. https://doi.org/10.1146/annurev.psych.58.110405.085654
3. Zhang N, Du S, Tang Z, Zheng M, Ma G (2017) Effect of water supplementation on cognitive performances and mood among male college students in Cangzhou, China: study protocol of a randomized controlled trial. Int J Environ Res Public Health 14(9):966. https://doi.org/10.3390/ijerph14090966
4. Wenger CB (1988) Human heat acclimatization. In: Pandolf KA, Burr MO (ed) Human performance physiology and environmental medicine at terrestrial extremes. pp 139–159

5. Bergeron MF, Waller JL, Marinik EL (2006) Voluntary fluid intake and core temperature responses in adolescent tennis players: sports beverage versus water. Br J Sports Med 40(5):406–410. https://doi.org/10.1136/bjsm.2005.023333
6. Buono MJ, Claros R, Deboer T, Wong J (2008) Na+ secretion rate increases proportionally more than the Na+ reabsorption rate with increases in sweat rate. J Appl Physiol (Bethesda, Md.: 1985) 105(4):1044–1048. https://doi.org/10.1152/japplphysiol.90503.2008
7. Jan KM, Chien S, Bigger JT Jr (1975) Observations on blood viscosity changes after acute myocardial infarction. Circulation 51(6):1079–1084. https://doi.org/10.1161/01.cir.51.6.1079
8. Van Nieuwenhoven MA, Brummer RM, Brouns F (2000) Gastrointestinal function during exercise: comparison of water, sports drink, and sports drink with caffeine. J Appl Physiol (Bethesda, Md.: 1985) 89(3):1079–1085. https://doi.org/10.1152/jappl.2000.89.3.1079
9. Chan J, Knutsen SF, Blix GG, Lee JW, Fraser GE (2002) Water, other fluids, and fatal coronary heart disease: the adventist health study. Am J Epidemiol 155(9):827–833. https://doi.org/10.1093/aje/155.9.827
10. Bongers CCWG, Alsady M, Nijenhuis T, Tulp ADM, Eijsvogels TMH, Deen PMT, Hopman MTE (2018) Impact of acute versus prolonged exercise and dehydration on kidney function and injury. Physiol Rep 6(11):e13734. https://doi.org/10.14814/phy2.13734
11. Lotan Y, Antonelli J, Jiménez IB et al (2017) The kidney stone and increased water intake trial in steel workers: results from a pilot study. Urolithiasis 45:177–183. https://doi.org/10.1007/s00240-016-0892-7
12. Rai V (2013) A study on physiological characteristics of national volleyball players. Res Hum Soc Sci 3(2). ISSN 2222–1719 (Paper) ISSN 2222–2863 (Online)
13. Nuccio RP, Barnes KA, Carter JM, Baker LB (2017) Fluid balance in team sport athletes and the effect of hypohydration on cognitive, technical, and physical performance. Sport Med (Auckland, N.Z.) 47(10):1951–1982. https://doi.org/10.1007/s40279-017-0738-7
14. Dube A, Gouws C, Breukelman G (2022) Effects of hypohydration and fluid balance in athletes' cognitive performance: a systematic review. Afr Health Sci 22(1):367–376. https://doi.org/10.4314/ahs.v22i1.45
15. Jain S, Dhiman S (2019) Analysis of reliability of audio-visual reaction time machine and ruler drop test. 6:693–697

Modeling of Cricket Ball Flight Using Navier–Stokes Equations

Rishi Kumar Pandey and Abhijeet K. Digalwar

Abstract This research study investigates the use of the Navier–Stokes equations to examine a cricket ball's flight. The trajectory of the ball is of paramount importance in the game, influencing the strategies employed by both bowlers and batsmen. Understanding the aerodynamics of the ball's flight is crucial for improving performance and developing new techniques. The Navier–Stokes equations, which govern fluid dynamics, offer a promising approach for studying and predicting ball trajectories. This paper discusses the significance of cricket ball flight and provides a concise explanation of the relevance of the Navier–Stokes equations. It emphasizes the need to investigate the application of these equations in analyzing cricket ball flight. By identifying gaps in the existing literature, the present study establishes a foundation for further investigation. The accuracy and limitations of the equations in modeling cricket ball flight are thoroughly examined and compare the results with existing theories and models. Overall, this research demonstrates the potential of the Navier–Stokes equations in analyzing cricket ball flight, contributing to the understanding of the game's aerodynamics, and providing valuable insights for enhancing performance and developing new techniques.

Keywords Navier–stokes equations · Numerical simulations · Mathematical modeling · Cricket ball · Trajectory

R. K. Pandey (✉)
Symbiosis Centre for Management Studies (SCMS), Symbiosis International (Deemed University) (SIU), Mouza-Wathoda, Nagpur, Maharastra, India
e-mail: rishi.research.pandey@gmail.com

A. K. Digalwar
Department of Mechanical Engineering, Birla Institute of Technology and Science, Pilani, India
e-mail: akd@pilani.bits-pilani.ac.in

S. Bhattacharyya and D. James (eds.), *Recent Trends in Sports Engineering*, Lecture Notes in Mechanical Engineering, https://doi.org/10.1007/978-981-97-8297-0_10

1 Introduction

The influence of fluid dynamics on sports equipment and athletic performance is significant and far-reaching. The behavior of various sports balls, such as cricket balls, soccer balls, tennis balls, and badminton shuttlecocks, during their flight is strongly influenced by the forces acting on them. Understanding the complex fluid dynamics involved in these sports is crucial for accurate analysis and prediction of ball trajectories. This research article focuses on the modeling of cricket ball flight using numerical simulations based on the Navier–Stokes equations.

Previous research efforts have aimed to investigate the flight characteristics of a cricket ball, with specific emphasis on conventional swing. Researchers have undertaken experimental investigations to examine the forces acting on the ball throughout its trajectory. Remarkably, the simulated results obtained using COMSOL Multiphysics® in this study align with the experimental findings of these previous researchers, validating the accuracy of the numerical approach [1].

In addition to conventional swing, this study delves into the variations in side and drag forces on the ball. By examining parameters such as ball velocity, seam angle, and backspin, a comprehensive understanding of the aerodynamic forces influencing the ball's flight is sought. It is important to keep in mind, nevertheless, that the k-ε turbulence model used in the simulations could not accurately represent the flow characteristics surrounding a cricket ball, necessitating more research and possibly model improvement [1].

Research endeavors in sports equipment and its aerodynamics extend beyond cricket. Force measurements have been carried out on a range of sports artifacts, including golf balls, field hockey balls, soccer balls, baseballs, tennis balls, cricket balls, volleyballs, and badminton shuttlecocks [2]. Interestingly, it was discovered that the artificial and duck-feather shuttlecock designs' coefficient of drag (CD) varied differently with Reynolds number (Re), emphasizing the need of taking particular sports equipment attributes into account in aerodynamic assessments [2].

Moreover, the flight characteristics of sports balls can be influenced by minute disturbances on their surfaces. The unpredictable trajectories of baseballs and cricket balls, tennis balls with felt covers, and volleyballs and soccer balls featuring patch seams stem from the distinctive characteristics of their respective designs [3]. These intricate surface details contribute to the complexity of ball flight dynamics and warrant thorough investigation.

Additionally, the use of inactive everlasting magnetic bearings has shown promise in improving perpendicular axis wind turbine efficiency. In particular, the arrangement of stacked rings of passive permanent magnetic bearings showed significant increases in rotational speed for a small cross-flow wind turbine with a vertical axis [4]. This suggests that magnetic bearing technology could be applied to optimize the performance of cricket balls or other sports equipment subjected to fluid forces.

Moreover, variations in the hydrodynamics of fluids arise from their distinct properties. Take water, for example, which exhibits a higher defenselessness to sound energy in contrast to extra fluids, whereas the impact of oil on sound energy proves more pronounced [5]. These characteristics of fluids can have implications for the flight dynamics of sports balls and should be considered when analyzing their trajectories.

In summary, comprehending the fluid dynamics governing the flight of a cricket ball is crucial for accurately analyzing its trajectory. This research article presents a numerical simulation approach utilizing the Navier–Stokes equations to investigate the flight characteristics of a cricket ball. By considering factors such as ball velocity, seam angle, backspin, surface disturbances, and the potential application of magnetic bearings, we aim to enhance our understanding of the intricate aerodynamic forces influencing the ball's flight path.

Large-scale wind tunnel experiments have shown that positional modifications of even a few degrees may have a substantial impact on the lift and drag areas of sporting goods [6]. To predict flow fields and evaluate aerodynamic forces in sporting applications, commercial computational fluid dynamics (CFD) simulation tools offer an alternative to wind tunnel measurements [6]. Notably, a human body model was developed using SOLIDWORKS® and imported into COMSOL Multiphysics® software to assess flow fields and aerodynamic forces for different angles and velocities in the context of ski jumping [6]. The improvement of Reynolds-Averaged Navier–Stokes (RANS) models is the aim of advancements in golf disc aerodynamics [7]. A comparative study of the comparing the k-ε shear stress transport model and the lag elliptic blending k-ε turbulence model improved accuracy and performance in CFD assessments of golf discs [7]. With implications for experts in sports and automotive engineering, our findings resolve difficulties in Reynolds Average Navier–Stokes turbulence models related to calculating turbulent kinetic energy and awaken size for bluff physiques [7].

Applying Navier–Stokes computational fluid dynamics simulations, efforts have been made to reduce aerodynamic drag and enhance yaw stability in golf drivers while ensuring compliance with USGA standards [8]. Employing the Reynolds-Averaged Navier–Stokes equation in conjunction with a time-domain flow solver has enabled the calculation of aerodynamic derivatives at transonic speeds, enriching insights into aerodynamic behavior [9].

A thorough analysis is given, encompassing the fundamentals of the RANS method as well as the main types of turbulence models that are used to calculate Reynolds stresses [10]. Furthermore, the Aeroelastic Prediction Workshop serves as a platform for computational fluid dynamicists to collaborate and establish the cutting-edge standards in computational aeroelasticity [11]. The solutions have been derived using the Implicit Reynolds-averaged NS solver (IMPRANS) using finite volume nodal point spatial discretization schemes [12].

The studies and methodologies discussed above highlight the significance of CFD simulations, turbulence modeling, and computational approaches in investigating and optimizing the aerodynamics of sports equipment. These advancements have implications for sports researchers, equipment designers, and engineers seeking to improve performance and efficiency while adhering to specific design constraints.

The authors report on a numerical simulation they ran to investigate the aerodynamic features of a non-slender BWB UCAV design through a 1.0 m span and a rounded leading edge [13]. Utilizing closure models based on simplifying assumptions, the study of high-Reynolds-number flows in computational fluid dynamics has made use of the NS equations [14].

Research has delved into the path of a cricket ball experiencing rectilinear drag force, involving the formulation of motion equations and the application of perturbation extension techniques to address the ensuing transcendental trajectory equation [15]. Investigation extends to the examination of fluid flow characteristics in a confined domain, considering stationary Stokes and NS equations alongside nonhomogeneous Navier boundary conditions [16]. The utilization of Computational Fluid Dynamics (CFD) software, particularly FLOTRAN, has facilitated the analysis of air flow patterns around a cricket ball, incorporating the creation of 3D spherical models [17].

In addition, drag, lift, and side force coefficients have been predicted using superficial covering meshing techniques and the RANS approach with the reachable k-ε turbulence model. This methodology enables the comparison of wake behavior and furnishes precise pressure distributions in close proximity to the stagnation point [18].

Cycling aerodynamics have been studied using both computational and experimental approaches, with particular attention to drag force variations and downstream movement construction with respect to crank angle changes [19].

Extensive analyses using commercial packages and Computational Fluid Dynamics (CFD) methods have examined the effects of surface roughness, bowling speed, and seam angle on cricket ball swing [20]. This study offers a thorough analysis of the aerodynamic characteristics connected to a cricket ball and offers scientific insights into the variables controlling cricket ball swing [21]. Moreover, innovative techniques for inducing swing in a cricket ball have been suggested, accompanied by compelling photographs, thus advancing the understanding and theory of swing and reverse swing generation [22].

The studies discussed in this article showcase the diverse applications of computational fluid dynamics, numerical simulations, and analytical techniques in understanding the aerodynamic characteristics and flight behavior of various objects, ranging from UCAV configurations to cricket balls. These investigations have contributed to the development of improved modeling approaches, shedding light on complex fluid flow phenomena and their impact on sports equipment and other engineering applications.

Overall, the research conducted in these studies has advanced our understanding of fluid dynamics and provided valuable insights for sports researchers, engineers, and designers, fostering innovation and optimization in the field of aerodynamics.

The variational iteration method (VIM) and its modified version, modified variational iteration technique (MVIM), are widely used in fluid dynamics and sports engineering to solve nonlinear partial differential equations. Prominent for its precision and effectiveness, the VIM has a simple implementation in contrast to substitutes such as the Adomian decomposition technique [23]. Similarly, the MVIM is a trustworthy and effective method that does away with the need for Adomian's polynomials for explaining singular fourth-order parabolic partial differential equations (PDEs) [24].

To broaden the convergence scope in dealing with nonlinear issues, the MVIM has been augmented with Pade approximants. This enhancement allows for the solution of equations such as Lax's equations, KdV, Burger's and mKdV [25]. Furthermore, the fusion of the Laplace transform (LT) and the MVIM has been successfully applied in resolving both linear and nonlinear partial differential equations (PDEs) across diverse fields of engineering and physics. This approach ensures simplicity, reliability, and a reliance solely on initial conditions [26].

The efficacy of the VIM in approximating analytical solutions has been demonstrated by investigations conducted on fractional-order partial integro-differential equations [27]. Additionally, the LVIM has been used as a potent method to solve fractional differential equations and nonlinear equations. Approximate analytical solutions can be achieved without discretization or restrictive assumptions by combining the Laplace transform approach with the variational iteration method [30].

It has been shown that the LVIM is useful for solving fractional integro-differential equations and linear and nonlinear PDEs [31, 32]. The results obtained from this technique demonstrate its effectiveness, accuracy, and resistance to round-off mistakes. By using this method, the reliance on numerical calculations is reduced, and results are obtained that are almost identical to the precise answer [33, 34].

This paper applies the VIM (particularly when combined with the LT) to solve nonlinear ordinary differential equations that are commonly found in boundary layer heat and mass transfer issues. Often, these equations have infinite boundary conditions [28]. Attractive features of the proposed method include its use of large integration time steps, strict enforcement of incompressibility, unified management of fluid and solid phases, and linear scaling. These characteristics increase its capacity to handle complex issues in sports engineering and fluid dynamics [29].

This research paper's following parts go into great detail on the suggested approaches for simulating cricket ball flight. In Sect. 2, basic ideas and notation norms are introduced, setting the stage for further investigations. In order to validate the effectiveness of the VIM in solving mathematical challenges, Sect. 3 focuses on applying it to functions that have two space variables. In Sect. 4, the general version of the NS equation for incompressible fluids is solved using a new fusion of the Laplace transform (LT) and the variational iteration method (VIM). Section 5 provides a detailed description of the technique and data generating procedure for numerical experiments. The results and a thorough analysis are shown in Sect. 6, which also assesses the usefulness and future prospects of the suggested techniques in fluid dynamics and sports engineering. Section 7 provides a summary of the paper's major discoveries, highlighting the importance of the study and outlining potential future prospects for additional developments in this area.

2 Initial Concepts and Notation Conventions

In this section, we will engage in an exhaustive discussion that covers the definitions and properties of LT. By delving into the fundamental notions and characteristics of LT, our goal is to establish a robust comprehension of this potent mathematical technique.

2.1 *Laplace Transform (LT)*

Let $W(t)$ be a function of t defined for all $t \geq 0$. Then the LT of $W(t)$, is defined as

$$\mathcal{L}\{W(t)\} = \int_0^\infty e^{-st} W(t)dt = \overline{W}(s),$$

if integral exists, given that "s" is the parameter,

Hence, $\mathcal{L}\{W(t)\} = \overline{W}(s)$,

i.e., $W(t) = \mathcal{L}^{-1}\{\overline{W}(s)\}$, where $\mathcal{L}^{-1}\{\overline{W}(s)\}$, represents the inverse of LT.

2.2 *Linear Behavior of LT*

The LT has a key property known as "linearity":

$$\mathcal{L}\{\mathcal{W}_1 a(t) + b\mathcal{W}_2(t)\} = a\mathcal{L}\{\mathcal{W}_1(t)\} + b\mathcal{L}\{\mathcal{W}_2(t)\}$$

where $\mathcal{W}_1(t)$ and $\mathcal{W}_2(t)$ are LT, and "a" and "b" are arbitrary constants. This property simplifies solving linear differential equations and analyzing linear systems.

2.3 *Linear Behavior of Inverse LT*

The linear behavior of the inverse LT asserts that the inverse LT of a linear combination of functions equals the same linear combination of their individual inverse LT. In mathematical terms, the linear behavior of the inverse LT can be articulated as follows:

$$\mathcal{L}^{-1}\{a\overline{\mathcal{W}_1}(s) + b\overline{\mathcal{W}_2}(s)\} = a\mathcal{L}^{-1}\{\overline{\mathcal{W}_1}(s)\} + b\mathcal{L}^{-1}\{\overline{\mathcal{W}_2}(s)\}$$

where:

- $\mathcal{L}^{-1}\{\}$ represents the inverse LT operator,
- $\mathcal{L}^{-1}\left\{\overline{\mathcal{W}_1}(s)\right\}$ and $\mathcal{L}^{-1}\left\{\overline{\mathcal{W}_2}(s)\right\}$ are the inverse LT of $\overline{\mathcal{W}_1}(s)$ and $\overline{\mathcal{W}_2}(s)$, respectively,
- a and b are fixed scalars.

In other words, if we have two LTs of $\overline{\mathcal{W}_1}(s)$ and $\overline{\mathcal{W}_2}(s)$, and we want to find the inverse LT of the linear combination $a\overline{\mathcal{W}}_1(s) + b\overline{\mathcal{W}}_2(s)$, we can do it by taking the inverse LT of each separate function and then multiplying each transform by its corresponding scalar constant and summing up the results.

This property is essential in solving linear differential equations and analyzing linear time-invariant systems, as it allows us to transform complex frequency-domain representations into time-domain functions, making it easier to work with the original signals or systems.

2.4 *Generalized LT of Derivatives*

The LT of $\mathcal{W}(t)$, representing the derivative of a function in a generalized form, is expressed as:

$$\mathcal{L}\left\{\frac{d^n\mathcal{W}(t)}{dt^n}\right\} = s^n\overline{\mathcal{W}}(s) - s^{n-1}\overline{\mathcal{W}}(0) - s^{n-2}\overline{\mathcal{W}}'(0) - \ldots - s\overline{\mathcal{W}}^{n-2}(0) - s\overline{\mathcal{W}}^{n-1}(0)$$

where:

- $\mathcal{L}\{\}$ represents the LT operator,
- $\frac{d^n\mathcal{W}(t)}{dt^n}$ symbolizes the nthnth rate of change of the function $\mathcal{W}(t)$ concerning time t,
- $\overline{\mathcal{W}}(s)$ is the LT of $\overline{\mathcal{W}}(s)$.

3 VIM for a Function with Three Spatial Variables

Addressing nonlinear terms through Variational Iterative Method (VIM): Examining the differential equation,

$$\mathcal{LW}(x, y, z, t) + \mathcal{NW}(x, y, z, t) = \mathcal{GW}(x, y, z, t) \tag{1}$$

Under the given Initial Conditions (ICs)

$$\mathcal{W}(x, y, z, 0) = \mathcal{H}(x, y, z) \tag{2}$$

Incorporating a linear operator "$\mathcal{L}$," nonlinear operator "$\mathcal{N}$," and nonhomogeneous function "$\mathcal{G}$": Utilizing the VIM to formulate the correction functional,

$$\mathcal{W}_{m+1} = \mathcal{W}_m + \int_0^t \lambda \Big[\mathcal{L}\mathcal{W}(x, y, z, t) + \mathcal{N}\mathcal{W}(x, y, z, t) - \mathcal{G}\hat{\mathcal{W}}(x, y, z, t) \Big] ds \tag{3}$$

Introducing Lagrange's Multiplier λ and m^{th} estimates for restricted function $\delta\hat{\mathcal{W}}_m = 0$: Deriving the solution $\mathcal{W}$ using λ and $\mathcal{W}_0$, with $\mathcal{W}_{m+1}$ as the iterative approximation is obtained as $\mathcal{W} = \lim_{m\to\infty} \mathcal{W}_m$,

where "s" is the complex variable used in the LT.

This formula permits us to find the LT of a higher-order derivative of a function based on its Laplace transform $\overline{\mathcal{W}}(s)$ and the initial values of the function and its derivatives at $t = 0$. It is a valuable tool in solving linear differential equations involving derivatives, as it enables us to convert differential equations into algebraic equations in the Laplace domain.

4 Utilizing a Fusion of LT and VIM to Address the Generalized Form of the Navier–Stokes Equation in Incompressible Fluids

In this segment, we demonstrate the application of a fusion approach involving LT and the modified VIM to solve the modeling of cricket ball flight using the NS equations. The analytical solution for this equation is derived in the form of a convergent series. We begin by considering the following representation of the NS equations:

$$\partial\mathcal{W}/\partial t + (\mathcal{W} \cdot \nabla)\mathcal{W} = -\frac{1}{\rho}\nabla P + \nu\nabla\mathcal{W} \tag{4}$$

where:

- $\mathcal{W}$ is the velocity vector of the fluid,
- $\partial\mathcal{W}/\partial t$ represents the partial derivative of $\mathcal{W}$ with respect to time,
- ∇ is the gradient operator,
- P is the pressure,
- ρ is the density of the fluid,
- ∇^2 is the Laplacian,
- ν is the kinematic viscosity of the fluid.

In Cartesian coordinates, the velocity vector $\mathcal{W}$ can be expressed as $\mathcal{W} = (w_x, w_y, w_z)$ with w_x, w_y and w_z denoting velocity along the x, y, and z axes, correspondingly.

Expanding the convective term $(w \cdot \nabla)w$, we have

$$\left(w_x\frac{\partial w_x}{\partial x} + w_y\frac{\partial w_x}{\partial y} + w_z\frac{\partial w_x}{\partial z}\right)i + \left(w_x\frac{\partial w_y}{\partial x} + w_y\frac{\partial w_y}{\partial y} + w_z\frac{\partial w_y}{\partial z}\right)j + \left(w_x\frac{\partial w_z}{\partial x} + w_y\frac{\partial w_z}{\partial y} + w_z\frac{\partial w_z}{\partial z}\right)k \tag{5}$$

Therefore, using (5), the Cartesian form of the NS equation becomes

$$\begin{aligned}
\frac{\partial w_x}{\partial t} + w_x\frac{\partial w_x}{\partial x} + w_y\frac{\partial w_x}{\partial y} + w_z\frac{\partial w_x}{\partial z} &= \frac{-1}{\rho}\frac{\partial P}{\partial x} + \nu\left(\frac{\partial w_x}{\partial x} + \frac{\partial w_x}{\partial y} + \frac{\partial w_x}{\partial z}\right)\\
\frac{\partial w_y}{\partial t} + w_x\frac{\partial w_y}{\partial x} + w_y\frac{\partial w_y}{\partial y} + w_z\frac{\partial w_y}{\partial z} &= \frac{-1}{\rho}\frac{\partial P}{\partial y} + \nu\left(\frac{\partial w_y}{\partial x} + \frac{\partial w_y}{\partial y} + \frac{\partial w_y}{\partial z}\right)\\
\frac{\partial w_z}{\partial t} + w_x\frac{\partial w_z}{\partial x} + w_y\frac{\partial w_z}{\partial y} + w_z\frac{\partial w_z}{\partial z} &= \frac{-1}{\rho}\frac{\partial P}{\partial z} + \nu\left(\frac{\partial w_z}{\partial x} + \frac{\partial w_z}{\partial y} + \frac{\partial w_z}{\partial z}\right)
\end{aligned} \tag{6}$$

The equations provided describe the temporal changes in the velocity components w_x, w_y, and w_z, as a function of their partial derivatives with respect to time and spatial coordinates. This $\frac{\nabla P}{\rho}$ represents the force per unit volume exerted by the fluid through pressure effects, while the term $(\nabla^2 w)$ accounts for the momentum's diffusion within the fluid. These formulas provide a basic picture of fluid dynamics and make it possible to comprehend the dynamics and behavior of the fluid medium in question in great detail. By analyzing the interplay of these terms, researchers can gain valuable insights into fluid flow patterns, pressures, and viscosity effects, crucial in various engineering, physics, and environmental studies.

With ICs,

$$\begin{aligned}
&w_x(x, y, z, 0) = f_1(x, y, z),\ w_y(x, y, z, 0) = f_2(x, y, z),\\
&w_z(x, y, z, 0) = f_3(x, y, z)
\end{aligned} \tag{7}$$

Upon employing the LT on both sides of Eq. (4), we obtain

$$\mathcal{L}\{\partial \mathcal{W}/\partial t + (\mathcal{W} \cdot \nabla)\mathcal{W}\} = \mathcal{L}\left\{-\frac{1}{\rho}\nabla P + \nu\nabla\mathcal{W}\right\} \tag{8}$$

$$s\mathcal{L}\{\mathcal{W}\} - f(x, y, z) + \mathcal{L}\{(\mathcal{W} \cdot \nabla)\mathcal{W}\} = \mathcal{L}\left\{-\frac{1}{\rho}\nabla P + \nu\nabla\mathcal{W}\right\} \tag{9}$$

where $f(x, y, z) = (f_1(x, y, z), f_2(x, y, z), f_3(x, y, z))$

$$\mathcal{L}\{\mathcal{W}\} = \frac{f(x, y, z)}{s} - \frac{1}{s}\mathcal{L}\left\{(\mathcal{W} \cdot \nabla)\mathcal{W} + \frac{1}{\rho}\nabla P - \nu \nabla \mathcal{W}\right\} \tag{10}$$

Utilizing Inverse LT on both sides of Eq. (10), yields

$$\mathcal{W} = \Theta(x, y, z) - \mathcal{L}^{-1}\left[\frac{1}{s}\mathcal{L}\{(\mathcal{W} \cdot \nabla)\mathcal{W} + 1/\rho \nabla P - \nu \nabla \mathcal{W}\}\right] \tag{11}$$

The term $\Theta(x, y, z)$ originates from the specified ICs. By incorporating the corrective functional of the VIM, we obtain

$$\mathcal{W}_{m+1} = \Theta(x, y, z) - \mathcal{L}^{-1}\left[\frac{1}{s}\mathcal{L}\{(\mathcal{W}_{m+1} \cdot \nabla)\mathcal{W}_{m+1} + 1/\rho \nabla P - \nu \nabla \mathcal{W}_{m+1}\}\right] \tag{12}$$

Equation (12) introduces the updated correctional functional, unveiling the solution

$$\mathcal{W} = \lim_{m \to \infty} \mathcal{W}_m \tag{13}$$

By using the Cartesian form of (6), the component solution of (13) in the directions of x, y, and z is determined.

5 Mathematical Simulation

Taking initial condition for the velocity of cricket flight at initial stage

$$\begin{aligned} &w_x(x, y, z, 0) = D \sin(\eta x)e^{-\alpha y}, \ w_y(x, y, z, 0) = E \cos(\eta y)e^{-\alpha z}, \\ &w_z(x, y, z, 0) = F \sin(\eta z)\, e^{-\alpha x} \end{aligned} \tag{14}$$

By decomposing the vector Eq. (12) into its three components along the x, y, and z axes and employing Eq. (6), we can deduce additional information. Furthermore, if the pressure term P remains constant, consequently, $\frac{\partial P}{\partial x} = 0$, $\frac{\partial P}{\partial y} = 0$, and $\frac{\partial P}{\partial z} = 0$.

$$\begin{aligned} (w_x)_{m+1} &= D \sin(\eta x)e^{-\alpha y} - \mathcal{L}^{-1}\left[\frac{1}{s}\mathcal{L}\left\{(w_x)_{m+1}\frac{\partial (w_x)_{m+1}}{\partial x} + \left(w_y\right)_{m+1}\frac{\partial (w_x)_{m+1}}{\partial y}\right.\right. \\ &\left.\left.+ (w_z)_{m+1}\frac{\partial (w_x)_{m+1}}{\partial z} - \nu\left(\frac{\partial (w_x)_{m+1}}{\partial x} + \frac{\partial (w_x)_{m+1}}{\partial y} + \frac{\partial (w_x)_{m+1}}{\partial z}\right)\right\}\right] \end{aligned} \tag{15}$$

$$\left(w_y\right)_{m+1} = E\cos(\eta y)e^{-\alpha z} - \mathcal{L}^{-1}\left[\frac{1}{s}\mathcal{L}\left\{(w_x)_{m+1}\frac{\partial\left(w_y\right)_{m+1}}{\partial x} + \left(w_y\right)_{m+1}\frac{\partial\left(w_y\right)_{m+1}}{\partial y}\right.\right.$$
$$\left.\left. + (w_z)_{m+1}\frac{\partial\left(w_y\right)_{m+1}}{\partial z} - \nu\left(\frac{\partial\left(w_y\right)_{m+1}}{\partial x} + \frac{\partial\left(w_y\right)_{m+1}}{\partial y} + \frac{\partial\left(w_y\right)_{m+1}}{\partial z}\right)\right\}\right] \quad (16)$$

$$\left(w_y\right)_{m+1} = F\sin(\eta z)\,e^{-\alpha x} - \mathcal{L}^{-1}\left[\frac{1}{s}\mathcal{L}\left\{(w_x)_{m+1}\frac{\partial(w_z)_{m+1}}{\partial x} + \left(w_y\right)_{m+1}\frac{\partial(w_z)_{m+1}}{\partial y}\right.\right.$$
$$\left.\left. + (w_z)_{m+1}\frac{\partial(w_z)_{m+1}}{\partial z} - \nu\left(\frac{\partial(w_z)_{m+1}}{\partial x} + \frac{\partial(w_z)_{m+1}}{\partial y} + \frac{\partial(w_z)_{m+1}}{\partial z}\right)\right\}\right] \quad (17)$$

In the process of solving the Navier–Stokes equation through the Laplace-Variational Iteration Method (LVIM), we initiate by applying the Laplace Transform (LT) to both sides of Eqs. (15), (16), and (17), incorporating (7), (8), (9), and (10). This application results in a transformed equation within the Laplace domain. Subsequently, we apply the inverse Laplace Transform (LT), using (11), to transition the outcomes derived from initial conditions (15) back to the time domain, establishing the foundation for the mathematical simulation.

Continuing with the solution process, the Variational Iteration Method (VIM) is employed, utilizing Eqs. (1), (2), and (3). We assume a solution in the form of a series, which is then substituted into the Laplace-transformed Eq. (12). This facilitates the determination of unknown coefficients through iterative procedures. After completing the variational iteration, we once again apply the inverse Laplace Transform to return the solution to the time domain.

The concluding steps involve a thorough verification and validation process, subjecting the obtained solution to rigorous testing against known conditions and physical principles. The primary objective is to ensure that the solution aligns with the specified initial conditions and boundary conditions. In Appendix I, the write-up provides a detailed account of the iterative values of the unknown coefficients, offering valuable insights into the convergence of the method and presenting a comprehensive overview of the entire solution process. This meticulous approach exemplifies the systematic methodology employed in solving the Navier–Stokes equation using the Laplace-Variational Iteration Method.

By focusing on the specific condition (17), we can derive the solution (4) by applying the simplifications outlined in Eq. (13) for each component along the x and y axes. This yields a simplified form of the solution, enabling a more efficient analysis of the system's behavior in each spatial direction.

$$w_x = -Dt\sin(\eta x)e^{-2\alpha y-\alpha z}\big(D\eta e^{\alpha z}\cos(\eta x) + \eta^2 v(-e^{\alpha y+\alpha z})$$
$$-\alpha e^{\alpha y+1}\cos(\eta y) + \alpha^2 v e^{\alpha y+\alpha z}\big) - t\big(D^2\eta e^{-2\alpha y}\sin(\eta x)\cos(\eta x)$$
$$-v\big(\alpha^2 D e^{\alpha(-y)}\sin(\eta x) - D\eta^2 e^{\alpha(-y)}\sin(\eta x)\big) - \alpha D\sin(\eta x)\cos(\eta y)e^{\alpha(-y)-\alpha z+1}\big)$$
$$-\frac{1}{2}De^{-2\alpha y-\alpha z}t^2 v\sin(\eta x)\big(\eta^4 v e^{\alpha y+\alpha z} - 2\alpha^2\eta^2 v e^{\alpha y+\alpha z} + \alpha^4 v e^{\alpha y+\alpha z}$$
$$-4D\eta^3 e^{\alpha z}\cos(\eta x) + \ldots$$

$$w_y = t\cos(\eta y)\big(-e^{\alpha(-x)-2\alpha z+1}\big)\big(-\alpha F e^{\alpha z}\sin(\eta z) + \eta^2 v(-e^{\alpha x+\alpha z})$$
$$-e^{\alpha x+1}\sin(\eta y) + \alpha^2 v e^{\alpha x+\alpha z}\big) - \frac{1}{2}vt^2\cos(\eta y)e^{\alpha(-x)-2\alpha z+1}$$
$$\big(2\alpha F\eta^2 e^{\alpha z}\sin(\eta z) - 2\alpha^3 F e^{\alpha z}\sin(\eta z) + 2\alpha^2 F\eta e^{\alpha z}\cos(\eta z)$$
$$+\eta^4 v e^{\alpha x+\alpha z} + 4\eta^3 e^{\alpha x+1}\sin(\eta y) - 2\alpha^2\eta^2 v e^{\alpha x+\alpha z}$$
$$-4\alpha^2\eta e^{\alpha x+1}\sin(\eta y) + \alpha^4 v e^{\alpha x+\alpha z}\big)$$
$$-\ldots$$

$$w_z = Ft\sin(\eta z)\big(-e^{-2\alpha x-\alpha y}\big)\big(-\alpha D e^{\alpha x}\sin(\eta x) + F\eta e^{\alpha y}\cos(\eta z)$$
$$+\eta^2 v(-e^{\alpha x+\alpha y}) + \alpha^2 v e^{\alpha x+\alpha y}\big) - \frac{1}{2}e^{-2\alpha x-\alpha y}Ft^2 v$$
$$\big(\eta^4 v e^{\alpha x+\alpha y} - 2\alpha^2\eta^2 v e^{\alpha x+\alpha y} + \alpha^4 v e^{\alpha x+\alpha y} + 2\alpha^2 D\eta e^{\alpha x}\cos(\eta x) -$$
$$+\ldots$$

6 Results and Analysis

In this research study, we investigate the behavior of the velocity components $\mathcal{W} = (w_x, w_y, w_z)$ of a cricket ball using the NS equation. We consider fixed values for key parameters, such as D $= 1m\,s^{-1}, E = 2m\,s^{-1}, F = 3m\,s^{-1}, k = \frac{\pi}{6}\,rad\,m^{-1}$, $\alpha = 0.05\,m^{-1}$, and $v = 1.5 \times 10^{-5}\,m^2\,s^{-1}$. Our aim is to gain insights into the variations and patterns of the velocity components in all directions.

The first visualization, shown in Fig. 1, presents a three-dimensional plot of the velocity profile in the x-direction, $w_x(x, t)$ at stable values of $y = z = 1$. We explore the variation of x from $-\pi$ to $+\pi$ and t from *0* to *5*, providing a comprehensive view of the x-directional velocity's evolution over time.

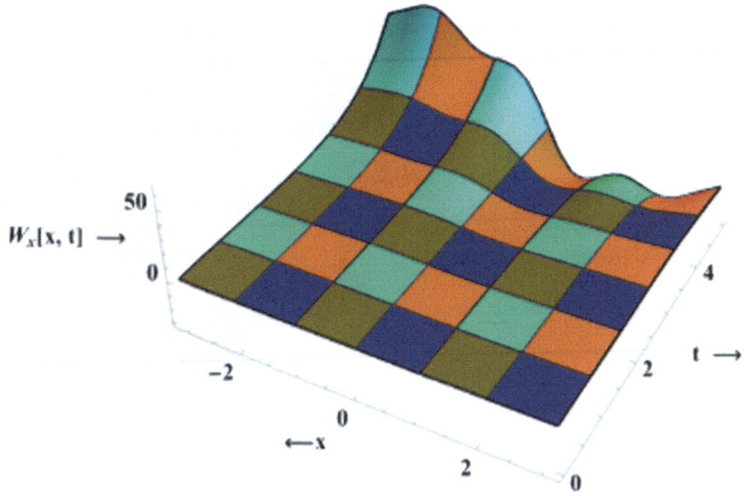

Fig. 1 Velocity profile of cricket ball w_x for the x-direction

Moving on to Fig. 2, we examine the three-dimensional plot of the velocity profile in the y-direction, $w_y(y, t)$ at stable values of $x = z = 1$. The range of y is set from $-\pi$ to $+\pi$, and the time span is from 0 to 5. This graph gives us valuable insights into the lateral movement of the cricket ball along the y-axis, with a slight impact observed on one extreme.

Lastly, in Fig. 3, we delve into the three-dimensional plot of the velocity profile in the z-direction, $w_z(z, t)$ at stable values of $x = y = 1$. Here, z ranges from $-\pi$ to $+\pi$,

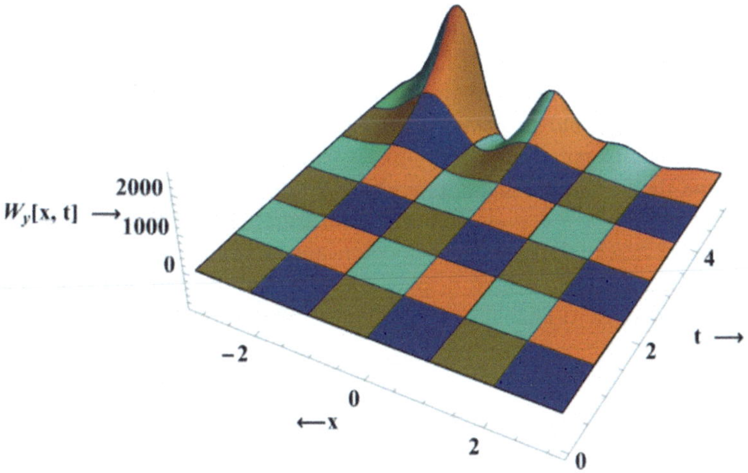

Fig. 2 Velocity profile of cricket ball for the y-direction

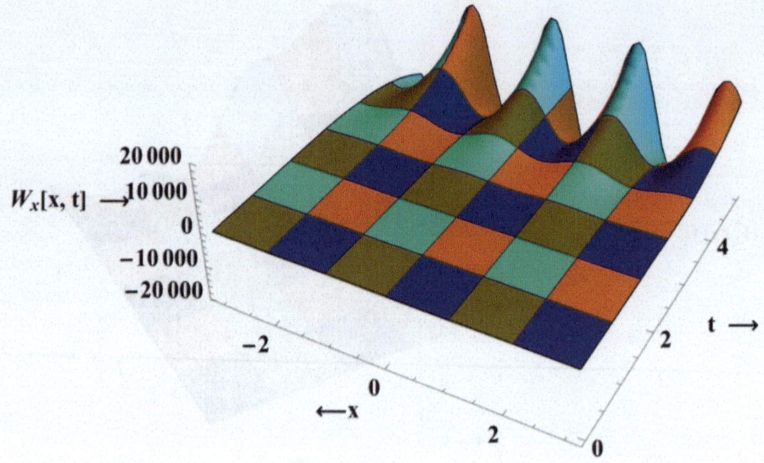

Fig. 3 Velocity profile of cricket ball for the *z*-direction

and t ranges from *0* to *5*. The graph provides a clear visualization of the *z*-directional velocity variation, revealing subtle impacts on one corner.

Through these graphical analyses, we gain a deeper understanding of the intricate velocity patterns and variations in different directions during the flight of a cricket ball. The insights obtained from these visualizations can be crucial for enhancing our knowledge of cricket ball dynamics, potentially leading to improvements in cricket strategies and techniques.

We investigate the velocity characteristics of a cricket ball during its flight using the NS equation, with constant values set at.

$D = 1 m s^{-1}, E = 2 m s^{-1}, F = 3 m s^{-1}, k = \frac{\pi}{6}\ rad\ m^{-1}, \alpha = 0.05\ m^{-1}$, and $\nu = 1.5 \times 10^{-5}\ m^2\ s^{-1}$. Our focus lies in understanding the intricate flow patterns of the ball's velocity in various planes.

Figures 4, 5, and 6 present vector field plots, illustrating the velocity distributions in the *xy*-, *yz*-, and *zx*-planes, respectively. In these plots, arrows represent the flow of velocity patterns. The xy-plane (Fig. 4) showcases how the velocity varies across the horizontal plane, providing insights into the lateral movement of the cricket ball. Figure 5 depicts the velocity distribution in the *yz*-plane, offering a comprehensive view of the ball's behavior along the vertical axis, with intriguing flow patterns observed. Finally, Fig. 6 displays the velocity distribution in the *zx*-plane, emphasizing the velocity changes along the longitudinal direction of the cricket ball.

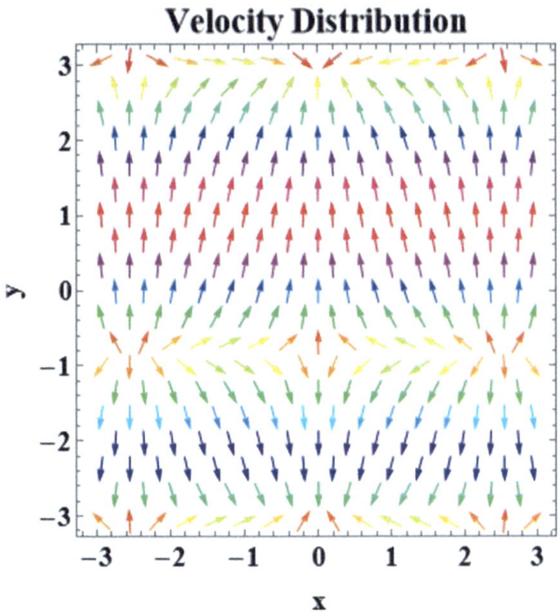

Fig. 4 Velocity distribution of cricket ball for the *xy*-plane

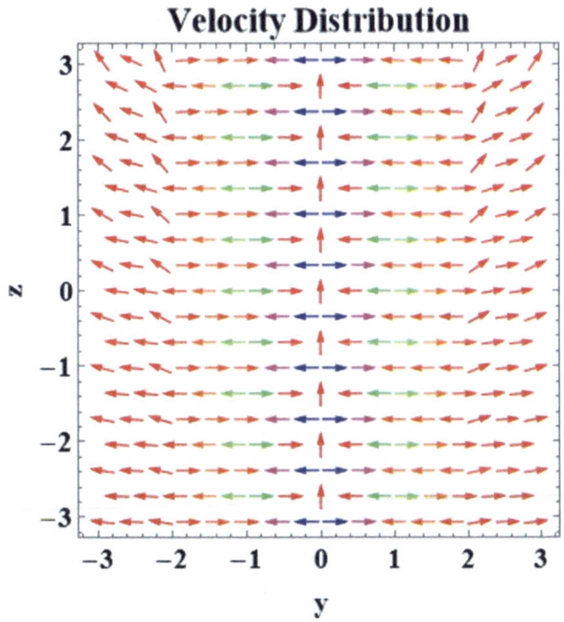

Fig. 5 Velocity distribution of cricket ball for the *zy*-plane

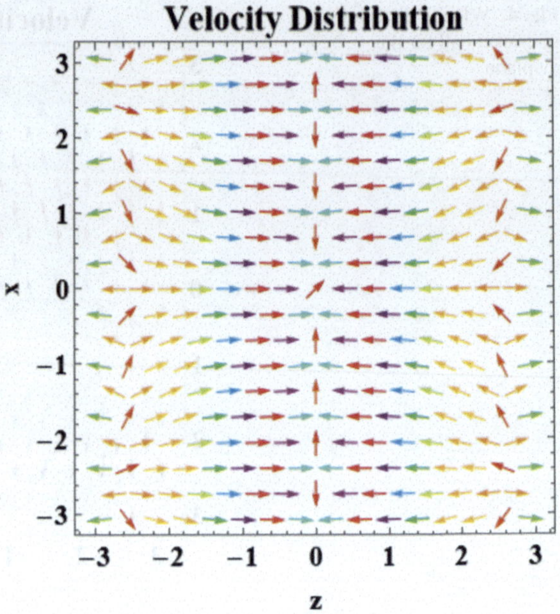

Fig. 6 Velocity distribution of cricket ball for the zx-plane

These vector field plots provide valuable visualizations of the complex velocity patterns during the cricket ball's flight. By analyzing the flow characteristics in different planes, we enhance our understanding of the ball's behavior, enabling potential improvements in cricket strategies and techniques. These findings contribute to the broader knowledge of cricket dynamics and can pave the way for further advancements in the field of sports aerodynamics.

7 Conclusion

In conclusion, our research on the Modeling of Cricket Ball Flight using NS Equations has provided valuable insights into the ball's velocity dynamics. The semi-analytical method, integrating Laplace transform and modified variational iteration, proved effective in analyzing the intricate flow patterns in different spatial planes. This technique holds promise for future applications in sports science and aerodynamics, enabling further advancements in cricket ball flight modeling. The semi-analytical method's versatility can be explored in other sports scenarios, contributing to a deeper understanding of fluid dynamics in various athletic disciplines. Moreover, this research lays the foundation for optimizing player performance, refining cricket strategies, and designing sports equipment for enhanced gameplay. Future studies could extend the method to incorporate additional factors such as spin, wind effects, and ball-material interactions, facilitating a more comprehensive analysis of cricket ball behavior. These avenues of exploration would propel the field of sports aerodynamics, benefiting athletes and sports enthusiasts alike.

Appendix-I

$$(w_x)_0 = D\sin(\eta x)e^{-\alpha y}$$

$$\left(w_y\right)_0 = E\cos(\eta y)e^{-\alpha z}$$

$$(w_z)_0 = F\sin(\eta z)\,e^{-\alpha x}$$

$$(w_x)_1 = -t\Big(D^2\eta\,e^{-2\alpha y}\sin(\eta x)\cos(\eta x) - \nu\Big(\alpha^2 De^{\alpha(-y)}\sin(\eta x) - Dk^2e^{\alpha(-y)}\sin(\eta x)\Big)$$
$$-\alpha D\sin(\eta x)\cos(\eta\,y)e^{\alpha(-y)-\alpha z+1}\Big) - De^{\alpha(-y)}\sin(\eta x)$$

$$\left(w_y\right)_1 = -t\Big(-\alpha F\cos(\eta y)\sin(\eta z)e^{\alpha(-x)-\alpha z+1} - \nu\Big(\alpha^2e^{1-\alpha z}\cos(\eta y) - \eta^2e^{1-\alpha z}\cos(\eta y)\Big)$$
$$-\eta\,e^{2-2\alpha z}\sin(\eta y)\cos(\eta y)\Big) - e^{1-\alpha z}\cos(\eta y)$$

$$(w_z)_1 = F\Big(-e^{\alpha(-x)}\Big)\sin(\eta z) - t\Big(-\alpha DF\sin(\eta x)\sin(\eta z)e^{\alpha(-x)-\alpha y}$$
$$+F^2\eta\,e^{-2\alpha x}\sin(\eta z)\cos(\eta z)$$
$$-\nu\Big(\alpha^2Fe^{\alpha(-x)}\sin(\eta z) - F\eta^2e^{\alpha(-x)}\sin(\eta z)\Big)\Big)$$

$$(w_x)_2 = \frac{1}{2}e^{\alpha(-x)-3\alpha y-2\alpha z}t^2\Big(-D\eta^4\nu^2\sin(\eta x)e^{\alpha x+2\alpha y+2\alpha z}$$
$$+2\alpha^2D\eta^2\nu^2\sin(\eta x)e^{\alpha x+2\alpha y+2\alpha z}$$
$$-\alpha^4D\nu^2\sin(\eta x)e^{\alpha x+2\alpha y+2\alpha z} - 2D^2\eta^3\nu\sin(\eta x)\cos(\eta x)e^{\alpha x+\alpha y+2\alpha z}$$
$$+2\alpha^2D^2\eta\nu\sin(\eta x)\cos(\eta x)e^{\alpha x+\alpha y+2\alpha z}$$
$$+2D^3\eta^2\sin(\eta x)\cos^2(\eta x)e^{\alpha x+2\alpha z} - 4\alpha D^2\eta\sin(\eta x)\cos(\eta x)\cos(\eta y)e^{\alpha x+\alpha y+\alpha z+1}$$
$$+\alpha^2D\sin(\eta x)\cos^2(\eta y)e^{\alpha x+2\alpha y+2} - D^3\eta^2\sin^3(\eta x)e^{\alpha x+2\alpha z}$$
$$-2\alpha^2DF\sin(\eta x)\cos(\eta y)\sin(\eta z)e^{2\alpha y+\alpha z+1}$$
$$+2\alpha D\eta\sin(\eta x)\sin(\eta y)\cos(\eta y)e^{\alpha x+2\alpha y+2}$$
$$+2\alpha^2D\eta\,\nu\sin(\eta x)\sin(y)e^{\alpha x+2\alpha y+\alpha z+1}\Big)$$
$$+\frac{1}{3}e^{-2\alpha x-4\alpha y-3\alpha z}t^3\Big(-D^2\eta^5\nu^2\sin(\eta x)\cos(\eta x)e^{2\alpha x+2\alpha y+3\alpha z}$$
$$+2\alpha^2D^2\eta^3\nu^2\sin(\eta x)\cos(\eta x)e^{2\alpha x+2\alpha y+3\alpha z}$$
$$-\alpha^4D^2\eta\nu^2\sin(\eta x)\cos(\eta x)e^{2\alpha x+2\alpha y+3\alpha z} - 2D^3\eta^4\nu\sin(\eta x)\cos^2(\eta x)e^{2\alpha x+\alpha y+3\alpha z}$$
$$+2\alpha^2D^3\eta^2\nu\sin(\eta x)\cos^2(\eta x)e^{2\alpha x+\alpha y+3\alpha z} - D^4\eta^3\sin(\eta\,x)\cos^3(\eta x)e^{2\alpha x+3\alpha z}$$

$$+\alpha D\eta^4 \nu^2 \sin(\eta x)\cos(\eta y)e^{2\alpha x+3\alpha y+2\alpha z+1}$$
$$-2\alpha^3 D\eta^2 \nu^2 \sin(\eta x)\cos(\eta y)e^{2\alpha x+3\alpha y+2\alpha z+1}$$
$$+\alpha^5 D\nu^2 \sin(\eta x)\cos(\eta y)e^{2\alpha x+3\alpha y+2\alpha z+1}$$
$$+4\alpha D^2\eta^3 \nu \sin(\eta x)\cos(\eta x)\cos(\eta y)e^{2\alpha x+2\alpha y+2\alpha z+1}$$
$$-4\alpha^3 D^2\eta\, \nu \sin(\eta x)\cos(\eta x)\cos(\eta y)e^{2\alpha x+2\alpha y+2\alpha z+1}$$
$$+2\alpha D^3\eta^2 \sin(\eta x)\cos^2(\eta x)\cos(\eta y)e^{2\alpha x+\alpha y+2\alpha z+1}$$
$$-\alpha^2 D\eta^2 \nu \sin(\eta x)\cos^2(\eta y)e^{2\alpha x+3\alpha y+\alpha z+2}$$
$$+\alpha^4 D\nu \sin(\eta x)\cos^2(\eta y)e^{2\alpha x+3\alpha y+\alpha z+2}-$$
$$\alpha^2 D^2\eta \sin(\eta x)\cos(\eta x)\cos^2(\eta y)e^{2\alpha x+2\alpha y+\alpha z+2}$$
$$+D^3\eta^4 \nu \sin^3(\eta x)e^{2\alpha x+\alpha y+3\alpha z}$$
$$-\alpha^2 D^3\eta^2 \nu \sin^3(\eta x)e^{2\alpha x+\alpha y+3\alpha z}+D^4\eta^3 \sin^3(\eta x)\cos(\eta x)e^{2\alpha x+3\alpha z}$$
$$-\alpha D^3\eta^2 \sin^3(\eta x)\cos(\eta y)e^{2\alpha x+\alpha y+2\alpha z+1}$$
$$-2\alpha D\eta^3 \nu \sin(\eta x)\sin(\eta y)\cos(\eta y)e^{2\alpha x+3\alpha y+\alpha z+2}$$
$$+2\alpha^3 D\eta\, \nu \sin(\eta x)\sin(\eta y)\cos(\eta y)e^{2\alpha x+3\alpha y+\alpha z+2}$$
$$-2\alpha D^2\eta^2 \sin(\eta x)\cos(\eta x)\sin(\eta y)\cos(\eta y)e^{2\alpha x+2\alpha y+\alpha z+2}$$
$$+\alpha^2 D\eta \sin(\eta x)\sin(\eta y)\cos^2(\eta y)e^{2\alpha x+3\alpha y+3}$$
$$+\alpha D\eta^2 \sin(\eta x)\sin^2(\eta y)\cos(\eta y)e^{2\alpha x+3\alpha y+3}$$
$$2\alpha^4 DF\nu \sin(\eta x)\cos(\eta y)\sin(\eta z)e^{\alpha x+3\alpha y+2\alpha z+1}$$
$$-2\alpha^2 D^2F\eta \sin(\eta x)\cos(\eta x)\cos(\eta y)\sin(\eta z)e^{\alpha x+2\alpha y+2\alpha z+1}$$
$$+\alpha^3 DF \sin(\eta x)\cos^2(\eta y)\sin(\eta z)e^{\alpha x+3\alpha y+\alpha z+2}$$
$$-\alpha^2 DF^2\eta \sin(\eta x)\cos(\eta y)\sin(\eta z)\cos(\eta z)e^{3\alpha y+2\alpha z+1}$$
$$+-2\alpha^2 DF\eta^2 \nu \sin(\eta x)\cos(\eta y)\sin(\eta z)e^{\alpha x+3\alpha y+2\alpha z+1}$$
$$+\alpha^3 D^2F \sin^2(\eta x)\cos(\eta y)\sin(\eta z)e^{\alpha x+2\alpha y+2\alpha z+1}$$
$$+\alpha^2 DF\eta \sin(\eta x)\sin(\eta y)\cos(\eta y)\sin(\eta z)e^{\alpha x+3\alpha y+\alpha z+2}\Big)$$

$$(w_y)_2 = t\cos(\eta y)\left(-e^{\alpha(-x)-2\alpha z+1}\right)\left(-\alpha F e^{\alpha z}\sin(\eta z) + \eta^2 \nu\left(-e^{\alpha x+\alpha z}\right)\right.$$
$$\left.-\eta e^{\alpha x+1}\sin(\eta y) + \alpha^2 \nu e^{\alpha x+\alpha z}\right)$$
$$-\frac{1}{2}e^{-2\alpha x-\alpha y-3\alpha z}t^2\left(\eta^4\nu^2\cos(\eta y)\left(-e^{2\alpha x+\alpha y+2\alpha z+1}\right) + 2\alpha^2\eta^2\nu^2\cos(\eta y)e^{2\alpha x+\alpha y+2\alpha z+1}\right.$$
$$-\alpha^4\nu^2\cos(\eta y)e^{2\alpha x+\alpha y+2\alpha z+1} - \eta^2\cos^3(\eta y)e^{2\alpha x+\alpha y+3}$$
$$+2\alpha^2 F\eta\nu\cos(\eta y)\cos(\eta z)e^{\alpha x+\alpha y+2\alpha z+1} + 2\eta^3\nu\sin(\eta y)\cos(\eta y)e^{2\alpha x+\alpha y+\alpha z+2} -$$
$$2\alpha^2\eta\nu\sin(\eta y)\cos(\eta y)e^{2\alpha x+\alpha y+\alpha z+2} + 2\eta^2\sin^2(\eta y)\cos(\eta y)e^{2\alpha x+\alpha y+3} +$$
$$2\alpha^2 DF\sin(\eta x)\cos(\eta y)\sin(\eta z)e^{\alpha x+2\alpha z+1} + \alpha^2F^2\cos(\eta y)\sin^2(\eta z)e^{\alpha y+2\alpha z+1}$$
$$\left.+4\alpha F\eta\sin(\eta y)\cos(\eta y)\sin(\eta z)e^{\alpha x+\alpha y+\alpha z+2}\right) - 2\alpha F^2\eta\cos(\eta y)\sin(\eta z)\cos(\eta z)e^{\alpha y+2\alpha z+1} +$$
$$\frac{1}{3}e^{-3\alpha x-2\alpha y-4\alpha z}t^3\left(\eta^4\nu\cos^3(\eta y)e^{3\alpha x+2\alpha y+\alpha z+3} - \alpha^2\eta^2\nu\cos^3(\eta y)e^{3\alpha x+2\alpha y+\alpha z+3}\right.$$
$$+\eta^5\nu^2\sin(\eta y)\cos(\eta y)e^{3\alpha x+2\alpha y+2\alpha z+2} - 2\alpha^2\eta^3\nu^2\sin(\eta y)\cos(\eta y)e^{3\alpha x+2\alpha y+2\alpha z+2}$$
$$+\alpha^4\eta\nu^2\sin(\eta y)\cos(\eta y)e^{3\alpha x+2\alpha y+2\alpha z+2} - \eta^3\sin(\eta y)\cos^3(\eta y)e^{3\alpha x+2\alpha y+4}$$
$$-2\eta^4\nu\sin^2(\eta y)\cos(\eta y)e^{3\alpha x+2\alpha y+\alpha z+3} + 2\alpha^2\eta^2\nu\sin^2(\eta y)\cos(\eta y)e^{3\alpha x+2\alpha y+\alpha z+3}$$
$$+\eta^3\sin^3(\eta y)\cos(\eta y)e^{3\alpha x+2\alpha y+4} + \alpha F\eta^4\nu^2\cos(\eta y)\sin(\eta z)e^{2\alpha x+2\alpha y+3\alpha z+1}$$
$$-2\alpha^3F\eta^2\nu^2\cos(\eta y)\sin(\eta z)e^{2\alpha x+2\alpha y+3\alpha z+1} + \alpha^5F\nu^2\cos(\eta y)\sin(\eta z)e^{2\alpha x+2\alpha y+3\alpha z+1}$$
$$-\alpha F\eta^2\cos^3(\eta y)\sin(\eta z)e^{2\alpha x+2\alpha y+\alpha z+3} + 2\alpha F^2\eta^3\nu\cos(\eta y)\sin(\eta z)\cos(\eta z)e^{\alpha x+2\alpha y+3\alpha z+1}$$
$$-2\alpha^3F^2\eta\nu\cos(\eta y)\sin(\eta z)\cos(\eta z)e^{\alpha x+2\alpha y+3\alpha z+1}$$
$$+\alpha F^3\eta^2\cos(\eta y)\sin(\eta z)\cos^2(\eta z)e^{2\alpha y+3\alpha z+1}$$
$$-2\alpha^2DF\eta^2\nu\sin(\eta x)\cos(\eta y)\sin(\eta z)e^{2\alpha x+\alpha y+3\alpha z+1}$$
$$+2\alpha^4DF\nu\sin(\eta x)\cos(\eta y)\sin(\eta z)e^{2\alpha x+\alpha y+3\alpha z+1}$$
$$-\alpha^2D^2F\eta\sin(\eta x)\cos(\eta x)\cos(\eta y)\sin(\eta z)e^{2\alpha x+3\alpha z+1}$$
$$+\alpha^3DF\sin(\eta x)\cos^2(\eta y)\sin(\eta z)e^{2\alpha x+\alpha y+2\alpha z+2}$$
$$-\alpha^2DF^2\eta\sin(\eta x)\cos(\eta y)\sin(\eta z)\cos(\eta z)e^{\alpha x+\alpha y+3\alpha z+1}$$
$$-4\alpha F\eta^3\nu\sin(\eta y)\cos(\eta y)\sin(\eta z)e^{2\alpha x+2\alpha y+2\alpha z+2}$$
$$+4\alpha^3F\eta\nu\sin(\eta y)\cos(\eta y)\sin(\eta z)e^{2\alpha x+2\alpha y+2\alpha z+2} -$$
$$\alpha^3DF^2\sin(\eta x)\cos(\eta y)\sin^2(k\eta)e^{\alpha x+\alpha y+3\alpha z+1}$$
$$+2\alpha^2DF\eta\sin(\eta x)\sin(\eta y)\cos(\eta y)\sin(\eta z)e^{2\alpha x+\alpha y+2\alpha z+2}$$
$$-\alpha^2F^3\eta\cos(\eta y)\sin^2(\eta z)\cos(\eta z)e^{2\alpha y+3\alpha z+1}$$
$$-\alpha^2F^2\eta^2\nu\cos(\eta y)\sin^2(\eta z)e^{\alpha x+2\alpha y+3\alpha z+1}$$
$$+2\alpha F^2\eta^2\sin(\eta y)\cos(\eta y)\sin(\eta z)\cos(\eta z)e^{\alpha x+2\alpha y+2\alpha z+2}$$
$$+\alpha^4F^2\nu\cos(\eta y)\sin^2(\eta z)e^{\alpha x+2\alpha y+3\alpha z+1}$$
$$+\alpha^2F^2\eta\sin(\eta y)\cos(\eta y)\sin^2(\eta z)e^{\alpha x+2\alpha y+2\alpha z+2}$$
$$\left.+2\alpha F\eta^2\sin^2(\eta y)\cos(\eta y)\sin(\eta z)e^{2\alpha x+2\alpha y+\alpha z+3}\right)$$

$$
\begin{aligned}
&-\alpha^4F^2\eta\nu^2\sin(\eta z)\cos(\eta z)e^{2\alpha x+3\alpha y+2\alpha z}\\
&-2F^3\eta^4\nu\sin(\eta z)\cos^2(\eta z)e^{\alpha x+3\alpha y+2\alpha z}\\
&+2\alpha^2F^3\eta^2\nu\sin(\eta z)\cos^2(\eta z)e^{\alpha x+3\alpha y+2\alpha z}\\
&-F^4\eta^3\sin(\eta z)\cos^3(\eta z)e^{3\alpha y+2\alpha z}\\
&+\alpha DF\eta^4\nu^2\sin(\eta x)\sin(\eta z)e^{3\alpha x+2\alpha y+2\alpha z}\\
&-2\alpha^3DF\eta^2\nu^2\sin(\eta x)\sin(\eta z)e^{3\alpha x+2\alpha y+2\alpha z}\\
&+\alpha^5DF\nu^2\sin(\eta x)\sin(\eta z)e^{3\alpha x+2\alpha y+2\alpha z}\\
&+2\alpha D^2F\eta^3\nu\sin(\eta x)\cos(\eta x)\sin(\eta z)e^{3\alpha x+\alpha y+2\alpha z}\\
&-2\alpha^3D^2F\eta\nu\sin(\eta x)\cos(\eta x)\sin(\eta z)e^{3\alpha x+\alpha y+2\alpha z}\\
&+\alpha D^3F\eta^2\sin(\eta x)\cos^2(\eta x)\sin(\eta z)e^{3\alpha x+2\alpha z}\\
&-2\alpha^2DF\eta^2\nu\sin(\eta x)\cos(\eta y)\sin(\eta z)e^{3\alpha x+2\alpha y+\alpha z+1}\\
&+2\alpha^4DF\nu\sin(\eta x)\cos(\eta y)\sin(\eta z)e^{3\alpha x+2\alpha y+\alpha z+1}\\
&-\alpha^2D^2F\eta\sin(\eta x)\cos(\eta x)\cos(\eta y)\sin(\eta z)e^{3\alpha x+\alpha y+\alpha z+1}\\
&+4\alpha DF^2\eta^3\nu\sin(\eta x)\sin(\eta z)\cos(\eta z)e^{2\alpha x+2\alpha y+2\alpha z}\\
&-4\alpha^3DF^2\eta\nu\sin(\eta x)\sin(\eta z)\cos(\eta z)e^{2\alpha x+2\alpha y+2\alpha z}\\
&+2\alpha D^2F^2\eta^2\sin(\eta x)\cos(\eta x)\sin(\eta z)\cos(\eta z)e^{2\alpha x+\alpha y+2\alpha z}\\
&-2\alpha^2DF^2\eta\sin(\eta x)\cos(\eta y)\sin(\eta z)\cos(\eta z)e^{2\alpha x+2\alpha y+\alpha z+1}\\
&+2\alpha DF^3\eta^2\sin(\eta x)\sin(\eta z)\cos^2(\eta z)e^{\alpha x+2\alpha y+2\alpha z}\\
&-\alpha^2D^2F\eta^2\nu\sin^2(\eta x)\sin(\eta z)e^{3\alpha x+\alpha y+2\alpha z}\\
&+\alpha^4D^2F\nu\sin^2(\eta x)\sin(\eta z)e^{3\alpha x+\alpha y+2\alpha z}\\
&-\alpha^2D^3F\eta\sin^2(\eta x)\cos(\eta x)\sin(\eta z)e^{3\alpha x+2\alpha z}\\
&+\alpha^3D^2F\sin^2(\eta x)\cos(\eta y)\sin(\eta z)e^{3\alpha x+\alpha y+\alpha z+1}\\
&-\alpha^2D^2F^2\eta\sin^2(\eta x)\sin(\eta z)\cos(\eta z)e^{2\alpha x+\alpha y+2\alpha z}\\
&+\alpha^2DF\eta\sin(\eta x)\sin(\eta y)\cos(\eta y)\sin(\eta z)e^{3\alpha x+2\alpha y+2}\\
&+\alpha^3DF^2\sin(\eta x)\cos(\eta y)\sin^2(\eta z)e^{2\alpha x+2\alpha y+\alpha z+1}\\
&+F^3\eta^4\nu\sin^3(\eta z)e^{\alpha x+3\alpha y+2\alpha z}\\
&-\alpha DF^3\eta^2\sin(\eta x)\sin^3(\eta z)e^{\alpha x+2\alpha y+2\alpha z}\\
&+F^4\eta^3\sin^3(\eta z)\cos(\eta z)e^{3\alpha y+2\alpha z}+\alpha^2F^3\eta^2\nu\sin^3(\eta z)e^{\alpha x+3\alpha y+2\alpha z}\Big)
\end{aligned}
$$

and soon.

References

1. Latchman R, Pooransingh A (2015) Modeling conventional swing of a cricket ball using COMSOL Multiphysics® (2015)
2. Shah K, Shakya RK, Mittal S (2019) Aerodynamic forces on projectiles used in various sports. Phys Fluids
3. Pallis J (2004) Aerodynamics and hydrodynamics in sports. https://www.semanticscholar.org/paper/d5363d3a0f189269f41ce37211 0c04f9e4d37b06
4. Seralathan S, B V, L R, Macha YY, Pinninti BK, MichaPremkumar T, Venkatesan H (2018) Modelling and development of passive permanent magnetic bearing for a small cross flow vertical axis wind turbine. Int J Renew Energy Res 8:1868–1880
5. Abed MH, Wali WA, Alaziz M (2022) Moveable inspection ball simulation for pipe leakage detection using an acoustic sensor. In: 2022 Iraqi international conference on communication and information technologies (IICCIT). pp 262–267
6. Latchman R, Pooransingh A (2016) Aerodynamic analysis of a ski jumper: a CFD approach
7. Honeycutt D, Uddin M (2023) Prediction of disc golf drivers' aerodynamic characteristics using Reynolds averaged Navier stokes computational fluid dynamics. Sports Eng 26:1–11
8. Neitzel KJ, Hanquist KM (2017) Aerodynamic optimization of a golf driver using computational fluid dynamics. AIAA. https://doi.org/10.2514/6.2017-0724
9. Othman N (2017) Prediction of aerodynamic derivatives using computational fluid dynamics (CFD) at transonic speed. Int J Aerodyn 4:1
10. Michael A (2010) Leschziner: Reynolds-averaged Navier-stokes methods. Flow Turbul Combust. https://doi.org/10.1002/9780470686652.EAE054
11. Heeg J, Chwalowski P, Jennifer P, Florance, Carol D, Wieseman, David M, Schuster, Raleigh B, Perry (2013) Overview of the aeroelastic prediction workshop. AIAA. https://doi.org/10.2514/6.2013-783
12. Sajjan S, Kumar KS (2012) Reynolds-averaged Navier-stokes computations over a two-element Aerofoil
13. Jo Y-H, Chang K, Sheen D-J, Park SH (2015) Numerical simulation of aerodynamic characteristics of a BWB UCAV configuration with transition models. Int J Aeronaut Space Sci 16(1):8–18. https://doi.org/10.5139/IJASS.2015.16.1.8
14. Edeling WN, Schmelzer M, Dwight RP, Cinnella P (2018) Bayesian predictions of Reynolds-averaged Navier-stokes uncertainties using maximum a posteriori estimates. AIAA J 56(5):2018–2029. https://doi.org/10.2514/1.J056287
15. Coutis P (1998) Modelling the projectile motion of a cricket ball. Int J Math Educ Sci Technol 29(6):789–798. https://doi.org/10.1080/0020739980290601s
16. Acevedo P, Tapia, Chérif, Amrouche, Carlos, Conca A, Ghosh A (2021) Stokes and Navier-stokes equations with Navier boundary conditions. J Differ Equ 285:258–320. https://doi.org/10.1016/J.JDE.2021.02.045
17. Penrose JMT, Hose DR, Trowbridge EA (2020) Cricket ball swing: a preliminary analysis using computational fluid dynamics. In: Sports engineering. pp 11–19. https://doi.org/10.1201/9781003078098-3
18. Sarah B, Chin SB, Carré M (2009) Sports ball aerodynamics: a numerical study of the erratic motion of soccer balls. Computers & Fluids 38(6):1091–1100. https://doi.org/10.1016/J.COMPFLUID.2008.11.001
19. Griffith MD, Crouch T, Thompson MC, Burton D, Sheridan J, Brown NAT (2014) Computational fluid dynamics study of the effect of leg position on cyclist aerodynamic drag. J Fluids Eng-Trans Asme 136(10):01105. https://doi.org/10.1115/1.4027428
20. Ashraf CM, Saifullah SMT (2013) The swing of a cricket ball-a computational study
21. Mehta RD (2005) An overview of cricket ball swing. Sport Eng 8(4):181–192. https://doi.org/10.1007/BF02844161
22. La-Brooy R, Alam F, Watmuff J (2009) New methods of initiating reverse swing on a cricket ball. pp 281–285

23. Darvishi MT, Khani F, Hamedi-Nezhad S (2010) Solution of some systems of nonlinear partial differential equations by variational iteration method. J Algorithms & Comput Technol 4(1):1–14. https://doi.org/10.1260/1748-3018.4.1.1
24. Noor MA, Noor KI, Mohyud-Din ST (2009) Modified variational iteration technique for solving singular fourth-order parabolic partial differential equations. Nonlinear Anal-Theory Methods & Appl 71(12). https://doi.org/10.1016/J.NA.2008.11.011
25. Abassy TA, El-Tawil MA, El Zoheiry H (2007) Solving nonlinear partial differential equations using the modified variational iteration Padé technique. J Comput Appl Math 207(1):73–91. https://doi.org/10.1016/J.CAM.2006.07.024
26. Hilal EMA, Elzaki TM (2014) Solution of nonlinear partial differential equations by new laplace variational iteration method. J Funct Spaces Appl 2014:1–5. https://doi.org/10.1155/2014/790714
27. Kassim A, Hussain N, Rusli FS, Fadhel, Yahya ZR (2016) Solution of one-dimensional fractional order partial integro-differential equations using variational iteration method. 1775(1):030096. https://doi.org/10.1063/1.4965216
28. Ganji DD, Sajjadi H (2011) New analytical solution for natural convection of Darcian fluid in porous media prescribed surface heat flux. Therm Sci 15:221–227. https://doi.org/10.2298/TSCI100424001V
29. Mazhar H, Pazouki A, Rakhsha M, Jayakumar P, Negrut D (2018) A differential variational approach for handling fluid–solid interaction problems via smoothed particle hydrodynamics. J Comput Phys 371:92–110. https://doi.org/10.1016/J.JCP.2018.05.013
30. Hesameddini E, Latifizadeh H (2009) Reconstruction of variational iteration algorithms using the laplace transform. Int J Nonlinear Sci Numer Simul 10:11–12, 1377–1382. https://doi.org/10.1515/IJNSNS.2009.10.11-12.1377
31. Yépez-Martínez H, Gómez-Aguilar JF (2020) Laplace variational iteration method for modified fractional derivatives with non-singular kernel. Appl Comput Mech 6(3):684–698. https://doi.org/10.22055/JACM.2019.31099.1827
32. Matinfar M, Saeidy M, Ghasemi M (2013) The combined laplace-variational iteration method for partial differential equations. Int J Nonlinear Sci Numer Simul 14(2):93–101. https://doi.org/10.1515/IJNSNS-2011-0092
33. Wu GC, Wu GC, Baleanu D, Baleanu D (2013) Variational iteration method for fractional calculus-a universal approach by Laplace transform. Adv Differ Equ 2013(1):18. https://doi.org/10.1186/1687-1847-2013-18
34. Biala TA, Afolabi YO, Asim OO (2014) Laplace variational iteration method for integro-differential equations of fractional order. Int J Pure Appl Math 95(3):413–426. https://doi.org/10.12732/IJPAM.V95I3.9

Using MCDM Techniques for Talent Identification and Management in Sports

Jainil Dharmil Shah, Chaitanya Iyer, and Mani Sankar Dasgupta

Abstract Talent identification in sports is a process of identifying individuals with the potential to excel in a particular sport or athletic activity. It entails evaluating a range of psychological, technical, tactical, and physical characteristics in order to pinpoint those who have what is needed to excel in a given sport. Talent management involves strategic management, nurturing already existing talent, and providing support to optimize performance with a vision for long-term success. MCDM Techniques provide a solid framework for determining the best solution to a decision problem taking into account the alternatives and free of any bias. In this paper, we explore the use of MCDM techniques for Talent Identification and Management with a special focus on football. We compiled data of five different players from Europe's Top 5 leagues using WhoScored and applied AHP and TOPSIS methods to rank and order attackers, defenders, and midfielders on multiple facets of their game. AHP helps to establish a structured and objective framework for determining the criteria and sub-criteria. TOPSIS is then used to rank these players on the basis of these criteria, thus providing valuable insights to clubs for talent identification and player selection in football. With sophisticated football strategies, and with an increasing number of players to pick from, it is becoming harder to distinguish them based on numbers. This study presents a feasible solution to address the issue by objectively evaluating all criteria in a scientific manner, in order to select the most qualified individuals.

Keywords Talent identification · AHP · TOPSIS

1 Introduction

In the information age, more and more sports teams are investing into advanced analytical tools to gain any slightest edge over competitors. And one of the major areas where teams can gain an advantage is talent recognition. While there is no

J. D. Shah (✉) · C. Iyer · M. S. Dasgupta
Birla Institute of Technology and Science, Pilani, Rajasthan, India
e-mail: f20200504@pilani.bits-pilani.ac.in

S. Bhattacharyya and D. James (eds.), *Recent Trends in Sports Engineering*, Lecture Notes in Mechanical Engineering, https://doi.org/10.1007/978-981-97-8297-0_11

agreed upon definition of talent, we can consider talent to be the natural skill or ability to perform a given task better than average. It allows for identification of players who have an aptitude for a particular sport and identifies their weaknesses to train them more effectively. It is an indicator of the future potential for a player. Especially for a sport like football where teams spend billions of dollars on transfer fees for top talent, a robust talent identification framework can effectively save a lot on transfer costs and provide sporting success as well. Early talent identification is essential because it enables clubs and academies to invest in talented players and mentor them through the different phases of their football careers.

Talent in football can be grouped into three major aspects—physical, cognitive, or mental, and technical. While the physical aspect is easily measurable and there are training systems in place in major academies for mental strength and conditioning, we can use the technical aspect to predict performance by using analytical techniques on key data points collected from games.

Currently, football players are selected by coaches through subjective analyses, based on their potential to complement the club's playing style using factors like agility, creativity, strength, speed, and size. However, the personal preferences of the coach can have a great impact on the selection process and thus newer technology-enhanced methods can provide a more objective analysis and incorporate factors that might get ignored in the current method.

Used as a technique for decision-making, MCDM generates a solution that necessitates consensus by combining the performance of alternatives across various, contradictory qualitative and/or quantitative criteria [1, 2]. Successful and effective decision-making requires an understanding of MCDM approaches as well as a suitable grasp of the perspectives of decision-making process itself (players who are participating in the decision-making process). Numerous MCDM techniques are available, including TOPSIS, AHP, VIKOR, DEMATEL, etc.

2 Literature Review

Due to there being no single agreed definition of talent, identification frameworks and models differ vastly. Talent is a mastery of knowledge and skills that places an individual in the top 10% of their field [3]. Therefore, talent in sport means the natural ability to excel at the game. The process of identifying athletes who have the potential to become exceptional players might be referred to as talent identification [4]. This method comprises assessing technical skills along with sociological and physical characteristics to forecast performance over time.

Different approaches to talent development have been followed since the millennium. Their study identified that physical predictor variables have been most commonly used in these models. Apart from this scouts and coaches typically observe current match performance in an effort to find and choose young soccer players for development programs. This method can be biased since it is not strictly based on data and objective facts. Talented players share traits in common, like being taller, heavier, and having more developed skeletal development. They also scored higher on tests evaluating technical ability (long/short passing, ball possession, dribbling, and shooting), flexibility, coordination, agility, and speed [5].

There is extensive research on the different approaches to talent identification in sport. Abbott et al. (2005) provides a lot of insight into a few important aspects of talent identification. Firstly, performance is related to range of different variables. A lot of these variables are not just genetic therefore, there is a need to consider aspects apart from just the physical. Therefore, models need to identify weaknesses as well since many of the key performance determinants in sport can be developed with the appropriate training opportunities [6].

Also, single-factor models that identify talent on only one discrete characteristic measure are poor predictors of talent. Examples of this include the free throw shooting in basketball or sprint performance for athletes. Thus, there is a need for models that take into account multiple criteria and base the decision off that.

Crespo and Reid [7] identifies 4 major tests for identification of talent in sport, (i) Physical Test (ii) Physiological Test (iii) Psychological Test, and (iv) Technical Test. The technical test measures components such as accuracy in skills, the consistency of a player, and more. For our framework, we focus exclusively on this technical test and rank players based on this. A review on talent identification in male football states that great accomplishments in football are strongly correlated with exceptional technical skills like ball retention, long & short passes, dribbling, and shooting. However, very few studies have examined the importance of technical and tactical skills for talent discovery and development [8].

There is more evidence in support of technical skills being a good indicator of future performance. Kannekens et al. (2011) concluded that particularly for midfielders, positioning, and decision-making seemed to be the tactical skills that most accurately predict adult performance level [9]. In comparison to players who stayed amateurs, football players who went on to become professionals were, on average, 0.3 s faster in 30 m of peak dribbling performance and 1 s faster in 3×30 m of repeated dribbling performance during their teens [10].

The literature indicates that while there is a great amount of research has been made into physical indicators and developing psychological strength however there is no framework for specifically using technical indicators to predict talent.

3 Problem Definition

For the purpose of model development, we decided to select football as it provides a very clear distinction of the different positions on the pitch and their roles such as attack, defense and midfield. This paper uses Analytic Hierarchy Process (AHP) to establish a structured and objective framework for determining different criteria and sub-criteria and to employ the Technique for Order of Preference by Similarity to Ideal Solution (TOPSIS) method to rank players based on these criteria, providing valuable insights to clubs for talent identification and player selection in football.

The data that we have taken is of five actual players from Europe's top 5 leagues using WhoScored. We wouldn't like to mention the names of the players for privacy reasons and would like to keep them anonymous as players 1–5.

3.1 Objectives

1. To compile and analyze data of five different players, encompassing a wide range of attributes for attackers, defenders, and midfielders.
2. To employ AHP to ascertain the weights, criteria, and sub-criteria that influence the ranking of football players based on their attributes.
3. To utilize the TOPSIS method to rank players based on the identified criteria and sub-criteria, allowing for a refined talent assessment.

4 Methodology

As highlighted in the problem definition, we have integrated AHP and TOPSIS to rank football players based on various attributes and then make the best available choice.

AHP is a powerful method for solving challenging decision-making situations. In terms of hierarchical levels, any complex problem can be divided into a smaller number of various sub-problems, each of which corresponds to a set of criteria or attributes relevant to each sub-problem [11]. The AHP approach is a MCDM analysis technique that relies on an additive weighting mechanism to represent several relevant attributes depending on their relative value.

TOPSIS is used to determine the best alternative from a group of solutions by considering multiple criteria and comparing each alternative to an ideal solution. Using the Euclidean distance to calculate an alternative's relative proximity to the optimal solution, the concept is that the alternatives chosen must be the ones that are closest to the positive ideal solution and the farthest from the negative (anti) ideal solution from the geometrical point [12].

Step 1: Performing literature review for identifying the main criteria and sub-criteria for problem of football player selection in the sports industry.

Step 2: Collection of data for five players from WhoScored.com based on identified criteria and sub-criteria.

Step 3: Setting up the decision-making hierarchy for primary criteria, sub-criteria and players.

Step 4: Constructing a pairwise comparison matrix for every criterion, sub-criterion, and alternative.

Step 5: Utilizing AHP to determine the weights of primary criteria, sub-criteria, and the standard decision matrix.

Step 6: Using the weights and decision matrix obtained by using AHP for the TOPSIS method.

Step 7: Ranking the players by using TOPSIS, comparing the players, and then finally player selection.

5 Results and Discussion

After a careful literature review, we have considered three main criteria for player analysis, i.e., offense, passing, and defense. Each criterion is further divided into four sub-criteria as can be seen in the decision hierarchy matrix below. We have collected data from WhoScored.com for five players so as to aid in player selection by ranking the players based on various listed attributes (Table 1).

Consistency ratio for each developed pairwise comparison matrix (for every criterion, sub-criterion, and player alternative) has been verified and found to be <0.1 for the decision to be considered as acceptable. Each value in the "weight of sub-criteria" column has been multiplied by the corresponding value in the "criteria weight" column to determine the desirability index for each alternative. The results are then summed up. This gives us the final weights for every criterion, sub-criterion, and player alternative as shown in Fig. 1.

Weighted Normalized Distribution Matrix is used for applying TOPSIS as shown in (Fig. 2). Each of the sub-criteria considered in our model is beneficial for assessing the players, the best and worst values for each sub-criterion are noted, and then based on the TOPSIS principle of ranking on the shortest possible distance from an ideal positive solution, the final ranking of players is obtained (Table 2).

In TOPSIS, the more the closeness ratio, the better the solution. As seen in Table 2, player 1 (with a closeness ratio of 1) is ranked first and player 2 (with a closeness ratio of 0) is ranked fifth, i.e., last.

Table 1 Decision hierarchy matrix

Criteria	Offense/attacking	Passing	Defense
Sub criteria	Goals per game	Average number of passes	Tackles
	Assists per game	Pass success	Intercepts
	Shots per game	Crosses	Clearances
	Dibble	Long balls	Blocks
Alternatives	Player 1, Player 2, Player 3, Player 4, Player 5		

Criteria	Sub-Criteria	Criteria Weighs	Sub-Criteria Weight	Final Sub-Criteria Weights	Alternative Weights					Final Alternative Weights				
					P1	P2	P3	P4	P5	P1	P2	P3	P4	P5
Offense/Attacking	Goals per Game	0.647737535	0.648874472	0.420300351	0.34517	0.402	0.12844	0.08656	0.0378	0.14508	0.16896	0.05398	0.02361	0.01589
	Assists per Game	0.647737535	0.217820147	0.141090285	0.55104	0.2352	0.10442	0.06475	0.04457	0.07775	0.03318	0.01473	0.00199	0.00629
	Shots per Game	0.647737535	0.08811985	0.057078534	0.15792	0.26473	0.21861	0.17894	0.1798	0.00901	0.01511	0.01248	0.0009	0.01026
	Dribble	0.647737535	0.045185531	0.029268364	0.12345	0.23456	0.34567	0.14567	0.15067	0.00361	0.00687	0.01012	0.00019	0.00441
Passing	Average number of passes	0.219316603	0.636542769	0.139604398	0.08033	0.35936	0.1567	0.11732	0.2866	0.01121	0.05017	0.02188	0.01043	0.04001
	Pass Success	0.219316603	0.227431082	0.049879412	0.10951	0.26193	0.30054	0.24721	0.08082	0.00546	0.01306	0.01499	0.0028	0.00403
	Crosses	0.219316603	0.070372451	0.015433847	0.05597	0.35261	0.06754	0.19312	0.33076	0.00086	0.00544	0.00104	0.00021	0.0051
	Long Balls	0.219316603	0.065653698	0.014398946	0.10399	0.20383	0.41017	0.25943	0.02257	0.0015	0.00293	0.00591	0.00025	0.00032
Defence	Tackles	0.132945862	0.173859096	0.023113847	0.2122	0.00055	0.32503	0.37301	0.08921	0.0049	1.3E-05	0.00751	0.0015	0.00206
	Intercepts	0.132945862	0.42754605	0.056840478	0.24335	0.10524	0.29803	0.24402	0.10937	0.01383	0.00598	0.01694	0.00593	0.00622
	Clearances	0.132945862	0.359521941	0.047796954	0.39957	0.03805	0.28847	0.13094	0.14296	0.0191	0.00182	0.01379	0.00225	0.00683
	Blocks	0.132945862	0.039072913	0.005194582	0.18709	0.36242	0.15933	0.08514	0.20602	0.00097	0.00188	0.00083	1.7E-05	0.00107
Decision Index										0.29329	0.30543	0.17419	0.05007	0.1025
Ranking										2	1	3	5	4

Fig. 1 Final weights for criteria, sub-criteria, and player alternatives derived

Weighted Normalized Decision Matrix												
Criteria	Offense/Attacking				Passing				Defence			
SubCriteria	Goals per Game	Assists per Game	Shots per Game	Dribble	Average Number of Passes	Pass Success Rate	Crosses	Long Balls	Tackles	Intercepts	Clearances	Blocks
Player 1	0.4018873	0.078695	0.0470835	0.01616867	0.0456421	0.0222922	0.0027889	0.0035426	0.0007024	0.00718981	0	0
Player 2	0.0967508	0.051154	0.0261575	0.01270396	0.0272419	0.0207074	0.0006972	0.0015183	0.0021071	0.00359491	0.01092948	0
Player 3	0.0175893	0.074402	0.0146482	0.00057745	0.089492	0.0235913	0.0139446	0.0116399	0.0140471	0.03235417	0.01366185	0.0021207
Player 4	0.0703611	0.018599	0.0094167	0.01154905	0.0697775	0.0221363	0.0055778	0.0073382	0.0161542	0.03594907	0.03005607	0.0042414
Player 5	0.0227635	0.072217	0.0073241	0.01732358	0.0615332	0.0227079	0.0020917	0.0017713	0.0084283	0.02875926	0.03278844	0.0021207
Classification	Beneficial	Beneficial	Beneficial	Beneficial	Beneficial	Beneficial	Beneficial	Beneficial	Beneficial	Beneficial	Beneficial	Beneficial
Ideal Best	0.4018873	0.078695	0.0470835	0.01732358	0.089492	0.0235913	0.0139446	0.0116399	0.0161542	0.03594907	0.03278844	0.0042414
Ideal Worst	0.0175893	0.018599	0.0073241	0.00057745	0.0272419	0.0207074	0.0006972	0.0015183	0.0007024	0.00359491	0	0

Fig. 2 Weighted normalized decision matrix derived for applying TOPSIS

Table 2 Final ranking of players after TOPSIS

Alternatives	S_i^*	S_i^-	P_i	Rank
Player 1	0	0.060095	1	1
Player 2	0.3051365	0.032554	0.0964025	4
Player 3	0.384298	0.055803	0.1267951	2
Player 4	0.3315263	0	0	5
Player 5	0.3791238	0.053617	0.1239016	3

6 Conclusion and Future Work

This approach incorporates all essential factors for talent identification, evaluation, and choice. For field implementation in new situations, questionnaires can be distributed to collect expert opinions for the purpose of constructing pairwise comparison matrices in AHP. These matrices are used to assess the consistency of the weights allocated to criteria, sub-criteria, and player options, depending on the responses provided by experts. The integration of AHP and TOPSIS with Fuzzy can be achieved by utilizing FAHP and FTOPSIS methodologies. This integration allows for more effective management of uncertainty and imprecision present in the acquired data. This methodology can be extended and utilized by any nation, team, or organization to assess and choose several potential players from a certain group, according to the desired qualities.

References

1. Taherdoost H, Madanchian M (2023) Multi-criteria decision making (MCDM) methods and concepts. Encyclopedia, January 9, 2023. https://doi.org/10.3390/encyclopedia3010006
2. Singh S, Dasgupta MS (2015) Evaluation of refrigerant options using TOPSIS-MADM technique. IJSTA Spec Edn NCTECT-2015 1(3). https://www.ijsta.com/archives_v1n3.html
3. Gagne F (2011) Academic talent development and equity issue in gifted education. Talent Dev Excellence 3(1):3–22
4. Alayode A (2014) Designing template for talent identification and development in sport, July 25. https://core.ac.uk/download/pdf/236305553.pdf
5. Williams AM, Ford PR, Drust B (2020) Talent identification and development in soccer since the millennium. J Sports Sci 38(11–12):1199–1210. https://doi.org/10.1080/02640414.2020.1766647
6. Abbott A, Button C, Pepping G, Collins D (2005) Unnatural selection: talent identification and development in sport. PubMed 9(1):61–88. https://pubmed.ncbi.nlm.nih.gov/15629068
7. Crespo M, Reid M (2009) Talent search and talent detection-introduction. ITF Coaching Education Series, USA
8. Sarmento H, Anguera MT, Pereira A, Araújo D (2018) Talent identification and development in male football: a systematic review. Sports Med 48(4):907–931. https://doi.org/10.1007/s40279-017-0851-7
9. Kannekens R, Elferink-Gemser M, Visscher C (2011) Positioning and deciding: key factors for talent development in soccer. Scand J Med Sci Sports 21(6):846–852
10. Huijgen B, Elferink-Gemser M, Post W et al (2010) Development of dribbling in talented youth soccer players aged 12–19 years: a longitudinal study. J Sports Sci 28(7):689–698
11. Tisha SN, Abed MT, Alam MGR (2021) Decision theory-based content selection and ranking for social media newsfeed using fuzzy analytic hierarchy process. In: 2021 IEEE region 10 symposium (TENSYMP). https://doi.org/10.1109/tensymp52854.2021.9550917
12. Repository home—University of Twente Student Theses (n.d.). https://essay.utwente.nl

Novel and Comprehensive Performance Metric for Identifying Sports Talent in IPL T20 Cricket Match

Karan Agrawal, Lajiya Aleena Saji, and K. Asha

Abstract Cricket, a professional sport with a rich history spanning 160 years, generates vast amounts of data that require effective analysis for accurate predictions. The conventional evaluation of batter performance in T20 cricket relies on strike rate and average, but these metrics fail to capture the intricacies and context of the game, resulting in a limited assessment of a batter's true value to the team's success. This research paper introduces two additional metrics, latency and acceleration, to provide a more comprehensive evaluation of a batter's performance in the T20 format. To demonstrate the efficacy of these metrics, the top 40 batters with the most runs scored in the Indian Premier League (IPL) are ranked using the proposed latency and acceleration metrics. The research identifies the best batter who consistently performed well, adapted to various match situations, and made significant contributions to the team's success. The incorporation of latency and acceleration addresses the limitations of relying solely on strike rate and average by considering contextual factors, providing a more comprehensive assessment of a batter's performance in T20 cricket matches. The research uniquely determines a player's caliber based on novel metrics not previously explored in other research findings, making it extremely useful. Furthermore, it enhances the understanding of the game, facilitates effective player selection, and empowers teams to make informed decisions based on a batter's true impact on team success.

Keywords Latency and acceleration · Machine learning · T20 batter performance

1 Introduction

The landscape of cricket has undergone a significant transformation with the emergence of Twenty20 (T20) cricket, demanding a dynamic and aggressive style of play to maximize scoring potential within limited overs. The research paper seeks to

K. Agrawal · L. A. Saji · K. Asha (✉)
Department of Advanced Computing, School of Information Technology, St. Joseph's University, Bengaluru, Karnataka, India
e-mail: ashak@sju.edu.in

S. Bhattacharyya and D. James (eds.), *Recent Trends in Sports Engineering*, Lecture Notes in Mechanical Engineering, https://doi.org/10.1007/978-981-97-8297-0_12

address the evolving demands of this fast-paced format by introducing two innovative metrics—"Latency" and "Acceleration". These metrics aim to provide a more comprehensive evaluation of T20 batter performance, surpassing the limitations of conventional methods reliant on strike rate and average. It focuses on the different phases of the innings such as powerplay, middle overs, and death overs, offering a holistic perspective on a batter's consistency, adaptability, and ability to impact crucial moments throughout the complete innings, rather than an overall assessment. This research endeavors to fill gaps in current evaluation techniques, presenting selectors, coaches, and analysts with valuable insights for identifying and nurturing T20-specific batting talents.

2 Literature Review

In recent real-time searches focused on cricket, only a limited number of articles have explored the prediction of players' performance, particularly within the Twenty-20 format. This literature review emphasizes this gap and stresses the importance of delving into metrics beyond traditional measures such as strike rate and batting average to capture nuanced aspects of player performance during different phases of a T20 innings. The review reveals a growing interest in anticipating players' performances within the dynamic T20 format and signals a significant shift towards more sophisticated methodologies. This includes the integration of machine learning and innovative performance indicators to gain a comprehensive understanding of player capabilities, addressing the need for a more nuanced approach in predicting player performance, especially in T20 matches.

Noteworthy research efforts have utilized an Artificial Neural Network (ANN) architecture to create a prediction model for batter's performance in the Indian Premier League (IPL) 2020. The model incorporated variables such as pitch conditions, bowling form, and batting form, demonstrating accurate predictions [1]. Another study proposed a regression model for forecasting a batter's total performance in the IPL, incorporating unique T20 cricket performance metrics [2]. Both studies highlighted the limitations of conventional metrics like strike rate and average and emphasized the need for more nuanced approaches in T20 cricket analysis.

A parallel study also revealed a research focus on machine learning algorithms to predict cricket match results, specifically in the IPL. These studies evaluated bowlers and batters based on effectiveness [3]. Additionally, the significance of multi-objective optimization in cricket team selection was discussed [4], stressing the importance of big data and machine learning techniques in advanced analytics [5].

A critical evaluation of batting metrics led to the recognition of strike rate as a key measure of a batter's efficiency in one-day cricket. The research was able to capture the acceleration aspect of batter's performance by using strike rate as one of the important metrics [6]. Another study proposed a player selection mechanism based on Shannon entropy, considering a player's effectiveness, dependability, and significance in team strategy [7]. In the T20 format, research aimed to assess the importance of different batting metrics, introducing the ordered weighted averaging (OWA) operator for player evaluation for a more nuanced assessment beyond traditional statistics [8].

Two research papers specifically focused on batter's performance. The first paper utilized regression models, including Linear Regression, Decision Tree Regression, Random Forest Regression, and Support Vector Regression, to predict batter's performance in One Day International (ODI) matches [9]. The other paper concentrated on efficient player prediction for IPL using the Random Forest Classification technique, incorporating various batting and bowling parameters such as matches played, strike rate, batting and bowling averages, wickets, runs, and ICC rankings to rank players [10].

Additionally, two other studies looked into batter's performance prediction. One introduces predictive analytics tools like "Weka" and "Dataiku" to analyze and predict batter's performance in cricket matches [11], while another proposes a clear method for evaluating players' performances during the 2012 T20 Cricket World Cup by exploring the measurement of bowlers' and batter's effectiveness using correlation matrices and principal component analysis [12].

Furthermore, one investigation explored the application of machine learning in cricket, focusing on the selection of T20 cricket teams and the evaluation of individual performance, introducing the best practice for selecting batters based on the ordered weighted averaging (OWA) operator [13]. Utilizing supervised machine learning models, another study concentrated on predicting Star Cricketers (SCs) in both batting and bowling domains, employing Bayesian rule, function, and decision-tree-based models for SC prediction [14].

Another research focus offered a thorough analysis of the uses of machine learning (ML) in the sport of cricket, emphasizing the significance of outcome prediction in cricket games and the use of various performance indicators and ML techniques for predicting game outcomes [15]. In a separate study, authors concentrated on developing the Deep Player Performance Indicator (DPPI), a novel performance evaluation metric for Twenty20 (T20) cricket players. DPPI captures a player's current performance and position on the team, facilitating comparisons between players on different teams and providing an approximation of a team's strength [16].

In conclusion, this literature review unfolds a trajectory in cricket analysis that is gradually steering away from the sole reliance on strike rate and averages. Despite the prevalence of these conventional metrics, emerging trends indicate a growing recognition of the need for a more nuanced exploration of the intricacies of a T20 innings. The introduction of metrics like latency and acceleration signifies a concerted effort to delve into different phases of an innings, providing a holistic approach to player performance evaluation in the dynamic context of T20 cricket.

3 Objectives and Methodologies

In the pursuit of a more nuanced evaluation of T20 batter's performance, the research introduces two innovative metrics: "Latency" and "Acceleration". These metrics aim to transcend the limitations of traditional measures such as strike rate and average, which, while valuable, often fall short in capturing the intricacies and contextual factors that define T20 cricket.

Traditional metrics like strike rate, representing the average number of runs scored per 100 balls faced, and average, denoting the mean runs scored per dismissal, provide essential information but may be very misleading in T20 cricket. This discrepancy becomes apparent, especially when considering instances of not-outs, where a batsman remains unbeaten—whether by playing down the order or throughout the entire innings. Such occurrences result in an accumulation of runs, significantly inflating the average. This distortion can be particularly misleading when directly compared with another batter who consistently scores runs.

In Tests and ODIs, where the leisure of available balls allows for a focus on average runs scored and the pace at which they are accumulated, a simplified evaluation through metrics like averages and strike rates proves insightful. However, the T20 format necessitates a more intricate approach. Both averages and strike rates in T20s lack the granularity needed to dissect a player's performance across different phases of the game. They fail to shed light on critical aspects such as a player's ability to construct innings, proficiency in hitting boundaries, and the consistency with which they score runs at regular intervals. Moreover, the impact of not-outs on the average further accentuates the inadequacy of this traditional metrics in capturing the unique dynamics of T20 cricket. Therefore, a more refined and tailored approach is imperative to ensure an accurate representation of a player's performance and their contribution to the team in the fast-paced and dynamic T20 format.

To address these limitations, the research emphasizes the need to move beyond overall statistics and delve into the intricate details of a T20 batter's innings. Traditional metrics fail to capture a player's strengths and playing style comprehensively. Hence, the objectives of this study are twofold: first, to highlight the inadequacies of traditional metrics, such as averages and strike rates, in T20 cricket, and second, to introduce and apply novel metrics, namely Latency and Acceleration, to conduct a more in-depth analysis of a batsman's performance.

The study aims to shift from a broad overview to a detailed examination, seeking to understand a batter's consistency through Latency and their boundary-hitting capability through Acceleration. By doing so, we aim to provide a more accurate and holistic representation of a T20 batter's prowess and contribution to the game.

3.1 Introduction to Latency and Acceleration Metrics

The primary focus of this study centered on crafting two pivotal performance metrics: "Latency" and "Acceleration", each specifically tailored to different innings phases—powerplay, middle overs, and death overs. These metrics serve as indispensable tools for assessing a batter's effectiveness in the T20 format, providing insights into their ability to consistently score runs and accelerate the scoring rate throughout each phase. Together, these metrics offer a holistic assessment of a batter's contribution to the team's success.

For a comprehensive evaluation of batter performance, an extensive dataset spanning IPL seasons from 2008 to 2023 was meticulously gathered from "Cricsheet". The top 40 batters, based on the highest runs scored, were subsequently ranked using the proposed latency and acceleration metrics. The dataset, initially in JSON format, underwent rigorous processing through custom-built R code to transform it into a structured, tabular format. A thorough data pre-processing phase followed, eliminating inconsistencies and anomalies to ensure the data's readiness for in-depth analysis.

Latency Metric. This critical metric quantifies a batter's ability to consistently score runs in every phase of the innings. It calculates the average number of runs scored by a batter in each three-ball interval throughout the match, providing insights into their adaptability and reliability. This analysis reveals how well batters can maintain the scoring rate across various game contexts, whether during the powerplay, middle overs, or death overs.

Acceleration Metric. This metric delves into assessing a batter's capability to rapidly increase the scoring rate through boundary-hitting. It measures the average number of boundaries scored by a batter in every three-ball interval, considering different innings phases—powerplay, middle overs, and death overs. Significantly, it accords more weightage to sixes than fours by applying a multiplier of 1.5 to sixes. This strategic adjustment underscores the impact of hitting sixes, acknowledging that two sixes are equivalent to scoring 12 runs, matching the value of three fours. Consequently, the 1.5 multiplier accentuates the influence of sixes on a batter's ability to effectively accelerate the scoring rate, particularly during critical moments of the game.

3.2 Detailed Analysis on Latency Metrics

Latency metrics offer insights into a batter's consistency of scoring runs in three-ball intervals, taking into account the various phases of the innings. These metrics are as follows:

Overall Latency. The metric measures a batter's ability to score runs consistently throughout an innings. It is calculated as the total runs scored by a batter in all intervals in a match, divided by the total number of intervals played. The average overall latency is determined by averaging this metric across matches for each batter. The formula is as follows:

$$\text{OL} = \frac{\sum_{i=1}^{n} \text{Runs } i}{\sum_{i=1}^{n} \text{Intervals } i}$$

where, n is the number of matches and Runs_i and Intervals_i represent runs scored and intervals played in the ith match.

Powerplay Latency. The metric evaluates a batter's performance during the powerplay phase. It is calculated as the total runs scored by a batter during powerplay intervals, divided by the total number of powerplay intervals played. An adjustment factor (0.9) is applied to account for the relative ease of scoring in powerplay due to fielding restrictions of only two fielders outside the circle. The average powerplay latency is obtained by averaging this metric across matches for each batter. The formula is as follows:

$$\text{PL} = \frac{\sum_{i=1}^{n} \text{PowerplayRuns } i}{\sum_{i=1}^{n} \text{PowerplayIntervals } i} \times 0.9$$

Middle Latency. The metric assesses a batter's performance during the middle overs. It is calculated as the total runs scored during middle overs intervals, divided by the total number of middle overs intervals played. No adjustment factor is applied, as this phase is deemed balanced, with only four fielders positioned outside the circle to guard the boundaries. The average middle latency is computed by averaging this metric across matches for each batter. The formula is as follows:

$$\text{ML} = \frac{\sum_{i=1}^{n} \text{MiddleRuns } i}{\sum_{i=1}^{n} \text{MiddleIntervals } i}$$

Death Latency. The metric gauges a batter's performance during the death overs. It is calculated as the total runs scored during death overs intervals, divided by the total number of death overs intervals played. An adjustment factor (1.1) is applied to account for the increased difficulty of scoring in this phase due to fielding restrictions increased to five fielders outside the circle. The average death latency is determined by averaging this metric across matches for each batter. The formula is as follows:

$$\text{DL} = \frac{\sum_{i=1}^{n} \text{DeathRuns } i}{\sum_{i=1}^{n} \text{DeathIntervals } i} \times 1.1$$

Weighted Latency. To provide a comprehensive measure of a batter's acceleration throughout an innings, weighted acceleration is introduced. This metric combines powerplay acceleration, middle acceleration, and death acceleration, with over-specific weights. The resulting metric provides an adjusted acceleration score for each batter across matches. The formula is as follows:

$$\text{WL} = \frac{6}{20} \times \text{PL} + \frac{9}{20} \times \text{ML} + \frac{5}{20} \times \text{DL}$$

Table 1 provides a detailed breakdown of the created latency metrics for the players, providing valuable insights into their performance in various phases of the innings.

3.3 *Detailed Analysis on Acceleration Metrics*

Acceleration metrics focus on a batter's ability to rapidly increase their scoring rate by hitting boundaries in three-ball intervals, taking into account the various phases of the innings. These metrics are as follows:

Overall Acceleration. The metric assesses a batter's ability to score boundaries consistently throughout an innings. It is calculated as the total boundaries scored by a batter in all intervals in a match, divided by the total number of intervals played. The average overall acceleration is determined by averaging the metric across matches for each batter. The formula is as follows:

$$\text{OA} = \frac{\sum_{i=1}^{n} \text{Boundaries } i}{\sum_{i=1}^{n} \text{Intervals } i}$$

where, n is the number of matches, and Boundaries_i and Intervals_i represent boundaries scored and intervals played in the ith match.

Powerplay Acceleration. The metric evaluates a batter's performance in scoring boundaries during the powerplay phase. It is calculated as the total boundaries scored during powerplay intervals, divided by the total number of powerplay intervals played. An adjustment factor (0.9) is applied to account for the relative ease of scoring boundaries in powerplay due to fielding restrictions of only 2 fielders outside the circle. The average powerplay acceleration is obtained by averaging the metric across matches for each batter. The formula is as follows:

$$\text{PA} = \frac{\sum_{i=1}^{n} \text{PowerplayBoundaries } i}{\sum_{i=1}^{n} \text{PowerplayIntervals } i} \times 0.9$$

Middle Acceleration. The metric assesses a batter's performance in scoring boundaries during the middle overs. It is calculated as the total boundaries scored during middle over intervals, divided by the total number of middles over intervals played.

Table 1 Latency metrics for the Top 40 batters in IPL seasons

Batsman	Overall latency	Powerplay latency	Middle latency	Death latency	Weighted latency	Role
CH Gayle	3.86	3.32	3.90	6.61	4.40	Opener
V Sehwag	4.01	3.55	4.08	5.10	4.18	Opener
DA Warner	3.65	3.22	3.66	5.42	3.97	Opener
JC Buttler	3.88	3.30	3.31	5.90	3.95	Opener
AB de Villiers	4.00	2.86	3.08	6.84	3.95	Middle
GJ Maxwell	3.88	3.39	3.41	5.48	3.92	Middle
S Marsh	3.49	2.75	3.86	5.06	3.83	Middle
RR Pant	4.00	2.89	3.59	5.35	3.82	Finisher
Q de Kock	3.50	3.20	3.22	5.57	3.80	Opener
SA Yadav	4.02	3.32	3.45	4.96	3.79	Middle
KL Rahul	3.62	3.13	3.50	5.07	3.78	Opener
SR Watson	3.58	2.74	3.61	5.22	3.75	Middle
SK Raina	3.56	3.18	3.40	5.00	3.73	Middle
N Rana	3.55	2.74	3.24	5.71	3.71	Middle
SV Samson	3.49	2.91	3.36	5.25	3.70	Middle
M Agarwal	3.43	2.92	3.25	5.40	3.69	Opener
F du Plessis	3.57	3.31	3.18	4.97	3.67	Opener
YK Pathan	3.77	3.42	3.09	4.81	3.62	Finisher
B McCullum	3.48	3.12	3.53	4.29	3.60	Opener
RV Uthappa	3.54	3.12	3.16	4.98	3.60	Opener
V Kohli	3.45	2.82	3.20	5.24	3.60	Middle
S Gill	3.60	3.22	3.33	4.46	3.58	Opener
S Dhawan	3.36	3.00	3.09	5.01	3.54	Opener
SPD Smith	3.64	2.93	3.21	4.76	3.51	Middle
SS Iyer	3.26	2.90	3.12	4.50	3.40	Middle
G Gambhir	3.20	2.98	3.05	4.52	3.40	Opener
DA Miller	3.75	2.61	2.97	5.08	3.39	Finisher
RG Sharma	3.34	2.70	2.88	5.12	3.39	Opener
M Vijay	3.10	2.80	2.88	4.97	3.38	Opener
AM Rahane	3.15	2.79	3.02	4.71	3.37	Opener
WP Saha	3.48	3.13	2.59	4.96	3.34	Middle
KA Pollard	4.07	2.60	2.85	5.08	3.33	Finisher
Yuvraj Singh	3.45	2.34	2.88	5.11	3.28	Finisher
KD Karthik	3.54	2.44	2.87	4.91	3.25	Finisher
AT Rayudu	3.45	2.39	2.88	4.77	3.21	Finisher
PA Patel	3.06	2.81	2.90	4.08	3.17	Opener

(continued)

Table 1 (continued)

Batsman	Overall latency	Powerplay latency	Middle latency	Death latency	Weighted latency	Role
MK Pandey	3.11	2.64	2.88	4.29	3.16	Middle
MS Dhoni	3.83	2.40	2.48	5.18	3.13	Finisher
RA Jadeja	3.92	2.57	2.73	4.49	3.12	Finisher
JH Kallis	2.75	2.54	2.48	4.52	3.01	Middle

No adjustment factor is applied, as this phase is deemed balanced, with only four fielders positioned outside the circle to guard the boundaries. The average middle acceleration is computed by averaging this metric across matches for each batter. The formula is as follows:

$$\text{MA} = \frac{\sum_{i=1}^{n} \text{MiddleBoundaries}\ i}{\sum_{i=1}^{n} \text{MiddleIntervals}\ i}$$

Death Acceleration. The metric gauges a batter's ability to score boundaries during the death overs. It is calculated as the total boundaries scored during death overs intervals, divided by the total number of death overs intervals played. An adjustment factor (1.1) is applied to account for the increased difficulty of scoring boundaries in this phase due to fielding restrictions increased to 5 fielders outside the circle. The average death acceleration is determined by averaging the metric across matches for each batter. The formula is as follows:

$$\text{DA} = \frac{\sum_{i=1}^{n} \text{DeathBoundaries}\ i}{\sum_{i=1}^{n} \text{DeathIntervals}\ i} \times 1.1$$

Weighted Acceleration. A comprehensive measure of a batter's acceleration is derived by considering the different phases of an innings. It combines powerplay, middle, and death acceleration, weighted by their respective over factors (6/20, 9/20, and 5/20). This metric provides an adjusted acceleration score for each batter across matches. The formula is as follows:

$$\text{WA} = \tfrac{6}{20} \times \text{PA} + \tfrac{9}{20} \times \text{MA} + \tfrac{5}{20} \times \text{DA}$$

Table 2 provides a detailed breakdown of the created acceleration metrics for the players, providing valuable insights into their performance in various phases of the innings.

Table 2 Acceleration metrics for the top 40 batters in IPL seasons

Batsman	Overall acceleration	Powerplay acceleration	Middle acceleration	Death acceleration	Weighted acceleration	Role
CH Gayle	0.74	0.67	0.71	1.31	0.85	Opener
V Sehwag	0.74	0.67	0.73	1.00	0.78	Opener
GJ Maxwell	0.64	0.60	0.56	0.98	0.68	Middle
SR Watson	0.60	0.48	0.63	0.88	0.65	Middle
JC Buttler	0.64	0.61	0.45	1.02	0.64	Opener
N Rana	0.58	0.49	0.51	1.04	0.64	Middle
RR Pant	0.63	0.45	0.55	1.00	0.63	Finisher
DA Warner	0.57	0.55	0.50	0.90	0.62	Opener
SA Yadav	0.64	0.58	0.50	0.87	0.62	Middle
AB de Villiers	0.62	0.47	0.38	1.25	0.62	Middle
SE Marsh	0.54	0.45	0.62	0.77	0.61	Middle
YK Pathan	0.56	0.65	0.46	0.74	0.59	Finisher
B McCullum	0.61	0.58	0.54	0.61	0.57	Opener
M Agarwal	0.52	0.46	0.43	0.96	0.57	Opener
SV Samson	0.51	0.46	0.46	0.88	0.57	Middle
Q de Kock	0.56	0.54	0.41	0.87	0.56	Opener
RV Uthappa	0.54	0.51	0.44	0.82	0.56	Opener
KL Rahul	0.55	0.50	0.47	0.77	0.55	Opener
SK Raina	0.51	0.55	0.46	0.70	0.55	Middle
Yuvraj Singh	0.53	0.40	0.44	0.88	0.54	Finisher
KA Pollard	0.64	0.44	0.44	0.84	0.54	Finisher
S Dhawan	0.50	0.49	0.38	0.78	0.51	Opener
RG Sharma	0.49	0.45	0.36	0.86	0.51	Opener
F du Plessis	0.51	0.55	0.35	0.73	0.50	Opener
V Kohli	0.47	0.42	0.37	0.80	0.49	Middle
DA Miller	0.49	0.42	0.36	0.77	0.48	Finisher
M Vijay	0.44	0.41	0.37	0.77	0.48	Opener
S Gill	0.52	0.52	0.40	0.59	0.48	Opener
KD Karthik	0.49	0.38	0.37	0.77	0.47	Finisher
AT Rayudu	0.48	0.36	0.38	0.75	0.47	Finisher
SPD Smith	0.48	0.45	0.35	0.68	0.46	Middle

(continued)

Table 2 (continued)

Batsman	Overall acceleration	Powerplay acceleration	Middle acceleration	Death acceleration	Weighted acceleration	Role
SS Iyer	0.45	0.41	0.38	0.65	0.46	Middle
WP Saha	0.50	0.55	0.25	0.67	0.45	Middle
MS Dhoni	0.52	0.44	0.26	0.80	0.45	Finisher
G Gambhir	0.44	0.46	0.34	0.59	0.44	Opener
AM Rahane	0.43	0.44	0.31	0.68	0.44	Opener
RA Jadeja	0.53	0.42	0.32	0.64	0.43	Finisher
PA Patel	0.46	0.46	0.34	0.46	0.41	Opener
JH Kallis	0.38	0.39	0.21	0.76	0.40	Middle
MK Pandey	0.39	0.40	0.32	0.52	0.39	Middle

3.4 Consolidation of Machine Learning

To ensure the accuracy and reliability of latency and acceleration metrics, a minimum threshold of 10 innings was established for each of the powerplay, middle, and death phases. Top 80 run-scorers in IPL were categorized into roles: "Opener", "Middle Order", and "Finisher" based on their batting order. Finishers typically had fewer than 10 matches in powerplay, and openers typically had fewer than 10 matches in the death overs, aligning with their roles. Machine learning was utilized to predict latency and acceleration for the top 40 batters with fewer than 10 matches in such specific phases. Linear regression models were trained using data from similar batters based on their batting roles, considering the performance in other phases. Specifically, the performance of death batters in powerplay was predicted using the performance in the middle overs, and the performance of opening batters in death overs was predicted using the performance in both powerplay and middle overs.

In emphasizing a pivotal aspect, it is also crucial to highlight that each linear regression model employed in this study demonstrated a commendable accuracy rate of approximately 93%. This significant precision underscores the reliability and efficacy of the machine learning approach in predicting latency and acceleration metrics for batters with fewer than 10 matches in specific phases. The robust accuracy can be attributed to the meticulous training of models with data from analogous batters, taking into account their roles and performance in comparable phases of the innings. This approach systematically mitigates potential biases, thereby ensuring more precise and dependable predictions.

3.5 Rating and Ranking

To rank the top 40 run-scorers in the IPL T20 format, we employed a comprehensive rating system that considered both weighted latency and weighted acceleration. The chosen method, common in econometrics, involves using a weighted average with the inverse of the standard deviations as weights. The two metrics were normalized using their standard deviations and then combined into a single rating score using a weighted formula that takes into account the standard deviations of both metrics for each batter. The rating score is computed using the following formula:

$$\text{Rating}(R) = \frac{(\text{WL} \times sd(\text{WA})) + (\text{WA} \times sd(\text{WL}))}{sd(\text{WA}) + sd(\text{WL})}$$

This rating system ensures a thorough assessment of a batter's performance, considering both the ability to score runs consistently and the capacity to accelerate the scoring rate by hitting boundaries throughout the innings. In simpler terms, this translates to multiplying the latency by a factor of almost 0.25 and acceleration by 0.75, then adding them to obtain the rating. A higher value indicates a better T20 batter in terms of consistency and scoring ability.

Table 3 presents the ranking of the top 40 run-getters in the entire IPL, combining both Weighted Latency and Weighted Acceleration. This table offers a clear comparison of these players' impact in IPL T20 cricket based on their novel performance metrics.

This rating table provides a comprehensive ranking of the top 40 run-scorers, showcasing their novel performance metrics in T20 cricket. Notably, Chris Gayle holds the top spot, and Virender Sehwag is the second-best, significantly ahead of all other IPL batters. Traditional metrics such as strike rate, average, or total runs scored fail to signify the impact these players have had on their teams throughout the years.

In cases where rating scores are tied, as seen with JC Buttler and DA Warner, both having a rating score of 1.44, priority is given to the player with higher acceleration. In this instance, JC Buttler takes precedence, ensuring that batters with superior abilities to accelerate the scoring rate are placed higher in the ranking. If both acceleration and rating are the same, the batter with the highest latency is then given the first preference over others with lower latency. The final rating table is ordered in descending order, ensuring that those with equal ratings are ranked according to the principles mentioned above.

It's crucial to highlight that these ratings can be computed separately for the first and second innings, enabling more refined comparisons of batters in various match situations. The individual metrics, encompassing powerplay overs, middle overs, and death overs Latency and Acceleration, play a pivotal role in evaluating a batter's performance across different phases of an innings. The combination of these individual metrics provides a comprehensive picture of a batter's ability and potential in diverse situations, facilitating a more in-depth and thorough analysis.

Table 3 The ratings for the top 40 batters in IPL seasons

Batsman	Weighted latency	Weighted acceleration	Rating
CH Gayle	4.40	0.85	1.71
V Sehwag	4.18	0.78	1.61
GJ Maxwell	3.92	0.68	1.47
JC Buttler	3.95	0.64	1.44
DA Warner	3.97	0.62	1.44
AB de Villiers	3.95	0.62	1.43
RR Pant	3.82	0.63	1.41
SR Watson	3.75	0.65	1.40
SA Yadav	3.79	0.62	1.39
SE Marsh	3.83	0.61	1.39
N Rana	3.71	0.64	1.38
Q de Kock	3.80	0.56	1.35
KL Rahul	3.78	0.55	1.34
YK Pathan	3.62	0.59	1.33
SV Samson	3.70	0.57	1.33
M Agarwal	3.69	0.57	1.33
SK Raina	3.73	0.55	1.33
B McCullum	3.60	0.57	1.31
RV Uthappa	3.60	0.56	1.30
F du Plessis	3.67	0.50	1.28
S Dhawan	3.54	0.51	1.25
V Kohli	3.60	0.49	1.25
S Gill	3.58	0.48	1.23
KA Pollard	3.33	0.54	1.22
RG Sharma	3.39	0.51	1.22
Yuvraj Singh	3.28	0.54	1.21
SPD Smith	3.51	0.46	1.21
DA Miller	3.39	0.48	1.19
M Vijay	3.38	0.48	1.18
SS Iyer	3.40	0.46	1.18
G Gambhir	3.40	0.44	1.16
KD Karthik	3.25	0.47	1.15
WP Saha	3.34	0.45	1.15
AM Rahane	3.37	0.44	1.15
AT Rayudu	3.21	0.47	1.14
MS Dhoni	3.13	0.45	1.10
RA Jadeja	3.12	0.43	1.09

(continued)

Table 3 (continued)

Batsman	Weighted latency	Weighted acceleration	Rating
PA Patel	3.17	0.41	1.08
MK Pandey	3.16	0.39	1.06
JH Kallis	3.01	0.40	1.03

Furthermore, it's important to underscore that this rating methodology is versatile and adaptable, providing a holistic understanding of a batter's proficiency in scoring runs consistently and their capability in hitting boundaries. This approach permits a nuanced comprehension of a batter's performance in each phase, offering insights into strengths and weaknesses in different contexts and enabling effective comparisons between players.

4 Findings and Interpretations

In this segment, we present the analysis of results through visualizations, charts, and data tables, providing a comprehensive view of player comparisons, insights into batter performances, and the identification of top performers in recent IPL seasons. The presentation includes comparisons with traditional metrics such as strike rate and average. Additionally, we delve into the effectiveness of these metrics, exploring practical applications, implications, and challenges.

4.1 Visualizing Key Metrics

To grasp the dynamic relationship between two key performance metrics, namely, weighted latency and weighted acceleration, visual representations are employed for a direct comparison. These plots act as a lens, allowing us to evaluate a batter's proficiency in consistently scoring runs (Latency) and their capability to enhance the scoring rate through boundary hits (Acceleration) across the innings.

4.2 Recent Performance Analysis—Last 5 IPL Seasons

In an effort to pinpoint the standout performers in the last five seasons of the IPL, we have ranked and identified the top 10 players based on the rating obtained by combining the weighted latency and acceleration metrics. These players have showcased remarkable consistency and acceleration in their batting performances, distinguishing themselves as standout performers in recent IPL T20 cricket. The rating results are presented in Table 4, while Fig. 4 illustrates the weighted latency and weighted acceleration for effective visualization and comparison.

The analysis offers a nuanced perspective on batters' performances, and the visualization of key metrics provides a clear understanding of their proficiency in scoring consistently and accelerating the run rate. In T20 cricket, an acceleration of 1 means that the batter averages one boundary hit every 3 balls, and an acceleration of 0.5 translates to an average of one boundary once in 6 balls. A latency below 3 means that the batter is averaging less than a run a ball, while anything above 4 is becoming the new standard.

Figure 4 represents the top 10 players with their respective latency and acceleration values. The players with higher consistency and acceleration are towards the top right side of the plot, suggesting the power hitting and run-scoring prowess of JC Buttler and GJ Maxwell in the last 5 years, indicating the strong ability to score more than 4 runs in every 3-ball interval with a high probability of scoring boundaries, which is highly commendable.

Batters clustered in the middle of the plot are often game-changers who win two to three games for their team almost on their own, while batters at the bottom left are typically players who play death roles very effectively and aren't looking to play long innings. A death over comparison would better showcase their prowess.

While analyzing the different plots, several noteworthy trends emerge. As seen in Fig. 1 of the Top 10 run getters in entire IPL, it is noticed that a batter like CH Gayle excels in both latency and acceleration, positioning him as a standout contributor to his team in any phase of the innings with a boundary almost every 3-ball interval and an average scoring of 4.4.

We also notice a player like AB De Villiers, despite being a middle-order batter, has a latency comparatively similar to that of Warner but holds better acceleration than Rohit Sharma, S Dhawan and Suresh Raina, who are openers and play most of their innings in the powerplay, suggesting the strong boundary hitting and run-scoring capabilities of AB De Villiers over the other players in any given phase.

Table 4 Sample rating table of last 5 IPL seasons (top 10 batters)

Batsman	Weighted latency	Weighted acceleration	Rating
JC Buttler	4.15	0.71	1.43
GJ Maxwell	4.06	0.69	1.39
CH Gayle	3.85	0.68	1.35
AD Russell	3.75	0.70	1.34
SA Yadav	3.89	0.64	1.33
SV Samson	3.88	0.64	1.33
YBK Jaiswal	3.86	0.65	1.32
F du Plessis	3.99	0.58	1.30
N Pooran	3.67	0.63	1.27
N Rana	3.66	0.63	1.27

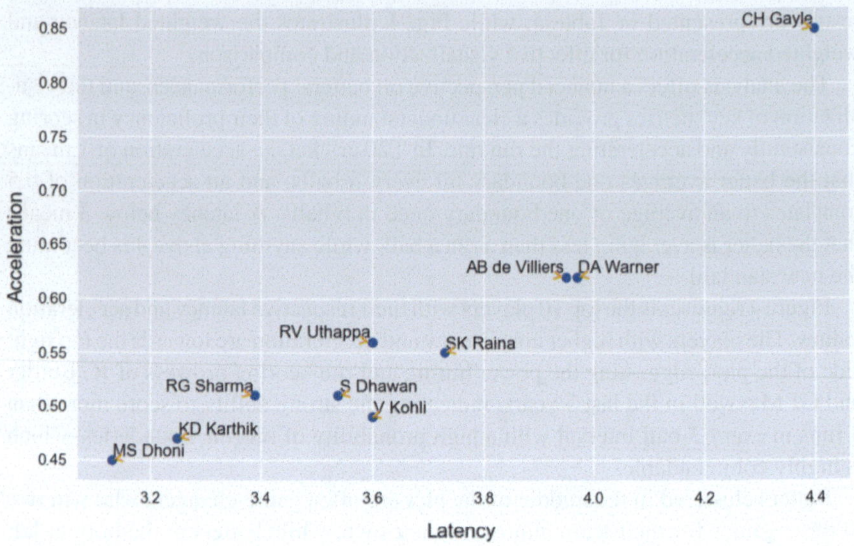

Fig. 1 Sample plot of weighted latency versus weighted acceleration for top 10 run getters of IPL

These plots also reveal key points of V Kohli's innings and reveal his slight inability to score boundaries consistently throughout his innings with an acceleration value of 0.5, suggesting a boundary scored in every 2–3-ball intervals, i.e., every 6 balls. However, on the other hand, he has a latency of 3.5 throughout his career, revealing his ability to score runs by constantly rotating strike between overs, which is better than other openers like S Dhawan and Rohit Sharma who do excel in acceleration but lag behind in scoring runs as consistently as compared to Virat Kohli. Teams may prioritize such players in various batting positions, depending on the match situation.

Figures 2 and 3 further delve into innings-wise performance, highlighting the batters who excel in Latency and Acceleration in specific innings. These plots enable better comparison of players according to match conditions of either while setting a target or while chasing a target. We notice while batters like CH Gayle and AB De Villiers stood out as main performers during the 1st innings while setting a target compared to other batters, the likes of CH Gayle, DA Warner, RV Uthappa, and Virat Kohli turned out to be better in the 2nd innings, who not only maintain a high scoring rate consistently but can also accelerate effectively by scoring boundaries, making them invaluable assets in T20 cricket during chases. CH Gayle turned out to be the only batter who topped the charts in both 1st and 2nd innings, suggesting his strong ability to score runs in any phase of the innings and match situation.

We can infer from the plots that higher Latency values might be more suitable for stabilizing the innings, especially during the powerplay and middle overs, while those with superior acceleration values are potential finishers, capable of quick runs in the death overs. However, weighted acceleration provides an overall quality of a batsman

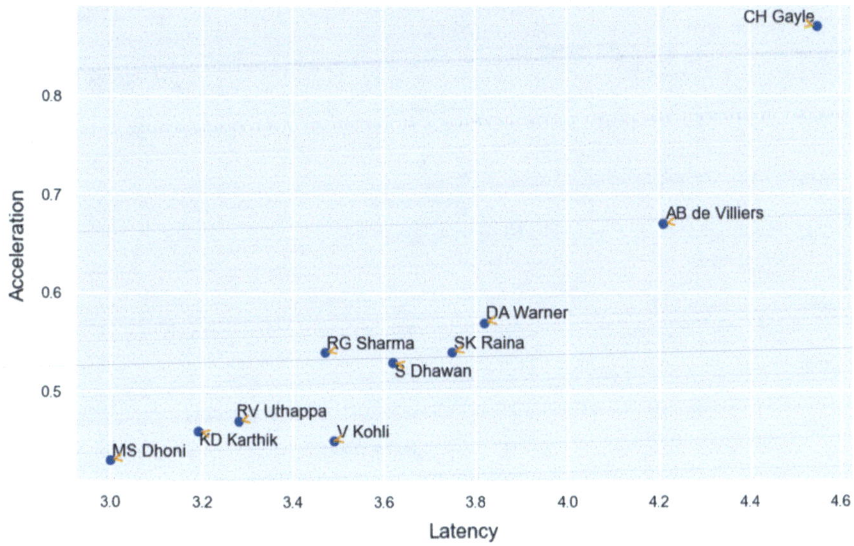

Fig. 2 Sample plot of weighted latency versus weighted acceleration in the 1st innings for top 10 run getters of IPL

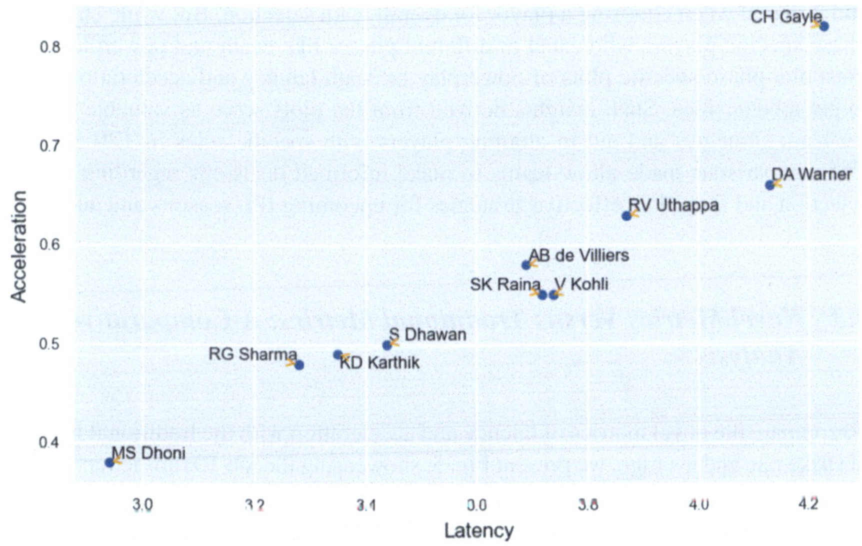

Fig. 3 Sample plot of weighted latency versus weighted acceleration in the 2nd innings for top 10 run getters of IPL

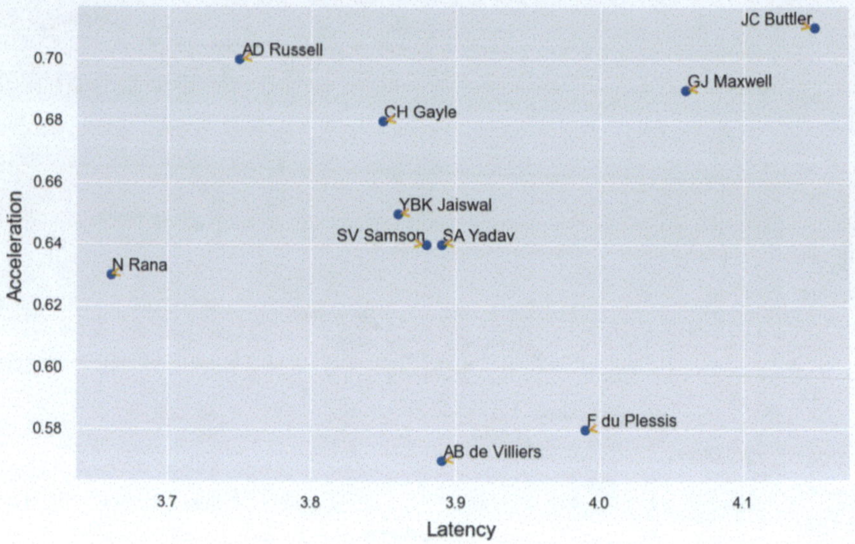

Fig. 4 Plot of weighted latency versus weighted acceleration for top 10 batter of last 5 IPL seasons

and is useful when choosing a player for overall team selection. But while choosing phase-specific players who excel in different phases like death and powerplay, one must plot phase-specific plots of powerplay or death latency and acceleration plots to get a better idea. Such insights derived from the plots serve as valuable guides for team strategies and aid in aligning players with specific roles in T20 cricket. The comparisons made allow teams to make informed decisions regarding player selection and formulate effective strategies for upcoming IPL seasons and auctions.

4.3 Novel Metrics Versus Traditional Metrics: A Comparative Analysis

To compare the novel metrics of latency and acceleration with the traditional metrics of strike rate and average, we present Fig. 5, showcasing the top 10 run-getters in IPL in terms of their strike rate and average. This visual representation aims to highlight the limitations and shortcomings associated with traditional metrics.

Analyzing the Fig. 5 plot, it becomes evident that over the entire IPL duration, players like MS Dhoni and AB De Villiers have maintained an impressive average of nearly 40. This seemingly robust average outshines that of players like Virat Kohli, Shikhar Dhawan, or Rohit Sharma, whose averages fall short. However, this comparison presents a limitation of using the average metric, as it considers average runs per dismissal rather than average runs scored per match. Batsmen like MS Dhoni and AB De Villiers, typically designated as finishers and batting towards the end of

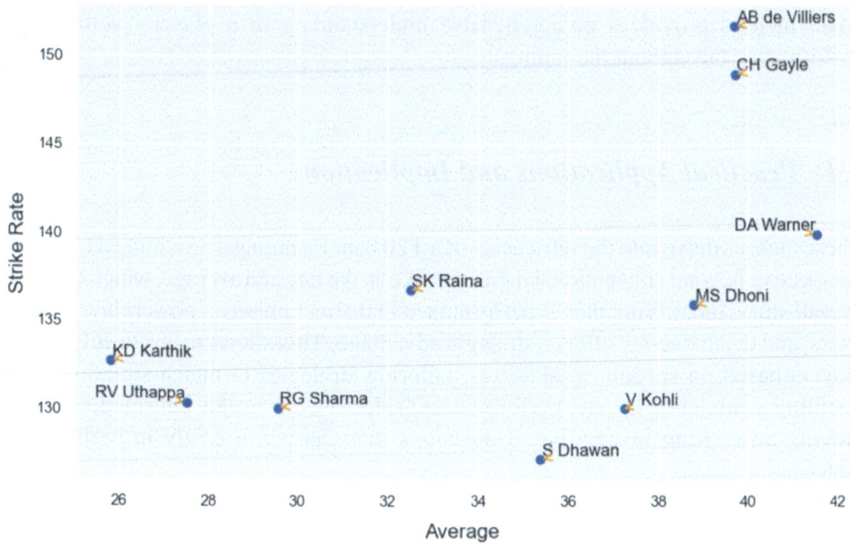

Fig. 5 Sample plot of strike rate versus average for the top 10 run getters of IPL

the innings, often accumulate higher not-out instances, inflating their averages. In contrast, players like Virat Kohli, Shikhar Dhawan, or Rohit Sharma, who bat in the early overs, have fewer chances to remain not out and adopt a more conservative approach, leading to a lower strike rate.

This discrepancy in not considering different phases of the innings leads to misleading direct comparisons. For instance, a quick glance at the traditional metrics might suggest that MS Dhoni is a superior run-scorer with better striking ability than Virat Kohli. However, this oversimplified comparison doesn't account for the nuances of player roles and the diverse phases of the innings they operate in, making it invalid. This is where the novel metrics come into play. Figure 1 illustrates the plot of Weighted Latency versus Weighted Acceleration, which factors in different innings phases. It evaluates a batter's proficiency in consistently scoring runs (Latency) and their capacity to enhance the scoring rate through boundary hits (Acceleration) across the innings. By combining the performance of the player from each phase, these novel metrics provide a more comprehensive basis for comparison.

The plot clearly highlights that Virat Kohli, contrary to initial impressions from strike rate and average metrics, boasts superior Latency and acceleration, indicating better run-scoring ability and boundary-hitting capability than MS Dhoni across all innings phases. This nuanced understanding of player performance is critical and not evident when relying solely on traditional metrics. Furthermore, the plot brings to light the striking strength of Chris Gayle, whose rapid scoring, evident in nearly scoring a boundary every three balls, wasn't adequately captured by traditional metrics. The shortcomings of relying solely on traditional metrics become apparent, offering a broad perspective that can be misleading, while the proposed novel metrics provide a comprehensive understanding of a player's ability and performance throughout the innings.

4.4 Practical Applications and Implications

These metrics delve into the intricacies of a T20 batter's innings, providing a nuanced perspective beyond conventional measures like strike rate and average, which offer an overall story. Identifying player performance in distinct phases—powerplay, middle overs, and death overs—offers a strategic advantage. This allows teams to strengthen players based on specific weaknesses, tailoring strategies to match situations. By instilling confidence in past performances, selectors can make informed choices, moving away from instinct-based decisions that can prove costly in high-stakes matches.

Player development is a key aspect, where these metrics contribute significantly. Guiding players to recognize their strengths and abilities becomes crucial, especially for those with high boundary-hitting capabilities. The use of latency scores aids in training players to maintain consistency. Understanding individual strengths, whether in maintaining consistency or providing acceleration, facilitates dynamic batting order decisions, aligning strategies with each player's unique abilities.

Additionally, these metrics simplify player comparisons across different phases, enabling more informed lineup decisions. Leagues like the IPL can leverage these metrics to identify impactful players for crucial match situations and while choosing the "Impact Player" based on match conditions. The approach not only helps in talent identification but also allows for direct and confident comparisons, overcoming the interpretational challenges often associated with traditional metrics like strike rate and average. This shift in focus towards the intricacies of building a T20 innings benefits both team strategies and precise decision-making in high-stakes T20 matches.

The insights derived from these metrics have tangible applications in the format of T20 cricket. Teams and selectors can leverage these performance metrics for more informed decisions about player selection and batting order. The nuanced view provided by these metrics assists teams in constructing a well-rounded batting lineup. Understanding a player's specific strengths, whether maintaining consistency or providing acceleration, enables teams to make dynamic batting order decisions, tailoring strategies to exploit unique player abilities. This data-driven approach supports both team strategies and precise decision-making in high-stakes matches.

4.5 Challenges and Limitations

While these metrics offer a comprehensive overview that surpasses traditional measures, it is essential to recognize their inherent limitations. Notably, these metrics do not account for external factors such as match pressure, situational nuances, pitch conditions, and boundary distances. The quantification of a batter's ability through Latency and Acceleration, while assessing consistency and aggressiveness across phases, may be influenced by the fall of wickets during an innings, impacting the natural game of the player. Comparing an opener with a batter who finishes matches may lead to faulty interpretations due to their inherent roles, limiting exposure to specific phases, for instance, an opener in the death overs.

Furthermore, the study treats every match equally, neglecting a player's lean patches. This limitation emphasizes the need for teams to selectively include matches where a player has performed well. This approach provides a more accurate understanding of a player's consistency and strengths in different phases, offering strategic advantages not achievable with traditional metrics like strike rate and average, which provide a limited overall narrative. It is crucial to acknowledge that the data collection and analysis are based on past performances, and T20 cricket is a dynamic and evolving format. Therefore, continuous updates and refinements to these metrics are imperative to keep pace with the changing dynamics of the sport.

The effectiveness of these metrics in phase-wise comparison underscores the necessity for careful consideration of external factors in specific match situations. This cautious approach aims to avoid misinterpretations and facilitate informed decisions, acknowledging the complexity of T20 cricket and the diverse challenges posed by different match scenarios.

5 Conclusion

In Conclusion, this research has successfully introduced innovative metrics, providing a profound understanding of player contributions in the dynamic T20 cricket format. The exploration of Weighted Latency versus Weighted Acceleration plots, particularly in the context of the last five IPL seasons, has yielded valuable insights into players' adaptability and reliability across diverse match situations. These metrics underscore the importance of identifying all-round contributors and emphasize the necessity for a mix of players with diverse Latency and Acceleration profiles.

As T20 cricket continues its evolution, these metrics are positioned to play a pivotal role in shaping a more informed and strategic cricketing experience. Their applicability extends beyond the IPL, encompassing a broader spectrum of T20 scenarios globally. Despite facing challenges related to external factors such as match pressure, situational nuances, and pitch conditions that may impact accuracy, the metrics prove effective in phase-wise comparisons, showcasing their potential impact on decision-making in various match scenarios.

Looking ahead, future research endeavors should explore additional factors influencing batter performance, such as situational awareness, strike rotation, and partnerships with other players. The incorporation of advanced factors like the form factor, considering lean patches in a player's performance, holds promise for enhancing the confidence in these metrics. Introducing a target factor to understand how specific targets impact scoring in each phase could aid in identifying players who excel under pressure. Considering factors like pitch conditions and boundary dimensions further adds depth to the analysis, allowing for more tailored player selections based on actual strengths and grounds.

While the current metrics primarily focus on batters, the underlying framework can be extended to derive bowlers' metrics, presenting a holistic approach to T20 cricket performance evaluation and talent identification. It's essential to underscore that these advancements can only be realized by leveraging the proposed metrics as foundational pillars, shaping the future trajectory of cricket analysis.

Acknowledgements We wish to thank Dr. Srinivas Bhogle, Honorary Scientist, CSIR-4PI, for his encouragement and many useful discussions.

References

1. Mody K, Malathi D, Jayaseeli JDD (2021) An artificial neural network approach for classifying cricket batter's performance by ad-am optimizer and prediction by derived attributes. Smart Technol Commun Robot (STCR) 2(5):99–110. https://doi.org/10.1109/stcr51658.2021.9588836
2. Santra A, Sinha A, Saha P, Das AK (2020) A novel regression based technique for batter evaluation in the Indian premier league. In: IEEE 1st international conference for convergence in engineering, vol 2(5), pp 99–110. https://doi.org/10.1109/icce50343.2020.9290569
3. Barot H, Kothari A, Bide P, Ahir B, Kankaria R (2020) Analysis and prediction for the Indian premier league. In: International conference for emerging technology, vol 2(5), pp 99–110. https://doi.org/10.1109/incet49848.2020.9153972
4. Chand S, Singh HK, Ray T (2018) Team selection using multi-/many-objective optimization with integer linear programming. In: IEEE congress on evolutionary computation, vol 2(5), pp 99–110. https://doi.org/10.1109/cec.2018.8477945
5. Awan MJ, Gilani SAH, Ramzan H, Nobanee H, Yasin A, Zain AM, Javed R (2021) Cricket match analytics using the big data approach. Electronics 10(19):2350. https://doi.org/10.3390/electronics10192350
6. Barr GDI, Kantor BS (2004) A criterion for comparing and selecting batsmen in limited overs cricket. J Oper Res Soc 55(12):1266–1274. https://doi.org/10.1057/palgrave.jors.2601800
7. Adhikari A, Majumdar A, Gupta G, Bisi A (2018) An innovative super-efficiency data envelopment analysis, semi-variance, and Shannon-entropy-based methodology for player selection: evidence from cricket. Ann Oper Res 284(1):1–32. https://doi.org/10.1007/s10479-018-3088-4
8. Amin GR, Sharma SK (2014) Measuring batting parameters in cricket: a two-stage regression-OWA method. Measurement 53:56–65. https://doi.org/10.1016/j.measurement.2014.03.029
9. Bhatambarekar G, Rai S (2022) Batsmen performance prediction using regression models. https://doi.org/10.1109/AIDE57180.2022.10060071
10. Ramalingam M, Gokul S, Mythravarshini LS, Harine KS (2022) Efficient player prediction and suggestion using machine learning for IPL tournament. https://doi.org/10.1109/mecon53876.2022.9752414
11. Passi K, Pandey N (2018) Increased prediction accuracy in the game of cricket using machine learning. Int J Data Min Knowl Manag Process 8(2):19–36. https://doi.org/10.5121/ijdkp.2018.8203

12. Manage ABW, Scariano SM, Hallum CR (2013) Performance analysis of T20-world cup cricket 2012. Sri Lankan J Appl Stat 14(1):1. https://doi.org/10.4038/sljastats
13. Sharma SK, Amin RG, Gattoufi S (2012) Choosing the best Twenty20 cricket batsmen using ordered weighted averaging. Int J Perform Anal Sport 12(3):614–628. https://doi.org/10.1080/24748668.2012.11868623
14. Ahmad H, Ahmad S, Asif M, Rehman M, Alharbi A, Ullah Z (2021) Evolution-based performance prediction of star cricketers. Comput Mater Contin 69(1):1215–1232. https://doi.org/10.32604/cmc.2021.016659
15. Wickramasinghe I (2022) Applications of Machine Learning in cricket: a systematic review. Mach Learn Appl 10:100435. https://doi.org/10.1016/j.mlwa.2022.100435
16. Deep Prakash C, Verma S (2022) A new in-form and role-based deep player performance index for player evaluation in T20 cricket. Decis Anal J 2:100025. https://doi.org/10.1016/j.dajour.2022.100025

Electromyographical Insights into Front-On Versus Side-On Bowling Actions

Jatin Bhosle and Yajuvendra Singh Rajpoot

Abstract This study compared muscle activity in 10 male medium-fast bowlers using electromyography during two different approaches: front-on and side-on. Five upper body muscles were selected for analysis: anterior deltoid, pectoralis major, latissimus dorsi, trapezius, and triceps brachii. Ten male medium-fast bowlers (aged 22.5 ± 3.19 years, height 176.78 ± 9.14 cm, weight 69.9 ± 8.7) were recruited for the study. Root Mean Square (RMS) values were calculated to measure muscle activation. The researchers compared the EMG signals of two groups using a t-test. While no statistically significant differences were found, trends emerged. Front-on bowlers exhibited higher RMS values for the anterior deltoid, trapezius, and triceps brachii, suggesting increased muscle activation. Conversely, the pectoralis major and latissimus dorsi displayed higher RMS values for side-on bowlers. The observed differences may be attributed to the range of motion required for each approach. The front-on approach involves shoulder alignment exceeding 200° during foot impacts, potentially necessitating greater muscle activation for accuracy. Despite the lack of statistical significance, the differences in muscle activation highlight the potential influence of the bowling approach on upper-body muscle activity. Further research is needed to confirm these results and explore their implications for bowling performance and injury prevention.

Keywords Medium-Fast bowling · Cricket · Electromyography

J. Bhosle (✉)
Department of Sports Biomechanics, Lakshmibai National Institute of Physical Education, Gwalior, Madhya Pradesh 474001, India
e-mail: jatinbhosle1@gmail.com

Y. S. Rajpoot
Department of Sports Management and Coaching, Lakshmibai National Institute of Physical Education, Gwalior, Madhya Pradesh 474001, India

S. Bhattacharyya and D. James (eds.), *Recent Trends in Sports Engineering*, Lecture Notes in Mechanical Engineering, https://doi.org/10.1007/978-981-97-8297-0_13

1 Introduction

Today, many people play cricket as a sport. The fascinating quality of this game has drawn viewers from all around the world. Although thousands of people attend cricket stadiums to see the game live, millions more watch it on television. An estimated 1.6 billion people worldwide watched the live coverage of the 2019 Men's Cricket World Cup, according to a media report issued by the International Cricket Council (ICC) in September 2019 [4]. In the game of cricket, there are three departments of skills, namely batting, bowling, and fielding.

Fast bowling, a crucial aspect of cricket, has led to injuries among young fast bowlers in the world recently that are practically epidemic in scope. According to reports, the cricketer most prone to severe or overuse injuries is the young fast bowler [11–13]. There have been more injuries recorded as a result of kids and teens participating in more organized sports and fitness programs [1, 5, 15]. These young athletes have also been reported to have recently developed injuries that are unique to particular sports.

When it comes to bowling action, fast bowling uses two distinct methods: mixed action and non-mixed action [14]. These actions are explained as follows upon simpler bifurcation: the additional explanation entails two different approaches, front-on and side-on (see Fig. 1). Front-on is defined as a shoulder alignment at the back or front foot impacts of more than 200 degrees, with the right hand horizontal defined as a line drawn through the leading shoulder and parallel to the pitch. On the other hand, side-on implies a shoulder alignment at the front foot or back foot impacts of less than 190 degrees [3]. A complete non-mixed action is produced by fewer bowlers throughout the world compared to non-mixed action [2].

The research aimed to compare electromyographic (EMG) recordings of upper body muscles in medium-fast bowlers. The objective was to assess the muscle activation of five selected upper body muscles during front-on and side-on bowling approaches. It was hypothesized that there would be no significant difference in electromyographical recordings of the selected upper body muscles between front-on and side-on medium-fast bowlers.

2 Materials and Methods

2.1 Study Participants

Ten male medium-fast bowlers (mean ± SD: aged 22.5 ± 3.19 years; height 176.78 ± 9.14 cm; weight 69.9 ± 8.7) were recruited for the study. They were categorized into two groups (n = 5 each) based on their bowling actions—Front-on and Side-on (refer to Table 1), who played cricket for at least 3 years and participated in at least one inter-university or state-level tournament. It was ensured that all participants were fit

(a) Front-on (b) Side-on

Fig. 1 The major difference is seen at the backfoot landing and chest position

Table 1 Descriptive statistics (bowling action-wise)

Bowling action	Parameters	N	Min	Max	Mean	Std. Dev
Front-on	Age (years)	5	18.3	24.1	21.42	2.27
	Height (cm)	5	166.3	182.1	171.12	6.55
	Weight (kg)	5	60.3	74.1	67.66	5.31
Side-on	Age (years)	5	18.4	25	21.96	2.45
	Height (cm)	5	166.4	185.2	174.24	7.88
	Weight (kg)	5	62	77.2	69.64	5.92

and healthy before the study commenced. A written consent form expressing their willingness to participate was completed by all participants.

The table above presents descriptive details of the study participants.

2.2 *Variables and Methods*

The selected muscles for investigation include Anterior Deltoid (AD), Pectoralis Major (PM), Latissimus Dorsi (LD), Trapezius (TPZS), and Triceps Brachii (TRCP) (refer to Fig. 2 for probe placement sites on selected muscles).

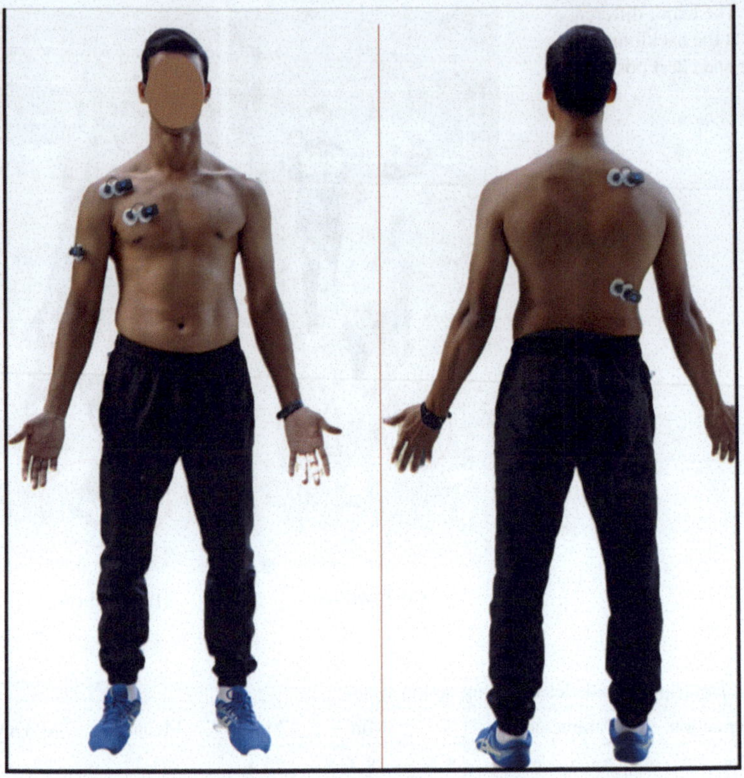

Fig. 2 Probe placement sites on selected muscles

Data collection was performed using the BTS Free EMG Analyzer, with the criterion measure being EMG signals (RMS values) as illustrated in Fig. 3. BTS Free EMG analyzer has been a trustworthy EMG signal recording device [10].

To ensure reliable surface electromyography (sEMG) data collection, several key considerations were addressed. First, proper equipment, including ECG electrodes and alcohol swabs, was gathered. Skin preparation was crucial, involving thorough cleaning and, if necessary, hair removal [6]. Adequate fixation of electrodes was necessary to account for muscle movement. Precise identification of probe placement sites and temperature control were essential. Vibrations were minimized, and data collection was conducted in controlled settings. Standardization of conditions, documentation of probe sequences and muscles, and clear protocol definition enhanced consistency and validity.

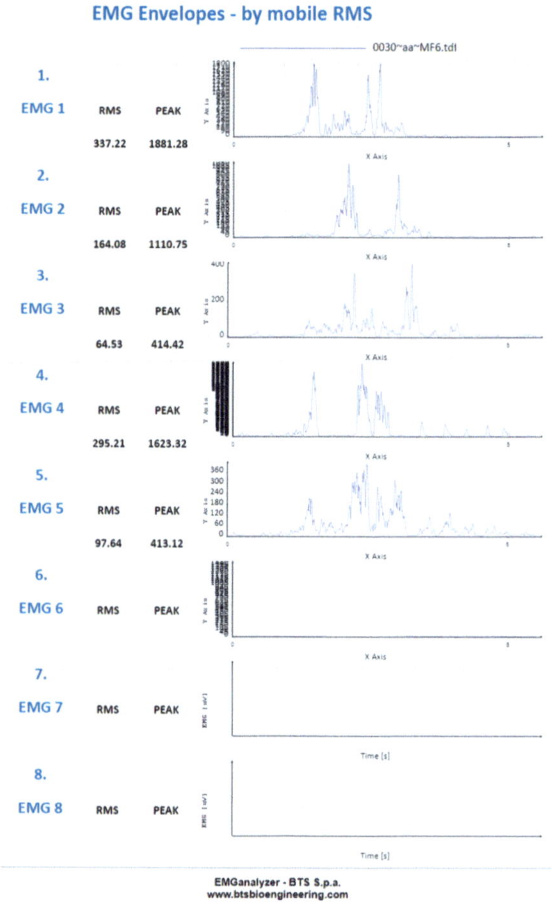

Fig. 3 EMG signal output of one of the participants

2.3 *Statistical Analysis*

Statistical analysis was conducted using IBM SPSS 26 software. Data comparison was performed using a two-sample independent t-test. Descriptive statistics, including mean and standard deviation, were utilized. The level of significance was set at $p < 0.05$.

3 Results

Muscle activation comparison revealed no statistically significant differences in any of the selected muscles (see Table 2). The null hypothesis for each muscle remains retained.

Table 2 Group statistics

Muscles	Bowing action	N	Min	Max	Mean	SD
Anterior deltoid	Front-on	5	102.21	337.22	210.72	23.69
	Side-on	5	89.01	343.49	197.1	50.51
Pectoralis major	Front-on	5	49.69	209.04	120.88	59.32
	Side-on	5	33.56	268.74	131.53	86.97
Latissimus dorsi	Front-on	5	20.2	93.07	52.88	26.79
	Side-on	5	26.7	218.14	71.36	48.51
Trapezius	Front-on	5	118.04	295.21	220.38	57.44
	Side-on	5	105.1	408.33	172.11	66.04
Triceps brachii	Front-on	5	30.11	228.61	106.6	44.67
	Side-on	5	40.65	322.65	75.38	26.87

Table 3 displays the significant values indicating the equality of variances in the RMS data for both bowling actions. All values exceeded the 0.05 level, signifying the equality of variances for both front-on and side-on bowlers. The results obtained from Table 3 revealed no significant difference in the RMS values of selected muscles' electromyography (EMG) between front-on and side-on medium-fast bowlers, as the p-values were > 0.05. It was observed that the TPZS and TRCP exhibited a higher mean difference between the root mean square values of both bowling actions (Fig. 4).

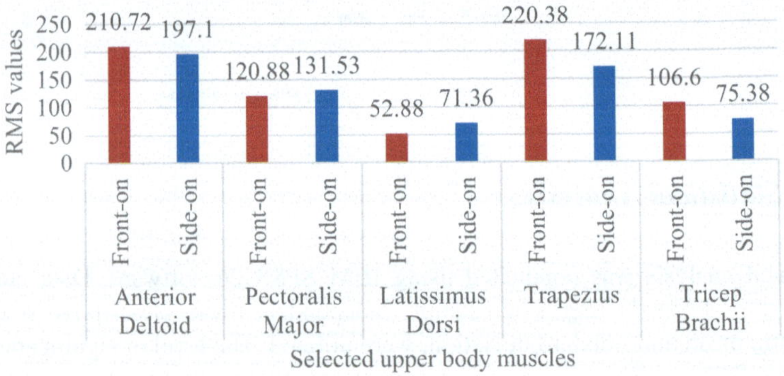

Fig. 4 RMS values chart depicting differences in the means of selected upper body muscles for both bowling actions

Table 3 Results of t-test on selected muscles, root mean square between front-on and side-on medium fast bowlers

Muscle	Levene's test		t-test for equality of means		
	Sig	T	df	Sig. (2-tailed)	MD
Anterior deltoid	0.038	0.546	8	0.6	13.61
Pectoralis major	0.695	0.226	8	0.827	10.64
Latissimus dorsi	0.047	0.746	8	0.477	18.48
Trapezius	0.999	1.233	8	0.253	48.26
Triceps brachii	0.382	1.339	8	0.217	31.21

t. obtained value of t; df. degree of freedom; Sig. (two-tailed). Significance level; MD. mean difference.

4 Discussion of Findings

The comparison of muscle activity between front-on and side-on bowling actions among medium-fast bowlers provided significant insights into the biomechanical demands of these techniques. While no statistically significant differences were detected in the electromyographic recordings of selected upper body muscles, the observed trends and the implications drawn from them merit further discussion and exploration.

The lack of statistical significance in the differences in muscle activation between front-on and side-on bowlers may suggest that both techniques engage similar musculature to a comparable extent [8]. However, it is essential to search deeper into the variations of these findings to understand the potential biomechanical implications for bowling performance and injury prevention.

The higher mean difference between the RMS values of TPZS and TRCP in both bowling actions suggests a potential area of interest for further investigation. This difference may be attributed to the differing demands placed on these muscles by the two bowling approaches. Specifically, the front-on technique, characterized by a shoulder alignment exceeding 200° during foot impacts, likely necessitates greater activation of the trapezius and triceps brachii to maintain accuracy and control throughout the delivery stride. Conversely, side-on bowlers, with a shoulder alignment under 190° during similar foot impacts, may distribute the workload differently among their upper body muscles, resulting in a lower mean difference in muscle activation for TPZS and TRCP.

These findings align with previous research by [3]), which highlighted the importance of shoulder alignment in determining the biomechanical demands of different bowling techniques. By exceeding the 200° threshold during front-on bowling, bowlers may experience increased stress on the trapezius and triceps brachii, potentially impacting performance and susceptibility to injury. Future studies could explore the relationship between shoulder alignment, muscle activation patterns, and bowling accuracy to provide a more comprehensive understanding of the biomechanics underlying different bowling techniques.

Moreover, while this study focused exclusively on electromyographical insights into bowling actions, it is crucial to consider other biomechanical variables that may influence performance and injury risk. Studies incorporating measures of linear and angular kinematics, anthropometrics, and dynamic joint loading could provide a more holistic understanding of the biomechanical factors contributing to bowling performance [7].

Additionally, the inclusion of female participants in similar studies would offer valuable insights into the biomechanical differences between male and female medium-fast bowlers. Given the potential variations in muscle activation patterns, joint kinematics, and injury mechanisms between genders, a comprehensive understanding of these differences is essential for developing targeted training programs and injury prevention strategies tailored to each demographic [9, 16].

Conclusively, the small sample size of this study warrants caution in generalizing the findings to a larger population of medium-fast bowlers. Future research endeavors should aim to replicate these findings with larger and more diverse participant cohorts to enhance the robustness and generalizability of the results.

Several limitations were encountered during the research process. Sweating among participants posed a significant challenge during data collection, making it difficult to adhere the probe stick to the selected muscle; consequently, sticky tapes were utilized to enhance probe adhesion. Additionally, the researcher lacked control over temperature conditions. Despite addressing all necessary limitations, the psychological states of the bowlers remained beyond the researchers' control, potentially impacting the study.

Understanding the muscle activation differences between front-on and side-on bowling approaches can aid in refining bowling techniques and injury prevention strategies. Coaches could tailor training regimes to target specific muscle groups highlighted in the study, potentially optimizing performance and reducing the risk of overuse injuries. Cricketers could adjust their bowling approach based on individual muscle activation patterns, aiming for a more balanced distribution of effort across relevant muscle groups for improved efficiency and injury resilience.

References

1. DiFiori JP, Benjamin HJ, Brenner JS, Gregory A, Jayanthi N, Landry GL, Luke A (2014) Overuse injuries and burnout in youth sports: A position statement from the American Medical Society for Sports Medicine. Br J Sports Med 48(4):287–288. https://doi.org/10.1136/bjsports-2013-093299
2. Ferdinands RED, Kersting UG, Marshall RN (2014) A new taxonomic system for the sub-classification of cricket bowling actions. Sports Technology 7(1–2):26–38. https://doi.org/10.1080/19346182.2014.893350
3. Foster D, John D, Elliott B, Ackland T, Fitch K (1989) Back injuries to fast bowlers in cricket: A prospective study. Br J Sports Med 23(3):150. https://doi.org/10.1136/bjsm.23.3.150
4. ICC Media Release (2019). 2019 Men's Cricket World Cup most watched ever. https://www.icc-cricket.com/media-releases/1346930

5. Micheli LJ (1983) Overuse injuries in children's sports: The growth factor. Orthop Clin North Am., 14(2), 337–357. PMID: 6843972
6. Farina D, Pozzo M, Merlo E, Bottin A, Merletti R (2004) Assessment of average muscle fiber conduction velocity from surface EMG signals during fatiguing dynamic contractions. IEEE Trans Biomed Eng 51(8).1383–1393
7. Glazier PS, Wheat JS (2014) An integrated approach to the biomechanics and motor control of cricket fast bowling techniques. Sports Med 44:25–36
8. Hilligan BK (2008) The relationship between core stability and bowling speed in asymptomatic male indoor action cricket bowlers
9. Landry SC, McKean KA, Hubley-Kozey CL, Stanish WD, Deluzio KJ (2007) Neuromuscular and lower limb biomechanical differences exist between male and female elite adolescent soccer players during an unanticipated side-cut maneuver. Am J Sports Med 35(11):1888–1900
10. Mishra R, Singh J, Bhosle J (2023) Effects of different racket coverings in table tennis forehand drive on selected upper body muscles. Ser Biomech 37(3):57–67. https://doi.org/10.7546/SB.08.03.2023
11. Payne W, Carlson J, Hoy G, Laussen SP (1987) What research tells the cricket coach. Sports Coach 10(4):17–22
12. Perrett C, Bussey M, Lamb P (2022) External workload intensity in cricket fast bowlers across maximal and submaximal intensities: Modifying PlayerLoad and IMU location. J Sports Sci 40(5):527–533. https://doi.org/10.1080/02640414.2021.2003570
13. Raghavendra Rao M, Srinivasan TM, Kumar Itagi R (2020) Epidemiology of annual musculoskeletal injuries among male cricket players in India. Indian J Community Health, 32(3), 590–593. https://doi.org/10.47203/IJCH.2020.v32i03.023
14. Ranson CA, Burnett AF, King M, Patel N, O'Sullivan PB (2008) The relationship between bowling action classification and three-dimensional lower trunk motion in fast bowlers in cricket. J Sports Sci 26(3):267–276. https://doi.org/10.1080/02640410701501671
15. Schneider KJ, Critchley ML, Anderson V, Davis GA, Debert CT, Feddermann-Demont N, Gagnon I, Guskiewicz KM, Hayden KA, Herring S, Johnstone C, Makdissi M, Master CL, Moser RS, Patricios JS, Register-Mihalik JK, Ronksley PE, Silverberg ND, Yeates KO (2023) Targeted interventions and their effect on recovery in children, adolescents and adults who have sustained a sport-related concussion: A systematic review. Br J Sports Med 57(12):771–779. https://doi.org/10.1136/bjsports-2022-106685
16. Zeller BL, McCrory JL, Ben Kibler W, Uhl TL (2003) Differences in kinematics and electromyographic activity between men and women during the single-legged squat. Am J Sports Med 31(3):449–456

Assessing Muscle Activation Variations in Table Tennis Forehand Smash with Differing Racket Coverings

Radhika Mishra and Joseph Singh

Abstract This study aimed to examine upper body muscle engagement during the forehand smash in table tennis using two different racket coverings—inverted and short-pimpled. Seven male university-level table tennis players (mean age 20.12 ± 2.38, height 164.9 ± 7.5 cm, weight 67.5 ± 5.2 kg) with adequate playing experience participated. Approval was obtained from the research committee, and participants were ensured to be physically fit. Electromyographical signals were recorded from four upper body muscles (Biceps Brachii, Triceps Brachii, Trapezius, and Anterior Deltoid). Root Mean Square (RMS), an EMG signal variable, was analyzed for maximum muscle involvement. Paired t-test method compared muscle activation with both racket coverings, using mean and standard deviation for descriptive statistics. While none of the muscles exhibited a significant difference in activation levels, Biceps Brachii and Trapezius showed higher RMS values with the Inverted covering, whereas Triceps Brachii and Anterior Deltoid had higher RMS values with the short-pip covering. Despite the forehand smash's offensive significance, its accuracy control remains challenging, particularly within a designated target area, as highlighted in previous studies citing the superior spin production of inverted rubber over pimpled rubber.

Keywords Muscle activation · Racket covering · Forehand smash · Electromyography

1 Introduction

Table tennis in the modern day has become a sport in which the development of high speed, power, endurance, strength, flexibility, quick reflexes, and dexterity is valued more than in the past. Prominent athletes typically focus on attacking plays or counterattacks [3, 6]. In the realm of table tennis, strength training, physical

R. Mishra (✉) · J. Singh
Department of Sports Biomechanics, Lakshmibai National Institute of Physical Education, Gwalior, Madhya Pradesh 474001, India
e-mail: mishraradhika722@gmail.com

S. Bhattacharyya and D. James (eds.), *Recent Trends in Sports Engineering*, Lecture Notes in Mechanical Engineering, https://doi.org/10.1007/978-981-97-8297-0_14

conditioning, and advanced fitness assessment techniques are essential components of the modern sports training program [5].

Table tennis is a physically demanding sport that requires a sophisticated understanding of strategic principles in addition to cognitive and physical training [1]. Similar to other sports, table tennis is susceptible to typical injuries sustained during play. However, modern sports sciences are essential for identifying and preventing injuries among table tennis players, especially when combined with knowledge from sports medicine. An extensive understanding of the functional anatomy and pathophysiology related to damage to different tissues is necessary for a thorough preventative approach. A thorough grasp of ideas like excessive load, load distribution, the mechanics behind sports injuries, and the biochemical reactions of body tissues to impact and overuse are all necessary for effective injury prevention [4].

Coaches put in a lot of work to improve player performance to dominate opponents. Among the various strategies that are used, one common one is the choice of particular table tennis racket coverings. These include inverted rubber surfaces, anti-spin, long, and short pip surfaces. Every variation adds unique qualities that impact ball spin, speed, and overall style of play. In the fast-paced world of table tennis, a player's ability to maximize their skills and obtain a competitive advantage depends heavily on their strategic choice of racket covering [6].

The objective of this study was to investigate the complexities of table tennis through a thorough examination of the activation of particular upper body muscles during a forehand smash. Understanding how the choice of racket coverings—inverted and short-pimpled, in particular—influenced the electromyographical patterns of the selected muscles involved in this dynamic table tennis movement was the main focus of the study. Through tackling the scientific subject of electromyographical variations in muscle activation, the study aimed to provide significant understanding into the biomechanics of playing techniques linked to these unique racket covers. This required a careful analysis of muscle activation, providing a better comprehension of the physiological difficulties imposed by using short-pimpled, inverted racket covers when performing a forehand smash.

2 Materials and Methods

2.1 Study Participants

Seven male table tennis players at the university level, with a mean age of 20.12 ± 2.38 years, an average height of 164.9 ± 7.5 cm, and a weight of 67.5 ± 5.2 kg, were carefully recruited for this study, as detailed in Table 1. Prior to their inclusion, stringent measures were taken to ensure the fitness and overall health of all participants. The selection process emphasized the voluntary nature of participation, and only those players who willingly agreed to take part in the study were included. The focus of this research involved the recording of electromyographical signals from specific upper body muscles. Electromyographic recordings were collected using the BTS Free EMG analyzer (Fig. 2). The chosen muscles for signal recordings included the Bicep Brachii, Triceps Brachii, Trapezius, and Anterior Deltoid (see Fig. 3).

Inverted: A racket with rubber featuring a flat surface (without pips) is called an inverted racket covering (see Fig. 1).

Table 1 Descriptive statistics of the study participants

Parameters	N	Min	Max	Mean	SD
Age (years)	7	19.3	23.5	20.12	2.38
Height (cm)	7	160.1	172	164.9	7.5
Weight (kg)	7	62.3	72.7	67.5	5.2

N. number of subjects; Min. minimum RMS recording observed; Max. maximum RMS recording observed; SD. standard deviation.

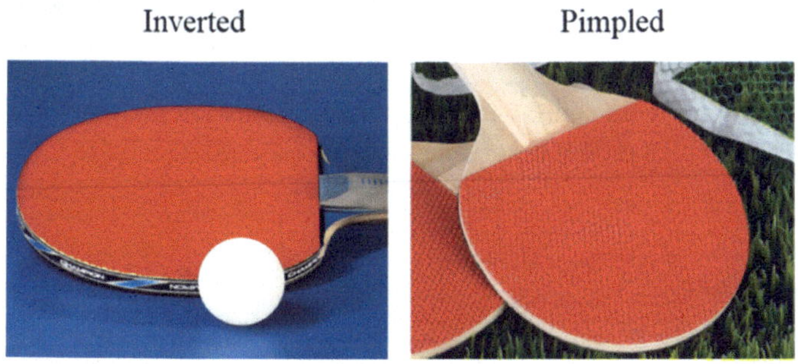

Fig. 1 Inverted and pimpled rubber coverings

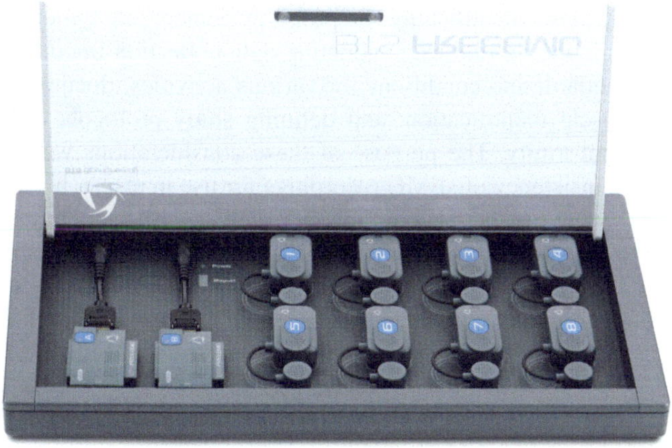

Fig. 2 BTS Free EMG analyzer device

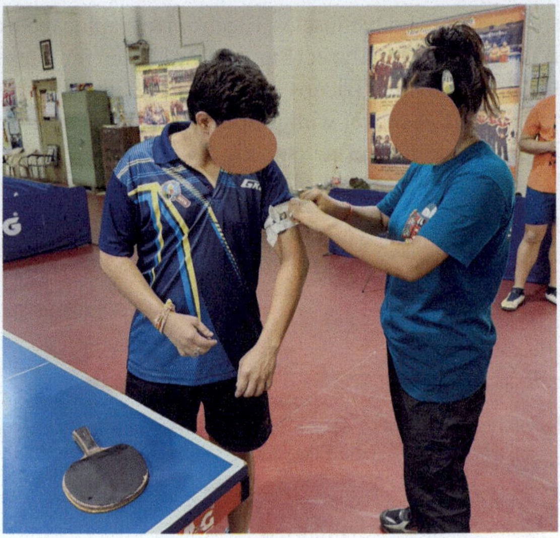

Fig. 3 Researcher placing the EMG probes on one of the participants' selected muscles

Pimpled: Rubber with short, wide pips that face outward and are tightly spaced is known as short pips rubber (see Fig. 1).

2.2 *Procedure*

Important factors for sEMG data collection included gathering the cotton pads, alcohol swabs, and ECG electrodes needed for electrode implantation and skin preparation (cleaning the skin at electrode sites and shaving off any hair). To ensure consistent and valid data collection, it was also addressed that muscle movement during activities, precise identification of probe placement sites, monitoring skin temperature, minimizing vibrations, conducting data collection in controlled laboratory settings, standardizing conditions for various activities, documenting probe sequences and muscle identification, and defining study protocols and variables with clarity and uniformity. The purpose of these considerations was to improve the accuracy and consistency of sEMG recordings for use in research settings.

All the players were instructed about the task well in advance. They were given sufficient time to warm up the targeted muscles and perform trial shots before collecting data. The Group of players performed the forehand smash stroke with Inverted and Pimpled racket coverings turn by turn. A target area of 55 × 55 cm (Fig. 4) was marked as per the drive ability test [7] as shown in Fig. 4.

Fig. 4 Target area as per drive ability test for maintaining accuracy

2.3 *Statistical Analysis*

The statistical analysis was conducted using IBM SPSS 26, employing a paired t-test for data comparison. Descriptive statistics, including mean and standard deviation, were utilized to characterize the data. The level of significance was set at $p < 0.05$. The criterion measure for the analysis was the Root Mean Square, recorded as the EMG signal.

3 Results

It was observed that the Anterior Deltoid exhibited a significantly higher mean difference (27.00) in RMS values when playing a forehand smash with both racket coverings, whereas the mean difference in other muscles was not as significant (Table 2).

Table 3 displays the t-values and two-tailed significance (Sig. two-tailed) values resulting from a paired t-test comparing the RMS data of Inverted versus Pimpled racket coverings. None of the calculated t-values exceeds the tabulated t-value (at $df_6 = 1.943$), leading to the retention of the null hypothesis. Across all pairs, the two-tailed significance values consistently exceed the significance level of $p > 0.05$, indicating a lack of significant difference between the paired observations.

Table 2 Descriptive statistics

Muscle	Racket covering	N	Mean	Mean difference	Standard deviation
Bicep brachii	Inverted	7	209.54	15.41	38.99
	Pimpled	7	194.13		35.96
Triceps brachii	Inverted	7	117.4	4.99	23.13
	Pimpled	7	122.39		23.22
Trapezius	Inverted	7	149.33	9.32	30.66
	Pimpled	7	140.01		29.89
Anterior deltoid	Inverted	7	175.15	27	21.66
	Pimpled	7	202.15		30.09

Table 3 Results of t-test on selected muscles RMS between inverted and pimpled racket coverings

Muscle	t	df	Sig. (two-tailed)
Bicep brachii (inverted-pimpled)	0.92	6	0.395
Triceps brachii (inverted-pimpled)	0.58	6	0.586
Trapezius (inverted-pimpled)	1.23	6	0.265
Anterior deltoid (inverted-pimpled)	1.71	6	0.138

t. obtained value of t; df. degree of freedom; Sig. (2-tailed). significance level.

4 Discussion of Findings

There were some limitations faced by the researchers during the time of data collection, such as the athletes sweating, which made it difficult to adhere the probes to the surface skin of the selected muscles. Additionally, all players used the same racket in order to reduce the heterogeneity of the equipment. However, since they preferred to play with their personal rackets, it sometimes took time to adjust to new equipment. Furthermore, sometimes the targeted area was not hit by the player, so retrials were conducted to ensure 100% accuracy. Some players achieved this in their first attempt, while others required second or third attempts, which researchers feared might have increased fatigue in the muscles to some extent.

In the conducted study, it was observed that there was no significant variance in muscle activation level across all muscles examined when playing with both racket coverings (refer to Fig. 5). However, upon comparing the means, it was noted that the BB and TPZS displayed higher RMS values when using the Inverted racket covering, whereas the TRCP and AD exhibited increased RMS values when using the short-pip racket covering. This suggests that different racket coverings may produce varied muscular responses during gameplay. Such findings underscore the importance of considering equipment variations in sports biomechanics research, particularly in activities involving specialized tools like rackets. Understanding how different racket characteristics affect muscle engagement can inform training regimens, equipment design, and injury prevention strategies in sports such as tennis and table tennis.

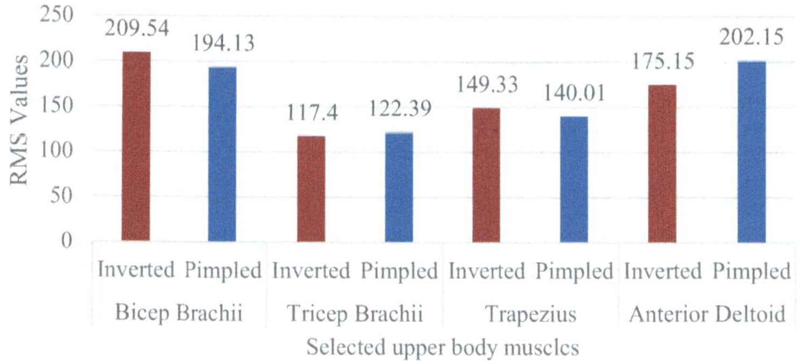

Fig. 5 RMS values chart depicting differences in the means of selected upper body muscles while playing smash with both racket coverings

Further investigations may dive into the fine interactions between racket properties and muscular activation to optimize performance and avoid injury risks.

In line with the findings of this study, previous research by Chow et al. (1999) has indeed emphasized the crucial role of AD in facilitating the forward movement of the arm and racket during racket sports. This muscle's higher activation, particularly with the inverted racket covering, suggests its importance in controlling the increased spin generated by this type of covering [9, 10]. Moreover, the differential activation patterns observed in the BB, TRCP, and TPZS muscles underscore the complex interactions between muscle groups during the execution of the forehand smash.

While the participants in this study were university-level table tennis players, it would be important to explore whether muscle activation patterns vary across skill levels and playing styles. For instance, elite players known for their aggressive playing style may exhibit distinct muscle activation profiles compared to intermediate or recreational players [8]. Additionally, investigating how muscle activation patterns adapt over time with training and experience could provide valuable insights into skill acquisition and performance optimization strategies.

The observed differences in muscle activation between racket coverings have practical implications for coaches and players seeking to optimize their training regimens and techniques. Coaches may consider tailoring drills and exercises to target specific muscle groups based on the player's preferred racket covering and playing style. Moreover, biomechanical analysis coupled with electromyography can inform the development of personalized training programs aimed at enhancing muscle strength, coordination, and endurance relevant to table tennis strokes [2].

While this study focused on the forehand smash, future research could explore muscle activation patterns during other table tennis strokes (e.g., backhand, serve) and defensive strokes. Comparative studies involving players of different skill levels and playing styles could provide a more comprehensive understanding of how muscle activation varies across diverse populations. Furthermore, investigating the impact of environmental factors (e.g., playing surface, lighting conditions) on muscle activation could explain additional determinants of performance in table tennis.

References

1. Bańkosz Z, Winiarski S (2017) The kinematics of table tennis racquet: Differences between topspin strokes. J Sport Med Phys Fit, 57(3). https://doi.org/10.23736/S0022-4707.16.06104-1
2. Ebrahim, H. (2010). The mechanical contribution of the arm movement during the performance of the back hand topspin table tennis players [Dissertation, University of Konstanz]. http://nbn-resolving.de/urn:nbn:de:bsz:352-opus-112139
3. Kahn J-F, Lees A, Maynard I (2004) Science and racket sports III: the proceedings of the eighth international table tennis. Routledge. http://public.eblib.com/choice/publicfullrecord.aspx?p=200710
4. Kondrič M, Furjan-Mandić G, Medved V (2006) Myoelectric comparison of table tennis forehand stroke using different ball sizes. 36(4):7
5. McCann PD, Bigliani LU (1994) Shoulder Pain in Tennis Players: Sports Medicine 17(1):53–64. https://doi.org/10.2165/00007256-199417010-00005
6. Mishra R, Singh J, Bhosle J (2023) Effects of different racket coverings in table tennis forehand drive on selected upper body muscles. Ser Biomech 37(3):57–67. https://doi.org/10.7546/SB.08.03.2023
7. Purashwani P, Datta AK, Purashwani M (2010) Construction of norms for skill test table tennis players. Int Table Tennis Fed Headquarter & Olymp Off 6:93–98
8. Reilly T (Ed) (1990) Physiology of sports (1st ed). E. & F.N. Spon
9. Varenberg M, Varenberg A (2012) Table Tennis rubber: tribological characterization. Tribol Lett 47(1):51–56. https://doi.org/10.1007/s11249-012-9961-4
10. Varenberg M, Varenberg A (2014) Table tennis: preliminary displacement in pimples-out rubber. Tribol Lett 53(1):101–105. https://doi.org/10.1007/s11249-013-0247-2

3D Numerical Simulations of a Tennis Ball Trajectory

Pasunuru Sai Vineeth and Mahesh V. Panchagnula

Abstract Tennis is one of the few ball sports where the playing surface is routinely shuffled to make the game engaging for both the fans and the players. These surfaces exhibit different characteristics and present their own challenges, thus demanding different playing strategies from the players. Another key element influencing the ball's trajectory is undoubtedly the imparted spin. Used appropriately, it can raise the shot's error margin while simultaneously affecting the event of bounce. The current study aims to visualize the trajectory of a tennis ball for various kinds of imparted spins on different playing surfaces. This needs an understanding of the various aerodynamic forces in operation while the ball is in flight and the court's behavior that manipulates the trajectory on impact. The numerous parameters in play can be collectively altered to vary the shot placement along and across the court, the bounce of the ball and the time the opponent player gets to return the shot. We investigate and elaborate on the effect each of the parameters has on the ball's trajectory. Understanding this physics can provide valuable insights to the players and coaches in terms of precise shot execution and playing strategies.

Keywords Sports science · Flight and bounce · Tennis ball

1 Introduction

A ball experiences aerodynamic forces in flight because of its interaction with the surrounding airflow [1]. The drag force occurs due to the separation of boundary layer and resists the relative motion between the ball and the air. Non-dimensional drag is defined as drag coefficient, and its value suddenly falls when the flow around the ball transitions to turbulence. It can be observed in the drag crisis plot [2], where the drag coefficient of a smooth sphere is plotted against the local Reynolds number. The transition of boundary layer to turbulence is accelerated when the sphere's surface is

P. Sai Vineeth (✉) · M. V. Panchagnula
Department of Applied Mechanics and Biomedical Engineering, Indian Institute of Technology Madras, Chennai 600036, India
e-mail: am21d201@smail.iitm.ac.in

S. Bhattacharyya and D. James (eds.), *Recent Trends in Sports Engineering*, Lecture Notes in Mechanical Engineering, https://doi.org/10.1007/978-981-97-8297-0_15

rough. Shah and Mittal found that for a tennis ball, the transition Reynolds number is 1.5×10^5 [3], and the drag coefficient remains fairly constant afterwards. The fuzz on a tennis ball results in a higher non-dimensional drag than that of a smooth sphere interpolated from the drag crisis plot. Additionally, if the ball spins, a lift force starts acting upon it. This arises because the boundary layer separation on the advancing and retreating sides of the ball is asymmetrical [4]. The magnitude of force is affected by both the velocity and the imparted spin [5]. These drag and lift forces are determined experimentally or numerically using computational fluid dynamics techniques.

The bounce process is difficult to formulate because the ball may undergo any of the following three interactions [6] before leaving the surface: sliding, sliding followed by rolling or sliding followed by gripping. Brody [7] presented a model for a tennis ball where the bounce was oblique. Using a torsion pendulum [8], he measured the moment of inertia of four types of tennis balls, whose value is vital to understanding the spin the ball acquires on impact. Cross [9] studied the case of a spinning ball incident perpendicular to the strings or the court but bouncing obliquely with modified spin. In another work [10], he found that the horizontal coefficient of restitution varies with the coefficient of sliding friction and the incident kinematic parameters. The restitution coefficient was observed to be between -0.51 and 0.24. Most of the studies in bounce literature have been restricted to two dimensions where the trajectories before and after the impact lie on the same plane.

2 Methodology

The trajectory of any sports ball can be split into two primary events: flight and bounce. The entire trajectory of the ball can be numerically simulated if we understand the underlying physics in both events. In this study, we will perform detailed three-dimensional mechanistic modeling of a tennis ball's flight and bounce. The output of the flight model becomes the input for the following bounce, whose output will then be the input for the succeeding flight and so on.

2.1 *Flight of the Ball*

An aerodynamic model is required to simulate a spinning ball's three-dimensional flight trajectory based on the imparted initial conditions, i.e. position, velocity and spin information. A tennis ball's flight is affected by three forces: drag, lift and gravity. The instantaneous acceleration of the ball is given by the sum of these three forces per unit mass of the ball [11].

$$\boldsymbol{F} = \boldsymbol{F}_{Grav} + \boldsymbol{F}_{Drag} + \boldsymbol{F}_{Lift} \tag{1}$$

The gravitational force always acts downward, perpendicular towards the ground and is given by $\boldsymbol{F}_{Grav} = m\boldsymbol{g}$ where m represents the mass of the ball and $\boldsymbol{g}$ corresponds to the acceleration due to gravity vector.

The drag force is given by the equation

$$\boldsymbol{F}_{Drag} = -\frac{1}{2}\rho V\boldsymbol{V}AC_D$$

where C_D is known as the coefficient of drag, ρ is the density of air, A is the frontal area and V is the velocity of the ball. It always acts opposite to the instantaneous velocity vector of the ball.

The lift force is given by the equation

$$\boldsymbol{F}_{Lift} = \frac{\rho V(\boldsymbol{\Omega} \times \boldsymbol{V})AC_L}{2\Omega}$$

where C_L is the coefficient of lift, and Ω is the angular velocity of the ball. It acts normally to the plane containing the instantaneous spin and velocity vectors, following Fleming's right-hand thumb rule.

The coefficients of drag and lift for a tennis ball moving in air depend on its instantaneous translational and rotational velocities. Robinson and Robinson [5] provided the empirical relationships as

$$C_D = 0.62 - 9.76 \times 10^{-4}(V - 50) + \left[1.03 \times 10^{-4} - 2.24 \times 10^{-6}(V - 50)\right]\Omega$$

and

$$C_L = \left[4.68 \times 10^{-4} - 2.10 \times 10^{-5}(V - 50)\right]\Omega$$

The tennis ball's spin vector, $\boldsymbol{\Omega}$ is assumed to be constant owing to the short duration of flight. The governing Eq. (1) is numerically integrated in MATLAB software using the Velocity Verlet algorithm to obtain the three-dimensional trajectory of the ball until it bounces.

2.2 Bounce of the Ball

At the point of bounce, the incoming spin and velocity of the ball are known from the flight model. The purpose of a bounce model is to extract the post-bounce kinematic parameters, i.e. spin and velocity of the ball immediately after the bounce, as a function of the pre-bounce kinematic parameters. Then, the flight model can be implemented again to obtain the airborne trajectory following the bounce.

Mencke et al. [12] utilized the concept of three-dimensional restitution for simulating the bounce. He assumed that the ball's deformation during the impact with the ground can be neglected compared to its radius hence considered a point contact. In the scenario of a ball bouncing on a surface, the vertical restitution coefficient e_y is well known, while the corresponding coefficients in two orthogonal horizontal directions, e_x and e_z, aren't widely used. If the three coefficients, e_x, e_y and e_z, corresponding to the impact of a tennis ball on a surface, are known, they can be employed to obtain the post-bounce kinematic parameters as a function of pre-bounce kinematic parameters. Generally, no difference is expected in surface behavior along two horizontal directions so $e_x = e_z$.

$$\begin{pmatrix} V_{x,2} \\ V_{y,2} \\ V_{z,2} \\ \Omega_{x,2} \\ \Omega_{y,2} \\ \Omega_{z,2} \end{pmatrix} = \begin{pmatrix} \frac{1-\alpha e_x}{\alpha+1} & 0 & 0 & 0 & 0 & \frac{-\alpha(1+e_x)R}{\alpha+1} \\ 0 & -e_y & 0 & 0 & 0 & 0 \\ 0 & 0 & \frac{1-\alpha e_z}{\alpha+1} & \frac{\alpha(1+e_z)R}{\alpha+1} & 0 & 0 \\ 0 & 0 & \frac{1+e_z}{R(\alpha+1)} & \frac{\alpha-e_z}{\alpha+1} & 0 & 0 \\ 0 & 0 & 0 & 0 & 1 & 0 \\ \frac{-(1+e_x)}{R(\alpha+1)} & 0 & 0 & 0 & 0 & \frac{\alpha-e_x}{\alpha+1} \end{pmatrix} \begin{pmatrix} V_{x,1} \\ V_{y,1} \\ V_{z,1} \\ \Omega_{x,1} \\ \Omega_{y,1} \\ \Omega_{z,1} \end{pmatrix}$$

where the kinematic parameters with subscripts 1 and 2 represent the corresponding values before and after the bounce, respectively. Here, R is the tennis ball's radius and α is the coefficient in the moment of inertia formulation, $I = \alpha m R^2$. The value of α in the current study is 2/3 as tennis ball is a hollow sphere.

3 Results and Discussion

3.1 *Validation and Sensitivity of the Numerical Model*

In the numerical integration of our flight model, we have achieved convergence of trajectory at a timestep of 0.001 s hence employed the same for all the simulations. This numerical model is validated against the model trajectories presented by Robinson and Robinson [5]. Their three-dimensional position data are obtained by digitization of two-dimensional plots presented in their work.

Figure 1 presents the trajectories from the current model and Robinson's model for the same initial conditions. The left sub-figure shows the trajectories straight down the center-line. Our model predicts the point of bounce within 1.08% of the Robinson's model range. The right sub-figure shows the cross-court trajectories with an imparted topspin. Our model predicts the point of bounce to be within 0.39% of the Robinson's model range along z-axis and 0.88% along x-axis. There haven't been any previous studies about the bounce outcome of a tennis ball hence couldn't be validated.

The six key input parameters in the trajectory of a tennis ball are three initial spin and three initial velocity components. We have considered a scenario of cross-court trajectory hit with topspin for the sensitivity analysis of the six said parameters. Each

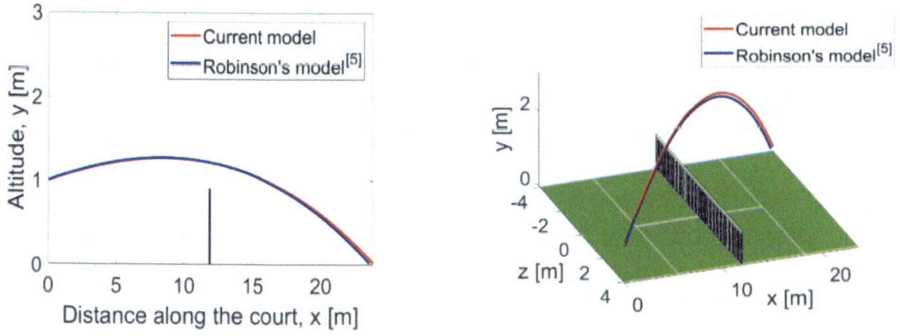

Fig. 1 Comparison of trajectories simulated by the current model and Robinson's model for center-line and cross-court (with topspin) shots

parameter, one at a time, is varied within –5% to + 5% of its starting value and the effect on the trajectory is quantified. The distance between the original and perturbed trajectories (in y–z plane) is calculated at every longitudinal (x coordinate) location on the original trajectory, and the root mean square of those distances is used as a measure for the deviation of perturbed trajectory.

Figure 2 presents the RMS change in the trajectory plotted against the relative change of each of the six parameters. The parameters can be classified into three classes based on sensitivity trend: As the velocity along the court (V_x) is the highest, a relative change in it is higher compared to the same change in other parameters thus affecting the trajectory the most, as seen in the left sub-figure. The dominant parameter V_x is removed from the right sub-figure for a better comparison of the remaining parameters. The next class has two spin components that are not perpendicular to the court ($\Omega_x$andΩ_z). The imparted high topspin decomposes into these two spin components hence they appear to deviate the trajectory more compared to the three remaining parameters (V_y, V_z andΩ_y).

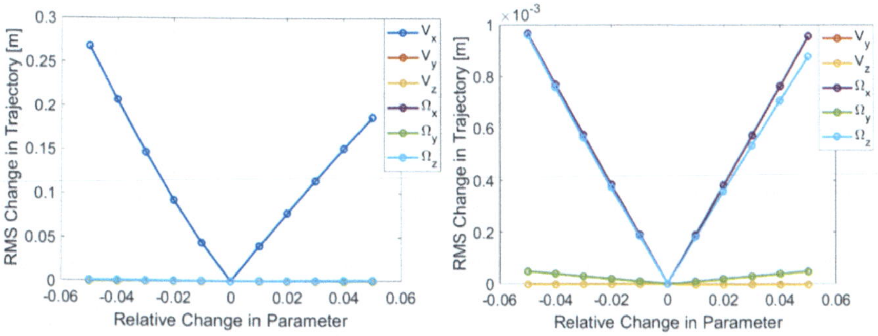

Fig. 2 Sensitivity analysis of the employed numerical model (original plot and the plot after removing V_x for comparison of other parameters)

3.2 *Study of Flight of the Ball Without Spin*

Let us first analyze only the ball's flight, which can be simulated using Eq. 1. Two extreme cases are the point of interest here. One case is the ball going over the highest part of the net, along the sidelines, termed hereon a down-the-line shot. The other is the ball going over the lowest part of the net diagonally across the court, termed hereon a cross-court shot. Assuming the ball is hit from the baseline in both scenarios, the maximum range it can cover without leaving the court's bounds is 23.77 m and 25.15 m, respectively.

Down-the-line simulations are represented in Fig. 3. When the ball is hit flat with no spin, it can be observed that the minimum speed required to reach the opposite baseline is 68.5 kmph. This can be achieved when the elevation angle is 40.16 deg about the ground. As we increase the velocity, it is easy to understand that the elevation angle needs to be reduced to maintain the same range. The ball can be hit at a maximum velocity of 164.5 kmph with an elevation angle of 2.36 deg, where it barely exceeds the net height. An interesting observation is that the fastest shot (0.69 s) takes less than one-thirds of the time taken by the slowest shot (2.28 s) to reach the opponent. This rushes the opponent into hitting a weaker shot or an error by giving them lesser time to return the incoming shot. It should be noted that there is no actual theoretical limit on the maximum velocity as it can be hit at elevation angles closing in on 90 deg but it is clearly impractical in a real game (Fig. 4).

Let us now look at Fig. 5 to understand the cross-court simulations. When hit with no spin, it can be observed that the minimum velocity required to reach the opposite diagonal point of the baseline is 71.7 kmph which is just a 4.6% increase in the velocity, compared to the slowest down-the-line shot. This can be achieved when the elevation angle is 39.75 deg above the ground. The ball can be hit at a maximum velocity of 206.7 kmph with an elevation angle of 0.97 deg. This corresponds to a 25.65% increase in the allowed maximum velocity. In this scenario, it can be

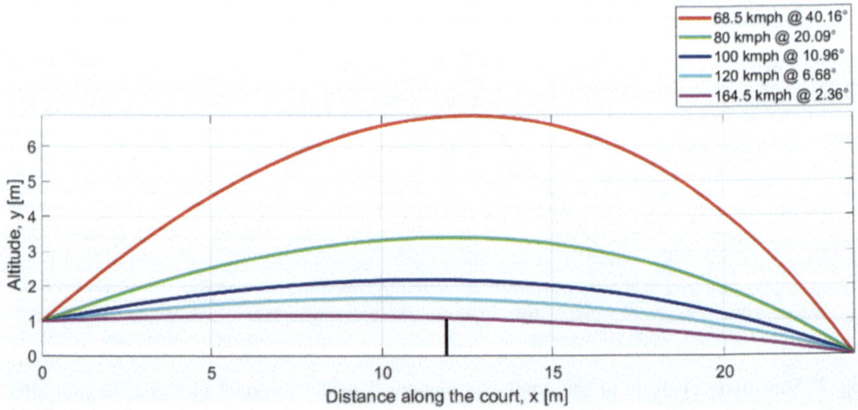

Fig. 3 The range of velocities and elevation angles for a down-the-line shot

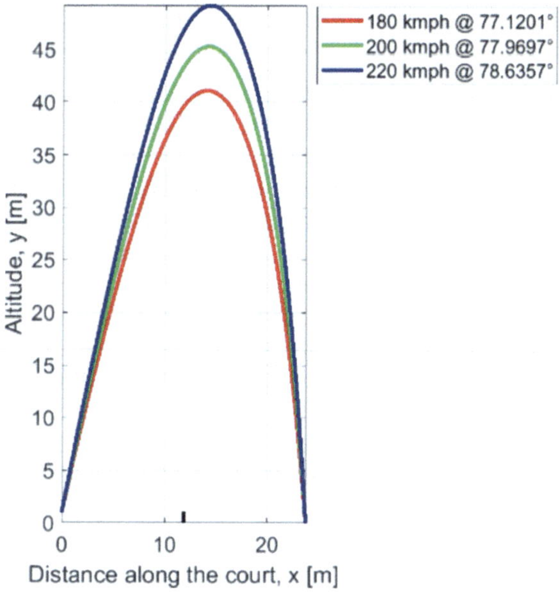

Fig. 4 Trajectories showing that there is no theoretical limit on the maximum velocity, though they aren't possible in a real game

understood that the lower height of net allows for lower elevation angle for the shots while a greater range allows for the ball to hit with a higher velocity.

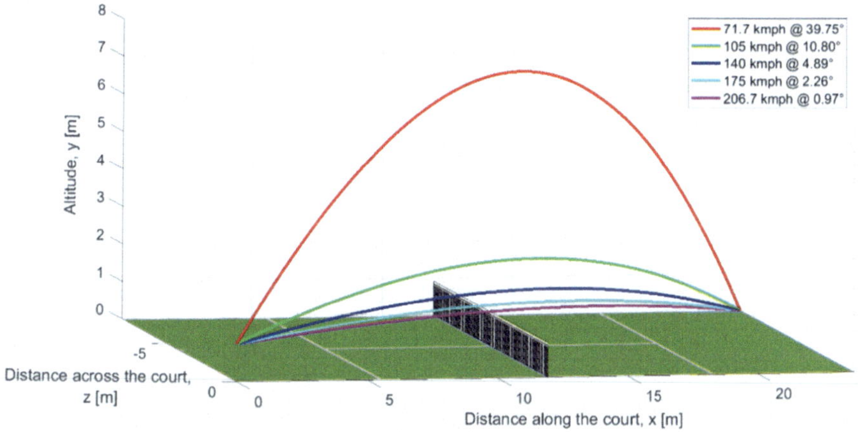

Fig. 5 The range of velocities and elevation angles for a cross-court shot

3.3 Study of Flight of the Spinning Ball

In general, any tennis shot is hit with some amount of spin, even the flatter looking ones. Most shots are imparted with significant topspin, resulting in a downward Magnus force that lets the ball bounce inside the court even at higher velocities. On the contrary, the application of backspin results in an upward Magnus force that allows the ball to stay afloat longer, thus forcing the player to hit it at lower velocities. Players use it for a variety in pace, from time to time in the rally. We can now perform an interesting exercise of fixing the elevation angle and observing the maximum velocity at which various magnitudes of topspin/backspin can be hit. The analysis for a down-the-line shot is shown in Fig. 6. Per the chosen sign convention, backspin is represented with a positive value and topspin with a negative value. At an elevation angle of 10 degrees, the maximum velocity with no spin is 103.55 kmph. As we increase the topspin to 4000 rpm, the maximum velocity increases by 58.93% to 164.57 kmph. If we impart a backspin of 4000 rpm, the maximum velocity reduces by 25.21% from the no-spin scenario, to 77.45 kmph. From Fig. 4, it can be understood that backspin results in a steeper trajectory than topspin despite the same elevation angle at the time of hit. Similar analysis was performed for a cross-court shot and the trajectories are shown in Fig. 7.

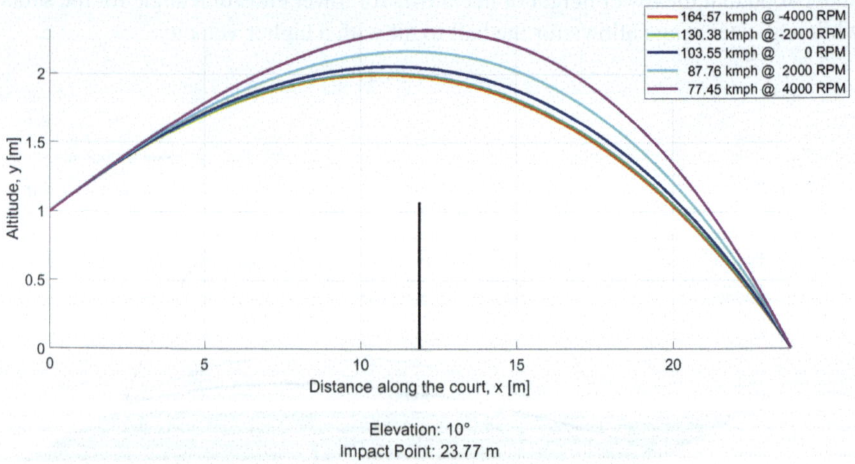

Fig. 6 The maximum initial velocity for a variety of spins for a down-the-line shot

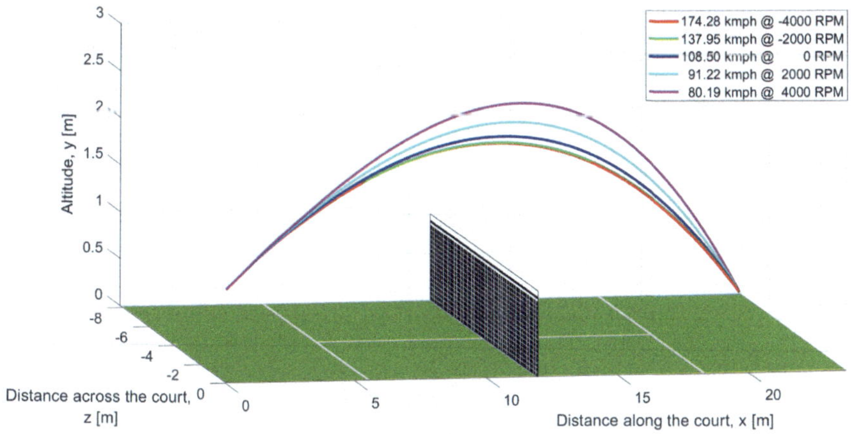

Fig. 7 The maximum velocity for a variety of spins for a cross-court shot

3.4 Study of Bounce and Subsequent Flight of the Spinning Ball

To understand the effect of spin on the bounce, the impact point can be moved back from 23.77 m (baseline) to 20 m (inside the court) to allow for flight after the bounce, within the range of the court. All the velocities are subsequently reduced to accommodate for the new range. It can be clearly observed from Fig. 8 that the trajectories post-bounce also get steeper as we move from topspin to backspin in the initialization. Given enough backspin, the ball can even reverse its direction and approach the net upon bounce.

Let us consider a realistic scenario where the ball is hit with a combination of topspin and sidespin. We will first fix the horizontal restitution coefficient ($e_{x,z}$) at 0.5 and vary the vertical restitution coefficient (e_y) to observe how the post-bounce trajectory is affected. The pre-bounce trajectory won't be affected, as restitution is just a bounce parameter that comes into play during the impact. It can be observed from Fig. 9 that the higher the value of e_y, the greater the range and maximum altitude of the ball. This is because a higher portion of the ball's energy is conserved.

In the next scenario (Fig. 10), e_y is fixed at 0.8 while $e_{x,z}$ varies. $e_{x,z} = -1$ corresponds to frictionless, gliding motion where the bounce doesn't alter the ball's horizontal motion. From the top view, it can be observed that the ball (red curve) moves as if the bounce never happened. $e_{x,z} = 1$ results in the maximum deviation from the pre-bounce trajectory as the impact corresponds to a perfectly elastic collision.

In reality, both the horizontal and vertical restitution coefficients vary from tournament to tournament, as no two courts are the same. The playing surfaces can be majorly classified as hard, clay and grass courts even though the characteristics would vary among each subset. Since published data aren't available regarding 3D

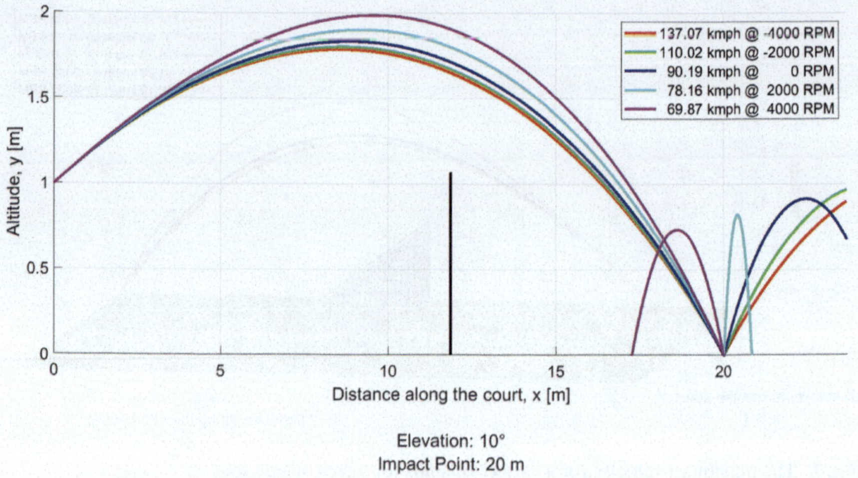

Fig. 8 The flight after the tennis ball's bounce for a variety of initial spins

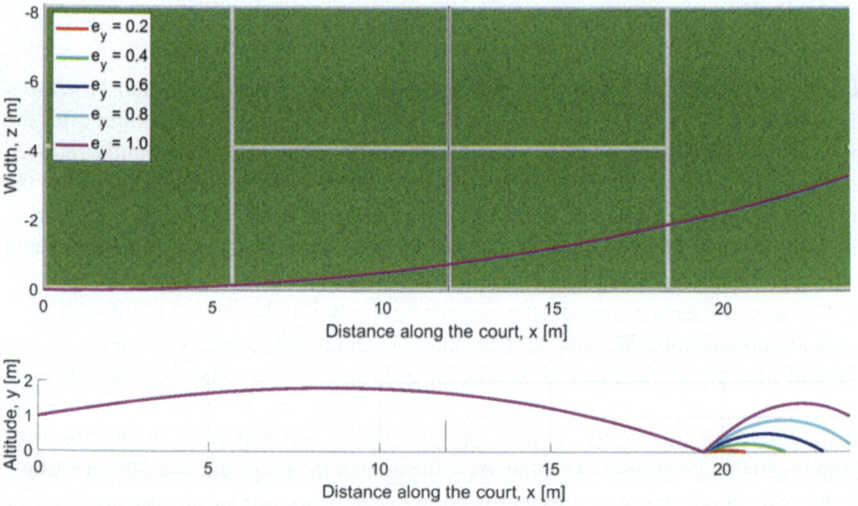

Fig. 9 The top and side views of a trajectory of a tennis ball hit with a combination of topspin and sidespin for various values of the vertical coefficient of restitution (e_y)

restitution coefficient between tennis balls and various courts, the following educated guess can be made based on their interaction.

	e_y	$e_x (= e_z)$
Grass	0.75	–0.50
Hard	0.80	0.20

(continued)

(continued)

	e_y	e_x (= e_z)
Clay	0.85	0.70

A negative value for e_x is considered on grass as the ball skids on impact and high positive value on clay is because the ball grips the surface. Figure 11 shows a visualization of how a tennis ball behaves differently on each of these surfaces.

4 Conclusions

This study aims to analyze the flight and bounce of a tennis ball and discuss the parameters that dictate the ball's trajectory. We first set the range of the shot so that the ball lands on the opponent's baseline. With a flexible elevation angle, we have observed the range of possible velocities without spin. The maximum velocity with which a ball can be hit cross-court is 25% greater than a down-the-line shot owing to lower net height and greater court length. In both these independent scenarios, the fastest ball took less than one-third of the time taken by the slowest shot to reach the opponent. In the subsequent simulations, spin is imparted during the hit, resulting in the lift force. For a fixed velocity and elevation angle, a ball hit with topspin experiences downward force and hence will have a lower range than a ball hit with backspin or no spin. This is why shots can be hit with greater velocity when topspin is imparted compared to backspin if the objective is a fixed horizontal range. Since the backspin shots float in the air compared to the dip observed with topspin, the former trajectories appear steeper than the latter for a constant elevation angle. The same behavior is noticed even in the second flight event, following the ball's bounce. Given enough backspin on the ball, the ball can change its direction on the bounce and move towards the net rather than the opponent. We have also looked at the effect of the coefficient of restitution on the flight trajectory following the bounce for shots imparted with a combination of topspin and sidespin, as usually done in professional matches. When the vertical restitution coefficient is varied, the horizontal deviation of the ball isn't affected because of the lack of coupling. However, the range and maximum altitude of the ball increased as the vertical coefficient increased. When only the horizontal restitution coefficient is varied, we observe difference in the horizontal movement. As the value of the horizontal coefficient increases from -1 to 1, the deviation after bounce increases too. Finally, we made an educated guess about the three-dimensional restitution for grass, hard and clay courts to visualize the impact each of the surfaces has on the ball's trajectory.

One of the biggest challenges in applying any numerical model would be the ever-changing condition of the ball which alters the physical characteristics as well as the drag and lift coefficients. Apart from that, the numerical flight model should work reasonably well in predicting an actual tennis ball trajectory. The current model would still be applicable in the presence of winds if the relative velocity of the ball

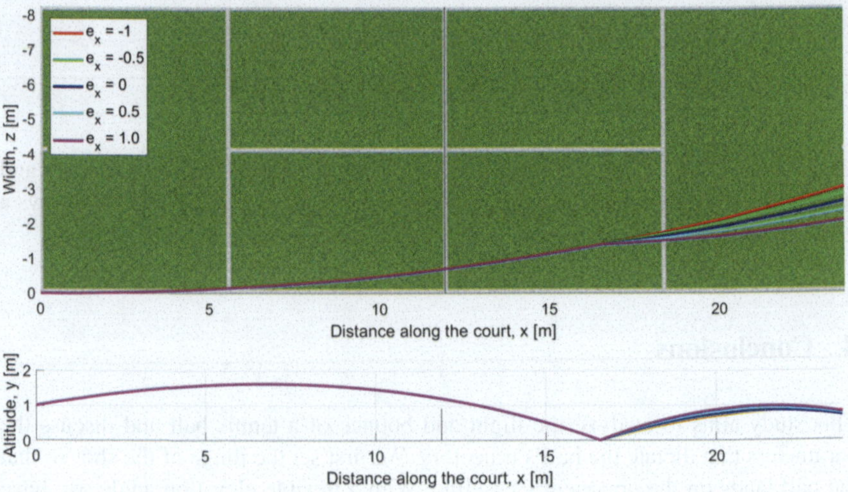

Fig. 10 The top and side views of a trajectory of a tennis ball hit with a combination of topspin and sidespin for various values of the horizontal restitution coefficient ($e_{x,z}$)

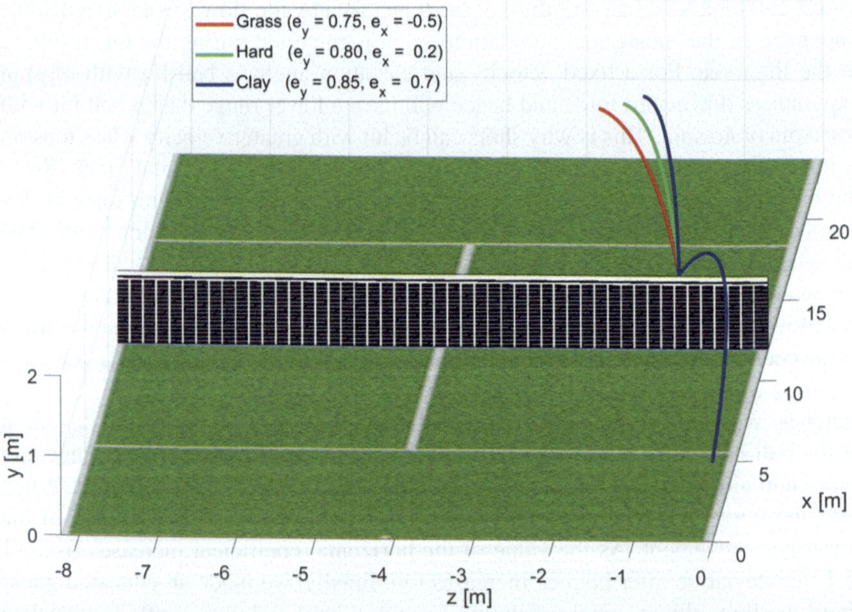

Fig. 11 The trajectory of a tennis ball simulated on three surfaces: grass, hard and clay

is considered in the equations of motion. One limitation of the bounce model is it intakes three-dimensional coefficient of restitution which has a working range for a ball-surface pair instead of a specific value, depending on the incident kinematic parameters. Due to the lack of three-dimensional bounce models in the literature, this existing yet incomplete model has been implemented. Currently, we are working on modeling the bounce of a spinning ball using experimental data.

Using the simulations, one can estimate the operating velocities with which players can hit the ball that would reach the baseline or just cross the net. The results have shown us that considerably faster shots can be hit by increasing the amount of topspin imparted. This can lead to improvement in string technology which allows imparting more spin on the ball resulting in an even faster game while not compromising on landing the ball in. Since the bounce characteristics vary from surface to surface, the players can take advantage of the court surface through simulations and employ different playing strategies. For example, hitting slices that generally skid would be advantageous on the fast grass courts whereas a heavy topspin play would be beneficial on slow, high-bouncing courts like clay. For example, on clay, simulations can be performed to detect the bounce point based on the returner's position, which will result in an uncomfortable height at which the balls reach them. The surface curators can alter the court composition and bring the restitution coefficient to the required limit that allows the game at the pace and bounce they desire.

References

1. Mehta RD (1985) Aerodynamics of sports balls. Annu Rev Fluid Mech 17(1):151–189
2. Achenbach E (1974) Vortex shedding from spheres. J Fluid Mech 62(2):209–221
3. Shah K, Shakya R, Mittal S (2019) Aerodynamic forces on projectiles used in various sports. Phys Fluids 31(1):015106
4. Barlow JB, Domanski MJ (2008) Lift on stationary and rotating spheres under varying flow and surface conditions. AIAA J 46(8):1932–1936
5. Robinson G, Robinson I (2018) Model trajectories for a spinning tennis ball: I. The service stroke. Physica Scripta 93(12):123002
6. Cross R (2002) Grip-slip behavior of a bouncing ball. Am J Phys 70(11):1093–1102
7. Howard Brody. That's how a ball bounces. Physics Teacher, 1984.
8. Brody H (2005) The moment of inertia of a tennis ball. Phys Teach 43:503–505
9. Cross R (2005) Bounce of a spinning ball near normal incidence. Am J Phys 73(10):914–920
10. Cross R (2002) Measurements of the horizontal coefficient of restitution for a superball and a tennis ball. Am J Phys 70(5):482–489
11. Clanet C (2015) Sports ballistics. Annu Rev Fluid Mech 47:455–478
12. Mencke JE, Salewski M, Trinhammer OL, Adler AT (2020) Flight and bounce of spinning sports balls. Am J Phys 88(11):934–947

Evaluation of Physical and Physiological Parameters Among Middle- and Long-Distance Runners

Chennakeshavulu Vemula, Tamali Mondal, Aishwarya Andugula, Maheshwar Mekam, and Suresh Challa

Abstract

Introduction Athletic is a very famous sport from school level to the international level. In this research, the investigators wanted to conduct the laboratory and field-based tests on athletes to know exactly about their body composition, physical abilities, and physiological variables to set and compare normative values and help them in improving the training strategies. **Aim**: To determine the physical characteristics, body composition, and physiological difference of middle- and long-distance runners and to compare physical, body composition, and physiological differences among national and state-level athletes. **Design**: A total of 41 (n = 41) sample size was taken from state and national level with the mean age of 24 ± 4. Cross-sectional study was carried out, the statistical analysis was based on descriptive statistics, and statistical analysis was done with the help of IBM SPSS Statistics 20 software. METHOD: Anthropometric measurements were carried out, to determine the body composition and physiological parameters among middle-distance and long-distance runners, the instruments used, namely, INBODY 770, CPET COSMED. Field tests like 12 min Cooper test were used. **Result**: There was no significant difference found in body composition and physiological variables except percentage of body fat, VO_2 max, maximum heart rate, recovery heart, respectively. There was no such significant difference seen except VO_2 max and maximum heart rate among state and national level long-distance runners. **Conclusion**: Body composition of middle- and long-distance runners is the same except percentage of body fat. VO_2max of long-distance runners is high because of more hours of endurance practice.

C. Vemula (✉) · T. Mondal · A. Andugula · M. Mekam
Department of Sports Science, Extension and Training Division, MYAS-NIN, Tarnaka, Hyderabad 500007, India
e-mail: chennalaxmi@gmail.com

C. Vemula
Department of Sports Sciences, ICMR-NIN, MYAS-NIN, Tarnaka, Hyderabad 500007, India

S. Challa
Department of Cell Biology, ICMR-NIN, Tarnaka, Hyderabad 500007, India

S. Bhattacharyya and D. James (eds.), *Recent Trends in Sports Engineering*, Lecture Notes in Mechanical Engineering, https://doi.org/10.1007/978-981-97-8297-0_16

Keywords INBODY 770 · CPET · Middle distance runners · Running performance

1 Introduction

Endurance running is one of the most popular sport practices in the world [1]. As athletics is a very famous sports from school level to the International level, so far independent India won one and only gold medal in the recent 2020 Summer Olympics in Japan. Indian athletes need to have athlete-centric coach-driven proper training plan with the much-needed support from sports science. Sports scientists should work at the grass route level for the talent selection and development in athletes to excel at international-level competitions. In this research, the investigators wanted to conduct laboratory and field-based tests on athletes to know exactly about their body composition, physical abilities, and physiological variables to set and compare normative values and help them in improving the training and nutritional strategies [6]. International Association of Athletic Federation (IAAF) conducts two types of events that include both middle- and long-distance running. Middle-distance running includes 800 m and 1,500 m [8].

Distance running athletic feats are the consequence of a complex interaction of physiological, biomechanical, psychological, environmental factors [7]. From a physiological standpoint, there are three important parameters those are largely hold performance, namely; maximum oxygen uptake (VO_2 max), running economy (RE), and fractional utilization (sustainable percentage of VO_2 max).

RE is the amount of oxygen that the body uses to maintain a certain pace [5]. There are several factors that affect running economy including metabolic rate, cardiorespiratory, biomechanical and neuromuscular, and training characteristics. For instance, metabolic rate indicates how much energy body uses in a certain time of frame, cardiorespiration refers to the strength of heart and lungs to coordinate with each other and supply oxygenated blood to the working muscles and remove deoxygenated blood from the body, biomechanical which bring up force, motion, momentum, levers and balance of an individual, and lastly neuromuscular which is a major adaptive system that promotes and control movement and stability in both skeletal and smooth muscle function.

The 800-m event originally came from the 880-yard run, which was rooted in the UK in the 1830s. Middle-distance running generally lasts longer than sprinting. In these events, the athlete has to run over a fixed distance over the minimum possible time frame. The 800 m is the shortest middle-distance running event where an athlete runs over two laps of the 400-m track. The 1500-m race is the second shortest middle-distance race held at the international level and places slightly higher attention on aerobic endurance and beneath anaerobic speed than 800-m races.

The aim of the study was to determine the physical as well physiological variables between middle- and long-distance runners. Performance in middle- and long-distance running depends on a variety of physiological factors, physical factors and also depends on different types of body composition of individuals [10]. Middle- and long-distance runners use different training methods that lead to different adaptations in the aspect of aerobic fitness.

Hypothesis It was hypothesized that middle-distance runners have different body composition, physical, and physiological parameters from long-distance runners.

2 Methodology

In this study, 41 male athletes took part as participants and some are middle-distance runners and some are long-distance runners. The state-level athletes were from Telangana state and national-level athletes were from army athletes of all the states of India.

2.1 Selection of Sample

To achieve the purpose of the study 41 (n = 41), middle and long-distance runners were selected from Telangana state preparing for state-level competition and national-level runners from army artillery center who were under off-season.

The athletes were aged between 18 and 28 years and all are males. Convenience sampling method was used for the study purpose.

2.2 Dependent Variables

Physiological variables

- Maximum heart rate
- Resting heart rate
- VO_2 max
- Recovery heart rate

Physical variables

- Anthropometric measurements [9]
- 12 min Cooper test [2, 3]

Body composition

- Fat mass % (BIA)
- Fat-free mass %
- Minerals % [4]
- Lower limbs
- Percentage of body fat %
- Visceral fat

The athletes were explained about the procedures and protocols of the tests which are being performed and a written consent was taken.

The inclusion criteria include:

- Aged between 18 and 28 years.
- Only male athletes were included in the study.
- Athletes having at least 2 years of regular practice.
- Body composition was assessed by BIA and anthropometric measurements were done with anthropometry kit and Harpedan skin fold caliper.

Cardiopulmonary Exercise Testing (CPET) was a non-invasive breath by breath indirect calometry maximal treadmill running method used to assess VO_2 max performance. All the athletes gave their consent and they were motivated to run their best. This NIN protocol was standardized with 12 min Cooper run test and 20 m PACER test.

This graded running test starts with 8 km/hr intensity and 2-grade elevation for entire test duration with 0.5 km/hr speed increments for every 2 min and after the completion of the test recovery heart responses were assessed.

Bioelectrical impedance analysis was done by using InBody 770 model for body composition. InBody 770 model has 2–4% error with gold standard DEXA test for body composition tests.

Descriptive statistic technique was used to note the differences of 41 samples of running athletes (middle distance and long distance). t-test was used among two independent samples in order to compare the classified group means, standard deviation, and *p* value. The statistical analysis was done with the help of IBM SPSS 20 software.

3 Results

RESULTS

MIDDLE DISTANCE RUNNERS (n=22)		LONG DISTANCE RUNNERS (n=19)			
VARIABLES	MEAN	SD	MEAN	SD	*p* VALUE
MINERALS	3.500	.597	3.526	.611	.890
SMM (%)	46.227	2.202	45.473	2.653	.326
PBF (%)	16.578	2.990	13.508	3.863	.017
LOWER LIMB(kg)	16	2.888	16.263	2.305	.716
VISCERAL FAT	38.318	15.923	41.368	13.626	.517

BODY COMPOSITION BETWEEN MIDDLE AND LONG DISTANCE RUNNERS

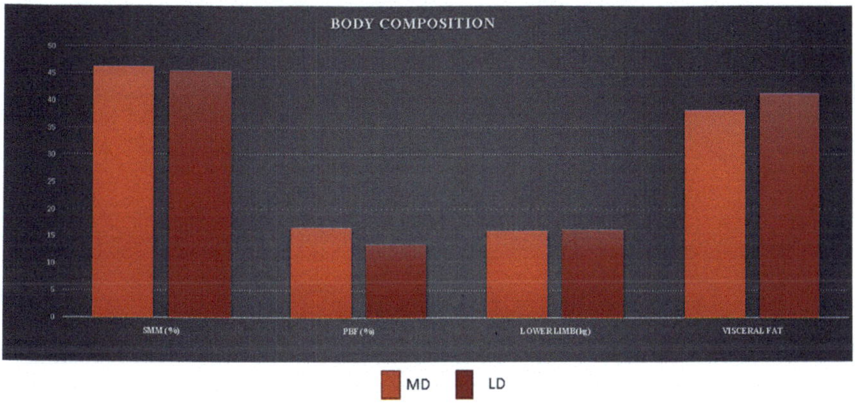

MIDDLE DISTANCE RUNNERS(n=22)			LONG DISTANCE RUNNERS (n=19)		
VARIABLES	MEAN	SD	MEAN	SD	p VALUE
RHR	60.45	5.61	61.63	5.53	.505
SPO2	97.90	.97	96	6.90	.206
R SYSTOLIC PRESSURE	112.31	10.34	117.68	13.06	.150
R DIASTOLIC PRESSURE	64.90	12.24	69.78	9.84	.172
BMR	1447.3	119.08	1451.3	118	.916
VO2 MAX COOPER	55.31	4.36	62	9.05	.004
VO2 MAX CPET	53	5.48	58	5.92	.007
MHR	200	21.93	172.94	29.18	.001
RHR (1 MIN)	183	19.52	164.47	27.65	.016
RSPO2 (1MIN)	88.50	9.22	88.68	8.27	.947
RHR (3MIN)	159.95	21.80	133.2	23.84	.001
RSPO2 (3MIN)	86.95	8.01	90.52	9.27	.194

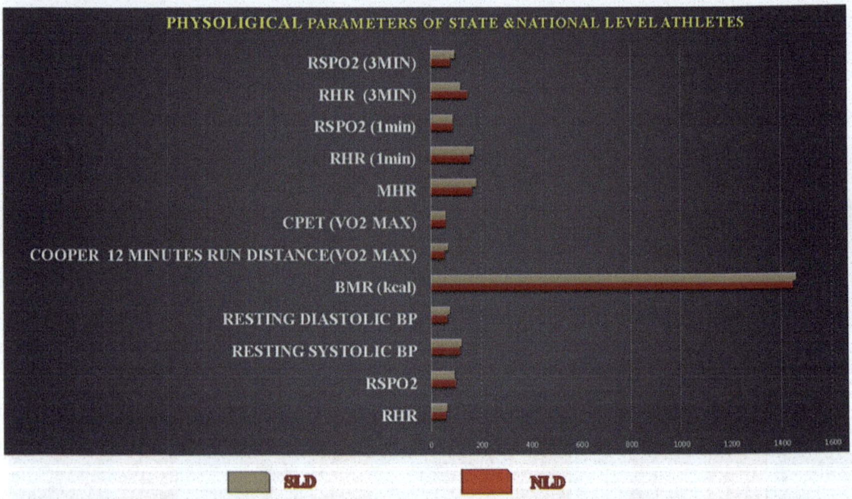

4 Discussion and Conclusion

The 41 subjects examined were well familiar with the testing procedures and performance was determined under controlled conditions. As per the data collected from the study and statistical analysis, there were no significant differences ($P > 0.05$) between most of the measured variables among middle- and long-distance runners because all national-level athletes were in off-season as they were just finished their National camp and there could be other reasons and regular assessment needs to be done in each phase of periodization. Percentage of body fat was higher in middle-distance runners compared to long-distance runners and it could be high volume of training. There was a significant difference in VO_2 max with the use of Cooper test as well as in CPET among middle- and long-distance runners. Resting heart rate as well as resting diastolic pressure is lower among long-distance runners compared to middle-distance runners. In Cooper test, VO_2 max is higher among long-distance runners compared to middle-distance runners. Recovery heart rate in 3 min is faster among middle-distance runners compared to long-distance runners. Resting heart rate was lesser among national level long-distance runners compared to state-level long-distance runners. Resting diastolic pressure and VO_2 max (Cooper) are more among national-level long-distance runners. Recovery heart rate (3 min) was more among state-level long-distance runners compared to national-level long-distance runners.

Within the limitations and delimitations of the study VO_2 max, recovery heart rate, body fat percentage, and maximum heart rate were significantly lower in long-distance runners, and all other parameters are similar in two groups of athletes. Middle- and long-distance athletes, coaches, physical educationists, trainers, sports scientists, strength, and conditioning need to assess and evaluate athletes' abilities and train accordingly.

References

1. Albouaini K, Egred M, Alahmar A, Wright DJ (2007) Cardiopulmonary exercise testing and its application. Postgrad Med J 83(985):675–682. https://doi.org/10.1136/hrt.2007.121558
2. Alvero-Cruz JR, Carnero EA, Giráldez García MA, Alacid F, Rosemann T, Nikolaidis PT, Knechtle B (2019) Cooper test provides better half-marathon performance prediction in recreational runners than laboratory tests. Front Physiol 10:1349. https://doi.org/10.3389/fphys.2019.01349
3. Apte S, Troxler S, Besson C, Gremeaux V, Aminian K (2022) Augmented cooper test: biomechanical contributions to endurance performance. Front Sports Active Living 4:935272. https://doi.org/10.3389/fspor.2022.935272
4. Bazzocchi A, Ponti F, Albisinni U, Battista G, Guglielmi G (2016) DXA: technical aspects and application. Eur J Radiol 85(8):1481–1492. https://doi.org/10.1016/j.ejrad.2016.04.004
5. Beck ON, Kipp S, Roby JM, Grabowski AM, Kram R, Ortega JD (2016) Older runners retain youthful running economy despite biomechanical differences. Med Sci Sports Exerc 48(4):697–704. https://doi.org/10.1249/MSS.0000000000000820
6. Damasceno MV, Pasqua LA, Lima-Silva AE, Bertuzzi R (2015) Energy system contribution in a maximal incremental test: correlations with pacing and overall performance in a 10-km running trial. Brazilian J Med Biol Res = Revistabrasileira de pesquisasmedicas e biologicas 48(11):1048–1054. https://doi.org/10.1590/1414-431X20154787
7. Denadai BS, Greco CC (2022) Could middle- and long-distance running performance of well-trained athletes be best predicted by the same aerobic parameters? Current Res Physiol 5:265–269. https://doi.org/10.1016/j.crphys.2022.06.006
8. Duffield R, Dawson B, Goodman C (2005) Energy system contribution to 400-metre and 800-metre track running. J Sports Sci 23(3):299–307. https://doi.org/10.1080/02640410410001730043
9. Fryar CD, Gu Q, Ogden CL, Flegal KM (2016) Anthropometric reference data for children and adults: United States, 2011–2014. Vital Health Stat 3 Anal Stud 2016(39):1–46. [PubMed]
10. Gabbett T, Georgieff B (2007) Physiological and anthropometric characteristics of Australian-junior national, state, and novice volleyball players. J Strength Cond Res 21(3):902–908. [PubMed]

An Analysis of Q-Angle in Relation to Anthropometry and Training Effects on Performance and Injury Risks Among Indian Female Athletes

Chennakeshavulu Vemula, Attaluri Kavya, Mekam Maheshwar, and Chella Suresh

Abstract The Q-angle, representing the angle formed between the patellar tendon and the quadriceps muscle, holds considerable significance in understanding knee biomechanics. This study focuses on the Q-angle in female athletes, recognizing their heightened vulnerability to knee injuries due to anatomical, joint laxity, and muscle strength differences. Specifically, the research targets female athletes who are engaged in volleyball, weightlifting, and boxing, comparing them to non-athletic counterparts ($N = 17$, age $= 19 \pm 3$). The investigation aims to correlate Q-angle values with anthropometry, body composition parameters, foot posture, and functional tests. Anthropometric variables such as height, weight, thigh circumference, limb length, and waist-hip ratio are measured alongside body composition parameters utilizing dual x-ray absorptiometry (DXA). The Foot Posture Index-6 (FPI-6) is employed to categorize foot types, encompassing criteria like talar head palpation and calcaneal angle. Bilateral differences in Q-angle among athletes are explored, with statistical analyses performed using independent T-tests and Pearson correlations. Surprisingly, the study reveals no significant difference in both right leg Q-angle ($p = 0.67$) and left leg Q-angle ($p = 0.30$) between athletes and non-athletes. Notably, negative correlations emerge between right-leg Q-angle and bone mineral content (BMC), total bone mineral density (BMD), and hip BMD in volleyball and boxing athletes, while weightlifters exhibit no such correlations. Pronated foot posture demonstrates a significant association with right leg Q-angle ($r = 0.508^*$) and left leg Q-angle ($r = 0.514^*$), additionally, negative correlations surface between right-leg Q-angle and functional tests. In conclusion, this study clarifies the link between knee biomechanics and athletic performance in female athletes by offering detailed insights into the delicate interplay of Q-angle with body composition, and postural parameters.

C. Vemula (✉) · A. Kavya · M. Maheshwar
Department of Sports Science, Extension and Training Division, MYAS—NIN, Hyderabad 500007, India
e-mail: chennalaxmi@gmail.com

C. Suresh
Department of Cell Biology, ICMR—NIN, Hyderabad 500007, India

S. Bhattacharyya and D. James (eds.), *Recent Trends in Sports Engineering*, Lecture Notes in Mechanical Engineering, https://doi.org/10.1007/978-981-97-8297-0_17

Keywords Q-angle · Lower limb functionality · Body composition · Foot posture · Anthropometry

1 Introduction

The Q-angle, commonly known as the quadricep angle, is created when two lines cross. Two lines were drawn, one from the tibial tuberosity to the patella's center and the other from the anterior superior iliac spine (ASIS) to the patella's (kneecap) center. Studies show that greater lateral stresses on the knee joint and modified patellar tracking can be caused by a raised Q-angle. Anterior cruciate ligament (ACL) injuries and patellofemoral pain syndrome are two disorders linked to these biomechanical abnormalities. A higher Q-angle, i.e. > 21° is often associated with patellofemoral pain syndrome, which is defined by anterior knee pain. In comparison to adults, children typically have a higher Q-angle; this anomaly is mainly explained by anatomical and developmental considerations. Changes in muscle mass and body fat distribution can impact the Q-angle by affecting the total musculoskeletal alignment. Through screening, athletes with abnormal Q-angles can be identified, allowing for focused conditioning and early intervention to lower the risk of knee injury. Anthropometry and body composition have a combined influence on Q-angle. Variations in body fat distribution and muscle mass can affect overall musculoskeletal alignment, potentially influencing the Q-angle.

In sports, the Q-angle is essential because it provides important information on knee joint mechanics and injury risk factors. It can be seen in both its static and dynamic forms. The Q-angle serves as a compass for athletes, helping them to avoid injuries and perform at their best. It influences their equipment selection, training, and measures for preventing injuries.

With an emphasis on Indian female athletes, this study explores Q-angle variations and their significant effects on lower limb functionality. Changes in the quantity and distribution of bone mineral density can affect the Q-angle and, in some cases, the way the bones are loaded. It is crucial to understand the minute details of Q-angle changes in this population because of possible anthropometric and cultural variables that could affect biomechanics.

2 Literature Review

One of the most explored areas in Q-angle research is its relationship with knee injury risk. A larger Q-angle is often associated with an increased risk of patellofemoral pain syndrome (PFPS). The measurement of Q-angle in a supine position, with the knees extended and the quadriceps relaxed, is considered the most conventional method. In closed kinetic chain (CKC) activities, the angle is often recorded along the motion trajectory in order to calculate the dynamic Q-angle. The most accurate method is to use three-dimensional (3D) kinematic analysis technologies, such as photogrammetry or camera motion capture systems with reflecting markers.

The Q-angle is considered multifactorial because it can be influenced by various anatomical and biomechanical factors for example gender, hip width, muscle imbalance, foot pronation, joint laxity, and injury history, etc. Sex differences in the Q-angle have been widely reported in the literature. Commonly, females are more vulnerable to having knee injuries because of the higher Q-angle than males. The primary reasons for this difference are thought to be related to variations in pelvic anatomy, including wider hips and increased femoral anteversion in females. According to athletes often have a Q-angle of less than 20°, with a mean that varies based on body side and gender between 6° and 13°. It has a mixed relationship with various forms of sports engagement.

Differences in physical activity, food, and ethnicity may also contribute to variances in body fat for the same BMI. Thus, disparities in Q-angle measurements across different regions may contribute to the variances in body characteristics. When compared to athletes in this group who did not have an ACL injury, there was a significant correlation between an ACL injury and a higher BMI. Higher BMI and weight have been reported to be strongly correlated with ACL injuries, particularly in men, as well as other intra-articular injuries that are discovered during ACL reconstructions.

3 Methodology

The primary objectives of the current study include:

- Assessing the Q-angle values of females who are involved in three sports (Volleyball, weightlifting, and boxing) and compare them with non-athletes.
- To determine the athletes' bilateral q-angle disparities.
- To establish a relationship between the Q-angle and several metrics in each particular sport, including anthropometry, foot posture, functional testing, and body composition.

Administration of Tests

Anthropometrical measurements are the basic measurements such as height, weight, etc., and these tests are conducted using Seca GmbH & Co manufactured equipment. Thigh circumference and limb lengths were measured by using Seca measuring body tape.

Body composition tests are conducted using the Dual X-ray absorptiometry (DXA). The equipment can measure fat mass, muscle mass, and bone mineral density. DXA offers a range of benefits, from accurate bone density measurements to comprehensive body composition analysis, enabling healthcare professionals to assess bone health, evaluate fracture risk, and monitor changes over time. The Q-angle was measured with a goniometer. It is a straightforward clinical procedure commonly used to assess the alignment of the quadriceps muscle, patella, and tibial tendon. Identify the anterior superior iliac spine (ASIS) on both sides of the pelvis. These bony protrusions are located at the front of the hip. With the patient in a supine position, place the center of the goniometer on the midpoint of the patella (kneecap). Perform the same procedure for the other leg. It's important to assess both legs to identify any asymmetry. Record the angles of both limbs.

The FPI-6 is commonly used in clinical settings, sports medicine, and research to assess foot posture and guide decisions related to treatment or interventions. FPI-6 takes into consideration both rearfoot and forefoot. Based on the scores obtained for each criterion, the examiner categorizes the overall foot posture into one of several classifications such as pronated, supinated, and neutral.

Selection of Sample

- Female athletes from volleyball, boxing and weight lifting (N=17, age=19±3).
- Non-athletic females (N=17, age=19±3).

Assessments

1. Laboratory based (Anthropometric, DEXA, Q-angle, Foot posture-FP16).
2. Field based (Triple hop test and Single leg squat test).

Statistical analysis

The statistical analysis was done with the help of IBM SPSS statistics 20 software. t-test was used among two independent samples in order to compare the classified groups to obtain p value and Pearson correlation was used to correlate variables in specific sports to obtain r value.

4 Results

The statistical results reveal that there is no significant difference in both right leg Q-angle ($p = 0.67$) and left leg Q-angle ($p = 0.30$) between athletes and non-athletes. From Table 1, though non-significant, athletes have slightly higher Q-angle values compared to non-athletic group.

Table 1 Comparison of the Q-angle values among athletes and non-athletes

Groups		Mean ± SD	Min	Max	Sig
Right leg Q-angle	Athletes (n = 17)	16.35° ± 3.14	11	22	0.67
	Sedentary (n = 17)	15.88° ± 3.38	11	23	
Left leg Q-angle	Athletes (n = 17)	16.41° ± 2.34	11	20	0.3
	Sedentary (n = 17)	15.58° ± 2.26	11	19	

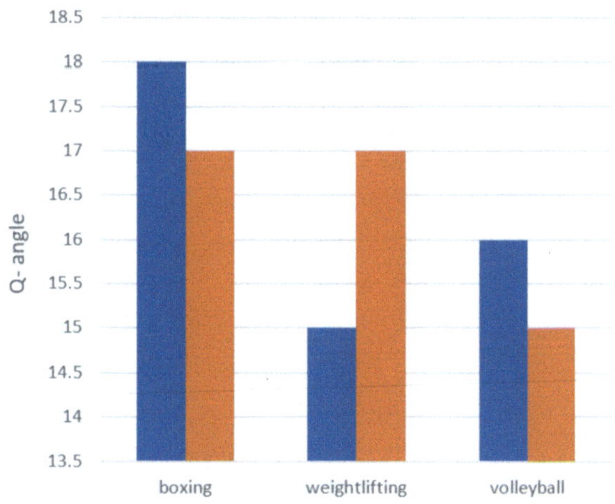

Fig. 1 Bilateral difference of q-angle among sports

Table 2 Average values of Q-angles of boxers, weightlifters, and volleyball players

Groups	Right Q-angle	Left Q-angle
Boxers (n = 4)	18.5° ± 1.29	17.5° ± 3.14
Weightlifting (n = 4)	15° ± 1.63	17.5° ± 3.10
Volleyball (n = 9)	16° ± 3.8	15.4° ± 2.18

A minor bilateral difference of the Q-angle was observed in each of the three sports. Upon analysis of the data from Fig. 1, boxers exhibit significantly higher Q-angles (right Q-angle = 18.5° ± 1.29 and left Q-angle = 17.5° ± 3.14) in comparison to athletes engaged in volleyball and weightlifting (Table 2).

Negative correlations emerge between right-leg Q-angle and bone mineral content (BMC), total bone mineral density (BMD), and hip BMD in volleyball and boxing athletes. Tables 3 and 5 illustrate the negative correlation between Q-angle and body composition parameters while Table 4 shows that weightlifters exhibit no such correlations.

Additionally, Table 6 shows that in female athletes, pronated foot posture has a strong correlation with both the left leg Q-angle (r = 0.514*) and the right leg Q-angle (r = 0.508*). This demonstrates how a foot that pronates excessively might alter the lower limb's alignment and perhaps cause an increase in the Q-angle.

And negative correlation exist between right—leg Q angle, triple hop test (r = −0.507*) and single leg squat test (r = −0.611**). Jumping and squatting are examples of functional tasks that may be impacted by altered lower limb biomechanics linked to an elevated Q-angle (Table 7).

Table 3 Correlation of Q-angles with training age, total BMD, BMD of hip, spine, limbs, and thigh circumference along with height, weight, limb length, waist/hip ratio, BMC of limbs, muscle mass, fat mass, and z-score in boxers (n = 4)

Boxers (N = 4)		TG	HGT	WGT	RL. L	LL. L	W/H	RT.C	Total BMD	Right leg BMC	Left leg BMC
Right Q−angle	r	0.944	−0.532	−0.701	−0.571	−0.541	−0.724	−0.405	−0.951*	−0.995*	−0.995*
	P	0.056	0.468	0.299	0.429	0.459	0.276	0.595	0.049	0.012	0.005
Left Q−angle	r	0.392	−0.296	−0.642	−0.416	−0.473	−0.065	−0.559	−0.55	−0.685	−0.628
	p	0.608	0.704	0.358	0.584	0.527	0.935	0.441	0.45	0.314	0.372

Boxers (N = 4)		Left fat mass	Right fat mass	Left muscle mass	Right muscle mass	z-score	Spine BMD	Left BMD	Right BMD	Hip BMD	LT. C
Right Q−angle	r	−0.647	−0.668	−0.756	0.588	−0.961*	−0.879	−0.981*	−0.314	−0.981*	−0.573
	P	0.353	0.332	0.244	0.412	0.039	0.121	0.019	0.686	0.019	0.427
Left Q−angle	r	−0.692	−0.8	−0.893	0.875	−0.587	−0.275	−0.665	0.326	−0.782	−0.665
	p	0.308	0.2	0.107	0.125	0.413	0.725	0.335	0.674	0.218	0.335

Table 4 Correlation of Q-angles with training age, total BMD, BMD of hip, spine, limbs, and thigh circumference along with height, weight, limb length, waist/hip ratio, BMC of limbs, muscle mass, fat mass, and z-score in weightlifters (n = 4)

Weightlifters (N = 4)		TG	Right thigh	Left thigh	RL. L	LL. L	WGT	Right BMC	Left BMC	Left fat mass	Right fat mass
Right Q-angle	r	−0.171	−0.12	−0.043	0.462	0.534	0.03	0.242	0.351	−0.225	−0.058
	p	0.829	0.88	0.957	0.538	0.466	0.97	0.758	0.649	0.775	0.942
Left Q−angle	r	0.42	−0.628	−0.581	−0.111	−0.027	−0.553	−0.56	−0.237	−0.715	−0.608
	p	0.58	0.372	0.419	0.889	0.973	0.447	0.648	0.763	0.285	0.392
Weightlifters (N = 4)		**Left muscle mass**	**Right muscle mass**	**z-score**	**Spine BMD**	**Left BMD**	**Right BMD**	**Hip BMD**	**Total BMD**	**W/H**	**HGT**
Right Q-angle	r	0.086	0.113	0.345	0.361	0.371	0.521	0.033	0.278	−0.126	0.415
	p	0.914	0.887	0.655	0.639	0.629	0.479	0.967	0.722	0.874	0.585
Left Q−angle	r	−0.496	−0.476	0.094	0.201	−0.118	0.114	−0.093	−0.224	−0.535	−0.186
	p	0.504	0.524	0.906	0.799	0.882	0.886	0.907	0.776	0.465	0.814

Table 5 Correlation of Q−angles with training age, total BMD, BMD of hip, spine, limbs, and thigh circumference along with height, weight, limb length, waist/hip ratio, android/gynoid ratio, muscle mass, fat mass, and z-score in volleyball players (n = 9)

Volleyball players (N = 9)		TG	Right thigh	Left thigh	RL. L	LL. L	WGT	HGT	Right BMC	Left BMC	Left Fat mass
Right Q-angle	r	0.062	−0.229	−0.077	0.039	0.053	−0.191	−0.539	−0.622	−0.592	−0.094
	p	0.875	0.553	0.843	0.921	0.892	0.623	0.134	0.074	0.093	0.81
Left Q−angle	r	−0.464	−0.265	−0.317	−0.188	−0.199	0.206	0.625	0.107	0.081	−0.175
	p	0.208	0.491	0.405	0.629	0.608	0.594	0.072	0.784	0.7836	0.653

Volleyball players (N = 9)		Right fat mass	Left muscle mass	Right muscle mass	z-score	Spine BMD	Left BMD	Right BMD	Hip BMD	Total BMD	W/H
Right Q-angle	r	−0.031	0.009	−0.094	−0.639	−0.444	−0.685*	−0.723*	−0.207	−0.646*	−0.169
	p	0.938	0.982	0.809	0.052	0.232	0.042	0.028	0.593	0.05	0.664
Left Q−angle	r	−0.295	0.122	0.089	−0.113	−0.146	−0.099	−0.006	−0.283	0.066	−0.232
	p	0.441	0.755	0.819	0.771	0.709	0.8	0.987	0.461	0.865	0.548

TG = training age, HGT = height, WGT = weight, RL. L = right limb length, LL. L = left limb length, W/H = waist-to-hip ratio, BMC = bone mineral content, BMD = bone mineral density

Table 6 Association of q-angle with foot posture

		Right leg Q-angle	Left leg Q-angle
FPI (foot posture index)	r	0.508*	0.514*
	p	0.037	0.035

Table 7 Association of q-angle with functional tests

		Right hop test	Left hop test	Left leg squat	Right leg squat
Right Q-angle	r	−0.507*	0.047	0.048	−0.611**
	p	0.038	0.858	0.856	0.009
	n	17	17	17	17
Left Q-angle	r	−0.221	−0.182	−0.016	0.245
	p	0.395	0.485	0.952	0.343
	n	17	17	17	17

5 Conclusion and Discussion

- The idea that athletic activity might affect Q-angle is called into question by the study's finding that there was no significant difference in Q-angle between athletes and non-athletes. This finding calls for a more thorough investigation of non-athletic variables that may also influence Q-angle variations. There may be a connection between Q-angle and bone health given the negative correlations observed in volleyball and boxing athletes' hip BMD, total BMD, and bone mineral content (BMC).
- This study begs the question of whether certain sports, with their particular movements or training regimes, can affect bone density as well as Q-angle. With its distinct biomechanical requirements, weightlifting could not show the same correlation between Q-angle and bone density as other sports, such as boxing and volleyball.
- There may be variations in functional performance among people with specific Q-angle features, as indicated by the negative connection observed between right-leg Q-angle and functional assessments. The strong link that was found between pronated foot posture and Q-angle highlights how various biomechanical aspects are interrelated. This correlation highlights the need to take into account various facets of lower limb alignment when evaluating foot posture and Q-angle.

Bibliography

1. Abulhasan JF, Grey MJ (2017) Anatomy and physiology of knee stability. J Funct Morphol Kinesiol 2(4). MDPI Multidisciplinary Digital Publishing Institute. https://doi.org/10.3390/jfmk2040034
2. Aggarwal N, Ved N (2020) Comparison of foot posture in runners (Sprinters) and non-runners in Indian population-an observational study. Indian J Public Health Res & Dev 11(6)
3. Bahadori S, Fatahi H, Ahmadpoor M (2020) The effect of TheraBand training on the Q angle and distance of ankle medial malleolus in individuals with Genu Valgum deformity. Phys Treat-Specif Phys Ther 10(3):117–126. https://doi.org/10.32598/ptj.10.3.304.3
4. Bartolomei S, Grillone G, Di Michele R, Cortesi M (2021) A comparison between male and female athletes in relative strength and power performances. J Funct Morphol Kinesiol 6(1). https://doi.org/10.3390/jfmk6010017
5. Basic Biomechanics of the Musculoskeletal System (n.d.)
6. Bélanger A (2010) Therapeutic electrophysical agents: evidence behind practice. Wolters Kluwer Health/Lippincott Williams & Wilkins
7. Bojicic KM, Beaulieu ML, Imaizumi Krieger DY, Ashton-Miller JA, Wojtys EM (2017) Association between lateral posterior tibial slope, body mass index, and ACL injury risk. Orthop J Sport Med 5(2). https://doi.org/10.1177/2325967116688664
8. Caylor D (1993) The Relationship between quadriceps angle and anterior knee pain syndrome. www.jospt.org
9. Chevidikunnan MF, Al Saif A, Pai KH, Mathias L (2015) Comparing goniometric and radiographic measurement of Q angle of the knee. Asian Biomedicine 9(5):631–636. https://doi.org/10.5372/1905-7415.0905.433
10. Clinical Biomechanics_20231002 (n.d.)
11. Dar G, Saposhnik A, Finestone AS, Ayalon M (2023) The effect of load carrying on gait kinetic and kinematic variables in soldiers with patellofemoral pain syndrome. Appl Sci (Switz) 13(4). https://doi.org/10.3390/app13042264
12. Dhakal I, Sharma S (2017) Study on quadriceps angle (Q Angle) in an adolescent and adult Nepali population with anterior knee pain. Int J Health Sci & Res (www.ijhsr.org) 7(9):9. www.ijhsr.org
13. El Gharib MH, El Tohamy AM, Mohamed NE (2021) Determining the relationship between the quadriceps and tibiofemoral angles among adolescents. J Taibah Univ Med Sci 16(1):70–76. https://doi.org/10.1016/j.jtumed.2020.10.003
14. Erdagi K, Altunok MK, Tüfekçi O, Özel R, Işik B, Ünüvar SB (2022) Quadriceps femoris angle of elite and non-elite athletes in olympic style weightlifting. Khyber Med Univ J 14(1):56–62. https://doi.org/10.35845/kmuj.2022.22016

Slip Risk Assessment of Commonly Employed Athletic Footwear Tread Patterns in the Indoor Sports Activity of Badminton

Subhodip Chatterjee, Binoy Krishna Biswas, Suvanjan Bhattacharyya, Shubham Gupta, and Arnab Chanda

Abstract Walking, jogging, or leaping can be done with exceptional plantar comfort thanks to modern athletic footwear. On the other hand, a perceptual illusion is produced when harmful plantar loads cause minimal felt plantar discomfort. This leads to insufficient impact-moderating behavior and subsequent injury. Sports footwear goods are in high demand due to the notable increase in physical activity. The footwear business has had to increase product variety, shorten timeframes for product development, and offer customized goods in order to stay competitive. To measure the functionalities of athletic footwear, evaluations are essential in product development, developing prototypes, biomechanics, and other domains. These include the mitigation of impact forces during motion, traction, and motion control. In this work, the level of optimum athletic performance was determined for certain Indian-based footwear brands. Five athletic footwears from some of the leading Indian footwear manufacturers were selected for slip risk assessment studies in an athletic activity flooring. Shore A hardness was measured with the help of a durometer for all the athletic footwear selected for the study. The heel tread contact area of all the athletic shoes was also determined prior to the questionnaire survey and athletic evaluation by the participants. The evaluation was performed on the flooring of a badminton court. The participants were asked to wear the five selected athletic footwears and then perform their practice session accordingly. Their responses were

S. Chatterjee · S. Gupta · A. Chanda
Centre for Biomedical Engineering, Indian Institute of Technology, Delhi, Hauz Khas, , New Delhi, India

S. Chatterjee · B. K. Biswas
Department of Applied Science and Humanities, B.P Poddar Institute of Management and Technology, Kolkata, India

S. Bhattacharyya (✉)
Department of Mechanical Engineering, BITS, Pilani, India
e-mail: suvanjan.bhattacharyya@pilani.bits-pilani.ac.in

A. Chanda
Department of Biomedical Engineering, All Indian Institute of Medical Science (AIIMS), New Delhi, India

S. Bhattacharyya and D. James (eds.), *Recent Trends in Sports Engineering*, Lecture Notes in Mechanical Engineering, https://doi.org/10.1007/978-981-97-8297-0_18

recorded which included the attributes from footwear comfort, slipping probability and ease of athletic effectiveness. On the basis of these feedback, the most optimum footwear outsole tread pattern was shortlisted.

1 Introduction

People are currently adopting more energy-saving practices in their daily lives. The same directions are being followed by exercise activities. These days, health is a bigger concern for people of all ages. Sports and other physical pursuits have grown in popularity, particularly among youth. Sports participation demands bodies that are flexible and appropriate foot placement. Every day, people utilize their feet to carry out a variety of tasks, including sitting, standing, walking, sprinting, and jumping. The quantity of weights used on each foot varies depending on the activity. In light of this, selecting the appropriate footwear for every game or activity is essential yet sometimes overlooked. Young users don't understand a product's capabilities unless they really use them.

Young adults typically focus more on the styles and designs of footwear and how they appear because of their individual wants and preferences. Athletic shoes should be picked with their purposes in mind to ensure that they are appropriate for the sports in mind, even though look is one of the most crucial considerations when making a purchase. A design for footwear must prioritize usefulness over look and identity, with comfort characteristics being one of the key components of the product's physical design. Therefore, in order to achieve a wide range of designing methods and a variety of looks and structures, footwear has currently been developed and evaluated by designers in the fields of athletic sciences and ergonomic disciplines. However, unless the product is actually used, young users are unaware of its capabilities. Making decisions during the purchasing process is mostly based on their personal experience with the designs, materials, and other elements that appeal to the individual buyers. Therefore, one of the most crucial elements influencing the appeal and market share of specific brands and designs aimed at younger consumers is appearance [1].

Playing sports is thought to improve one's health. Injury is a major public health issue and a likely consequence of participation [2, 3]. The frequency and distribution of injuries associated with sports differ depending on the sport, level of engagement, gender, and position of the athlete [4]. Injuries to cartilage, apophysis, and growth plates are among the numerous types of injuries that young athletes are more susceptible to than adults [5]. Playing sports at an early age promotes health, happiness, and professional growth. Young players exercise harder and longer when the competitive factor interferes, which causes sports-related injuries [6]. The presence of growth cartilage, which is less resistant to recurrent micro injury than its competent grown-up equivalent, skeletal infancy and forceful training, decreased flexibility due to pronounced growth spurt, a propensity for taking risks and experiments during sports, and finally a different physiological response to exercise make adolescents more prone to injury [7].

Adolescent athletes frequently sustain strains, sprains, development plate fractures, injuries from overuse, infections, epiphyseal injuries, stress fractures, and dislocations [7]. Sports injuries among boys are more common in football and basketball, whereas among girls, they are more common in field hockey, soccer, and basketball [8–10]. Sports injuries are caused by a multitude of variables that come into play at the scene of the injury rather than a single causative component. They could be extrinsic (environmental), intrinsic (personal, host), or the result of prior injury, which could injure up to 50–70% of people [6]. Adolescent athletes typically suffer lower extremity injuries when playing sports. Sport-related injuries, especially those to the knee and ankle, which affect more women than men, might change motor function and raise the chance of developing osteoarthritis in later life [11]. Additionally, it is projected that 8% of young people give up recreational sports each year due to injuries. This could result in less-than-ideal health in the road due to inactivity and cardiovascular disease and obesity [12].

It has been researched how often athletes between the ages of 11 and 19 sustained injuries in different nations [12–14]. There aren't many research like this in India; to the best of our knowledge, just one study has documented injuries in basketball and football, and another study has documented rates of injuries in other sports [15]. Therefore, it's important to understand the frequency, types, and contributing factors of injuries in sports like badminton, hockey, wrestling, weightlifting, and sprinting that are popular among athletes from rural India [16] (Tables 1 and 2).

The main objective of our study was to assess the effectiveness of five athletic footwears specifically intended for the sport of badminton. The five athletic footwears selected were from five different Indian brands and each of them having a different outsole tread pattern [17–19]. The evaluation of these footwear was done by supplying each type of footwear to 10 athletes. The athletes performed their regular training session activities while wearing these footwears and their responses were recorded. The variation in performance among the five athletic footwears in the badminton court was also recorded. On the basis of that the most optimum athletic footwear tread parameters for the sports activity of badminton was shortlisted.

Table 1 Competition of athletes at different levels [16]

Level of competition	Frequency	Percentage
School	256	55.5
University	27	5.8
District	120	26
State	46	9.9
National	12	2.6
Total	461	100

Table 2 Injuries in different age categories [16]

Age (years)	Frequency of injury/total	Percentage	Age wise (%)
11	9/23	4.2	39.1
12	16/57	7.5	28.1
13	24/73	11.4	32.8
14	52/95	24.5	54.7
15	46/82	21.6	56.1
16	20/42	9.4	47.6
17	26/46	12.4	56.5
18	4/15	1.8	26.6
19	15/28	7.2	53.5
Total	212/461	100	

2 Materials and Methods

The first part of the study was in shortlisting the athletic footwear which was generally used for badminton players. After a careful market survey, five such footwears were shortlisted. These athletic footwears were made by Indian footwear brands and were within the middle-income group range, i.e., <Rs.1000. It was also specifically noticed that no such selected footwear had the same outsole tread pattern. The Shore A hardness [20, 21] of the five different footwears was recorded with the help of a Shore A durometer (Precision Instruments, New Delhi). The variation in the Shore A hardness was noted. The outsole tread area was also calculated from the digital image of the footwear outsole by utilizing the ImageJ software. A point and shoot camera (Canon Powershot A400) was used to capture the images accordingly. The surface roughness of the flooring of the badminton court was measured with the help of a surface profilometer (Precise Equipment, Chennai, India). The average surface roughness was recorded to be 20.25 μm, respectively.

After the subsequent recording of the outsole tread geometry and hardness measurement, these selected footwears were given back to the badminton athletes. The usage time period of each of the five selected athletic footwear was also noted. Each footwear was worn by 10 badminton athletes and every single athlete who participated played an entire set, respectively. A questionnaire study was conducted after each athlete completed a game set with each of the five selected footwear. Various attributes ranging from footwear comfort to detection of footwear slips were recorded. The following Table 3 integrates all the findings from the questionnaire study accordingly (Figs. 1 and 2).

As we can see from Table 3, there is a wide variation among the athletic footwear selected. The variation is evident from the Shore A hardness to the tread contact area. Based on the questionnaire responses, the effect of these parameters was further assessed accordingly.

Table 3 Selected footwear parameters

Footwear designation	Footwear brand	Shore A hardness	Duration of usage (months)	Tread contact area (cm^2)
BS1	Paragon	55	6	148.417
BS2	Asian	42	3	159.348
BS3	Campus	35	2	166.013
BS4	Bata	45	4	130.364
BS5	Khadim's	52	12	125.322

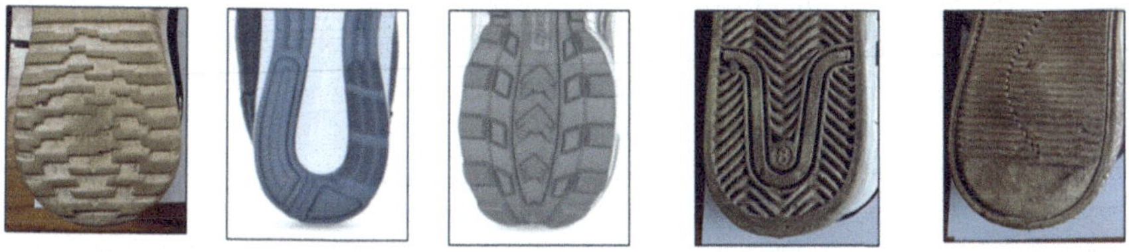

BS-1 **BS-2** **BS-3** **BS-4** **BS-5**

Fig. 1 Athletic footwear selected for study

Fig. 2 **a**, **b** Overall view of the badminton court, **c**, **d** badminton court flooring

On the basis of the responses from the athletes which is evident from Table 4, the athletic footwear BS2 and BS3 were the top choices in playing badminton.

3 Results and Discussions

The variation in Shore A hardness among the selected footwear ranged from 35 to 55. The footwear BS2 and BS3 exhibited the lowest Shore A hardness value of 35 and 42. This indicates the outsole is softer, which will result in greater comfort for the wearer and softer outsoles also increases the traction between the footwear and the flooring surface. Thus, lower shore A hardness was found to be a vital factor in selection of footwear among the participants.

The effect of the duration of usage of the footwear is also evident from the responses to the questionnaire responses. The footwear BS5 which had a usage duration of 12 months had the lowest performance feedback and had also the highest recordings of slipping. Comparatively new footwears such as BS2 and BS3, which had the usage duration below 4 months performed comparatively well both in terms of athletic performance and also on providing enhanced slip resistance in the event of rapid movements which are common in the activity of badminton.

As it is known from certain previous studies that the heel region of the footwear first makes contact with the flooring surface. More heel tread contact area will result in more contact between the outsole treads and the flooring surface which will help in enhanced traction performance. The athletic footwear designated as BS2 and BS3 had the highest heel tread area compared to the other footwear, for which its traction performance was high as it is evident from the questionnaire responses.

Thus, we can see that among Indian brand athletic footwear, the parameters of enhancing athletic performance are also present. Proper selection of athletic footwear will go a long way in improving the performance of the athletes specially in the rural part of India accordingly (Fig. 3).

4 Conclusion

The overall objective of this study was to identify the suitability of Indian brand athletic footwear in athletic activities such as badminton. The main factors influencing the selection of such footwears were identified and it ranged from Shore A hardness, duration of usage, and outsole tread contact area. The footwears which had the favorable combination of these parameters were found to be the most selected choice among the badminton athletes. This work serves as a guideline in identifying the proper footwear for the proper sports activities. If such selection can be made for other athletic activities such as hockey, wrestling, sprinting, etc., a lot of accidents can be reduced and the optimum level of performance can be reached for any athletic activities.

Table 4 Questionnaire study of the athletic footwear intended for playing badminton

Footwear used	Level of comfort	Ease of running	Ease of turning	Ease of rapid sprinting	Slips felt during running	Slips felt during turning	Slips felt during rapid sprinting	Difficulty in ankle movement	Overall traction	Level of selection
BS1	Medium	Medium	Low	Low	Medium	Medium	Medium	Medium	Medium	Medium
BS2	High	High	High	High	Low	Low	Low	Low	High	High
BS3	High	High	High	High	Low	Low	Low	Low	High	High
BS4	Medium	Medium	Medium	High	Medium	Medium	Medium	Medium	Medium	Medium
BS5	Low	Low	Low	Low	High	High	High	High	Low	Low

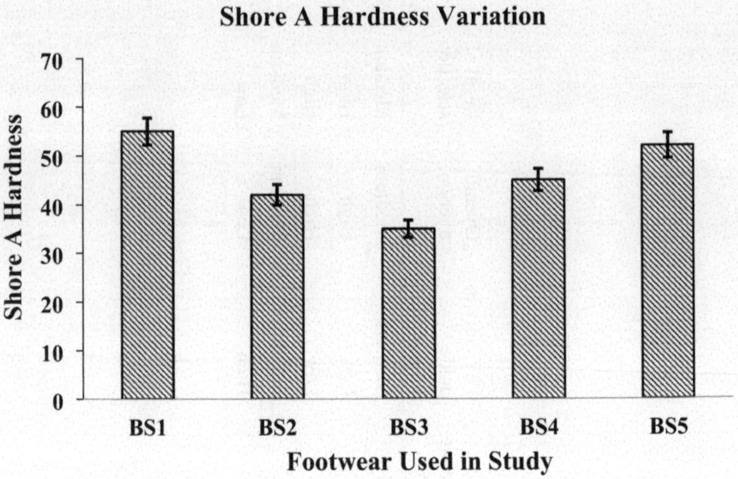

(a)

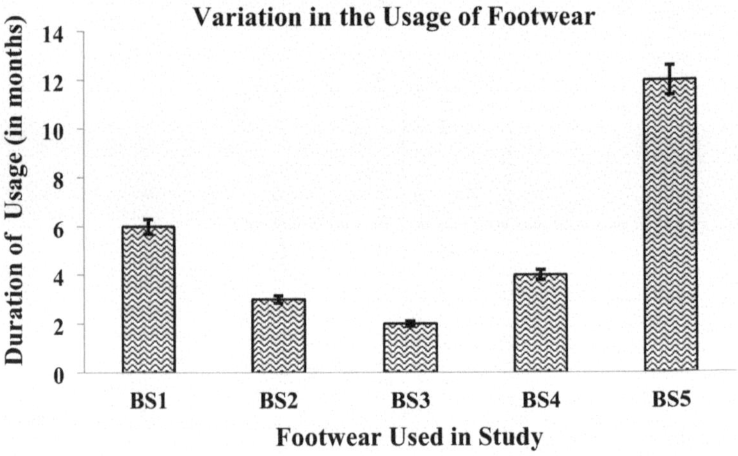

(b)

Fig. 3 **a** Variation in shore A hardness among the athletic footwear, **b** variation in the duration of usage among the athletic footwear, **c** variation in tread contact area among the athletic footwear

The economic issues faced by the athletes in the rural region were also addressed in this study as no high-cost footwear was selected in this study, and the availability of these footwear brands in rural regions is also evident. The proper selection among this existing footwear will boost the performance of the athletes from these rural regions accordingly.

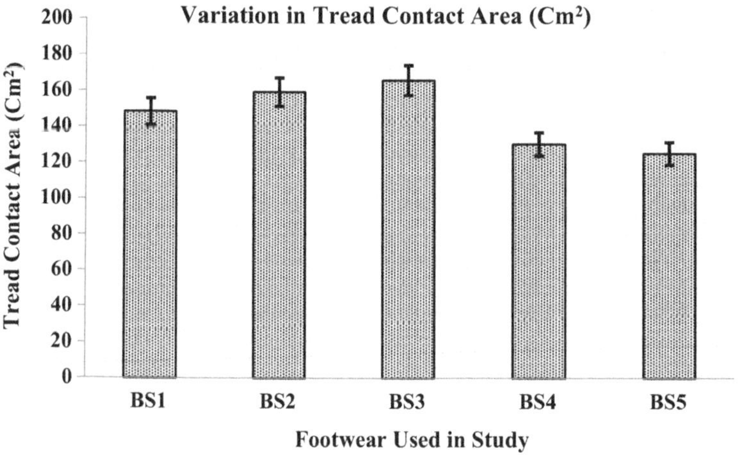

(c)

Fig. 3 (continued)

References

1. Promjun S, Sahachaisaeree N (2012) Factors determining athletic footwear design: a case of product appearance and functionality. Procedia—Soc Behav Sci 36:520–528. https://doi.org/10.1016/J.SBSPRO.2012.03.057
2. Eley RM, Hatt D, Baarbe S (2020) A retrospective observational study of presentations to an Australian emergency department for injuries sustained in exercise pursuits over 14 years. Ochsner J 20(3):261–266. https://doi.org/10.31486/TOJ.19.0059
3. Williams JM, Wright P, Currie CE, Beattie TF (1998) Sports related injuries in Scottish adolescents aged 11–15. Br J Sports Med 32(4):291–296. https://doi.org/10.1136/BJSM.32.4.291
4. Shanmugam C, Maffulli N (2008) Sports injuries in children. Br Med Bull 86(1):33–57. https://doi.org/10.1093/BMB/LDN001
5. Radelet MA, Lephart SM, Rubinstein EN, Myers JB (2002) Survey of the injury rate for children in community sports. Pediatrics 110(3). https://doi.org/10.1542/PEDS.110.3.E28
6. Emery CA, Tyreman H (2009) Sport participation, sport injury, risk factors and sport safety practices in Calgary and area junior high schools. Paediatr Child Health 14(7):439–444. https://doi.org/10.1093/PCH/14.7.439
7. Edwards T (2007) Clinical sports medicine, 3rd edn. Br J Sports Med 41(3):183. https://doi.org/10.1136/BJSM.2006.030312
8. Conn JM, Annest JL, Gilchrist J (2003) Sports and recreation related injury episodes in the US population, 1997–99. Inj Prev 9(2):117–123. https://doi.org/10.1136/IP.9.2.117
9. Frisch A, Croisier JL, Urhausen A, Seil R, Theisen D (2009) Injuries, risk factors and prevention initiatives in youth sport. Br Med Bull 92(1):95–121. https://doi.org/10.1093/BMB/LDP034
10. Powell JW, Barber-Foss KD (1999) Traumatic brain injury in high school athletes. JAMA 282(10):958–963. https://doi.org/10.1001/jama.282.10.958
11. Doherty C, Delahunt E, Caulfield B, Hertel J, Ryan J, Bleakley C (2014) The incidence and prevalence of ankle sprain injury: a systematic review and meta-analysis of prospective epidemiological studies. Sports Med 44(1):123–140. https://doi.org/10.1007/S40279-013-0102-5

12. Abernethy L, Bleakley C (2007) Strategies to prevent injury in adolescent sport: a systematic review. Br J Sports Med 41(10):627–638. https://doi.org/10.1136/BJSM.2007.035691
13. Habelt S, Hasler CC, Steinbrück K, Majewski M (2011) Sport injuries in adolescents. Orthop Rev (Pavia) 3(2). https://doi.org/10.4081/OR.2011.E18
14. Schneider S, Weidmann C, Seither B (2007) Epidemiology and risk factors of sports injuries—multivariate analyses using German national data. Int J Sports Med 28(3):247–252. https://doi.org/10.1055/s-2006-924293
15. Cromwell F, Walsh J, Gormley J (2000) A pilot study examining injuries in elite Gaelic footballers. Br J Sports Med 34(2):104–108. https://doi.org/10.1136/BJSM.34.2.104
16. Sreekaarini I, Eapen C, Cp Z (2014) Prevalence of sports injuries in adolescent athletes Charu Eapen KMC Mangalore. Manipl Academy of Higher Education Prevalence of Sports Injuries in Adolescent Athletes. https://doi.org/10.4172/2324-9080.1000168
17. Gupta S, Sidhu SS, Chatterjee S, Malviya A, Singh G, Chanda A (2022) Effect of floor coatings on slip-resistance of safety shoes. In: Coatings 2022, vol 12, p 1455, October 2022. https://doi.org/10.3390/COATINGS12101455
18. Gupta S, Chatterjee S, Malviya A, Chanda A (2022) Traction performance of common formal footwear on slippery surfaces. In: Surfaces 2022, vol 5, pp 489–503, November 2022. https://doi.org/10.3390/SURFACES5040035
19. Gupta S, Chatterjee S, Malviya A, Chanda A (2023) Frictional assessment of low-cost shoes in worn conditions across workplaces. J Bio-Tribo-Corros 9(1):1–13. https://doi.org/10.1007/S40735-023-00741-0/FIGURES/12
20. Gupta S, Chatterjee S, Malviya A, Singh G, Chanda A (2023) A novel computational model for traction performance characterization of footwear outsoles with horizontal tread channels. In: Computation 2023, vol 11, p 23, February 2023. https://doi.org/10.3390/COMPUTATION11020023
21. Gupta S, Chatterjee S, Chanda A (2023) Influence of vertically treaded outsoles on interfacial fluid pressure, mass flow rate, and shoe–floor traction during slips. In: Fluids 2023, vol 8, p 82, February 2023. https://doi.org/10.3390/FLUIDS8030082

Study on the Impact of Nutrients Intake on Athletic Performance by Using the Fibonacci Wavelet Method

R. Vidya Shree, B. Patil Mallikarjun, and Suvanjan Bhattacharyya

Abstract The current work investigates the impact of nutrients intake on athletic performance through the Fibonacci wavelet approach. The Athlete's performance is influenced by the nutrients intake in his diet. The governing nutrient, metabolism, and performance equations are transformed into algebraic equations. The Fibonacci wavelet method is used to solve achieved nonlinear coupled equations. The nature of the Athlete's performance is analyzed graphically for a scale of various values of physical parameters. The Fibonacci integral operational matrix is used to solve the achieved nonlinear equation and it has been used for the first time to solve problem related to sports engineering.

Keywords Fibonacci wavelet · Integral operational matrix · Nutrient intake · Athlete's performance

1 Introduction

A complete and well-balanced diet is a crucial component of the training regimen for reaching peak athletic performance. A negative energy balance is caused by consuming too little in comparison to what is expended, which severely restricts athletic performance [1]. For athletes, supplying calories is the most vital and fundamental nutritional goal [2]. Attaining energy balance serves as essential for ensuring the ideal conditions for lean tissue mass, the immunological system, the reproductive system, and maximum athletic performance. Sports performance is restricted and training benefits are reduced in the state when the calorie intake is less than energy expenditure. Moreover, persistently consuming low amounts of energy frequently

R. Vidya Shree · B. Patil Mallikarjun
Department of Studies and Research in Mathematics, Tumkur University, Tumkur-572103, Karnataka, India

S. Bhattacharyya (✉)
Department of Mechanical Engineering, BITS Pilani, Pilani Campus, Vidya Vihar, Pilani 333031, Rajasthan, India
e-mail: suvanjan.bhattacharyya@pilani.bits-pilani.ac.in

S. Bhattacharyya and D. James (eds.), *Recent Trends in Sports Engineering*, Lecture Notes in Mechanical Engineering, https://doi.org/10.1007/978-981-97-8297-0_19

results in nutritional deficiencies overall and micronutrient deficiencies specifically [3, 4]. Micronutrients, or vitamins and minerals, are important for the creation of energy even though they are not energy sources in and of themselves. They also help to maintain bone mass, improve hemoglobin synthesis, build and repair muscle tissue, boost immunity, and shield tissues and cells from oxidative damage [5]. As athletes have to perform for a long time, the body needs a lot of carbohydrates to fuel itself [6, 7]. To compensate for the wasted energy, the human body draws from its energy reserves.

A low carbohydrate diet is likely to hinder the resynthesis of muscle glycogen after exercise and cause premature muscle glycogen depletion during training or competition, which will ultimately limit athletic performance [8]. Maintaining fluid equilibrium is just as much of a struggle during ultra-endurance running as physical performance. The risk of not replacing fluids enough throughout a race has been highlighted by several writers, who state that doing so could result in dehydration, increased fatigue, cognitive impairment, and subpar performance [9, 10]. Aside from hydration, fuel availability is a crucial consideration when exercising, particularly in hot weather when it intensifies the oxidation of carbohydrates [11]. According to O'Brien et al. [12], carbohydrates are the main fuel used in marathons and should be ingested both before and during intense activity in order to reduce fatigue and improve performance. Aiming at a maximum carbohydrate oxidation rate of 1 g kg − 1 h − 1, the American College of Sports Medicine [13] recommends consuming 30–60 g h − 1 of carbohydrates during endurance exercise [14].

Signal processing has evolved dramatically during the last 10 years. Recent signal processing approaches, such as the Wavelet method, have enabled academics to tackle a number of difficult challenges. The wavelet approach has helped to tackle many physical issues in engineering and applied sciences. Wavelets are computational processes that have been widely used in digital signal processing in domains such as waveform representation and segmentation, nuclear physics, time–frequency analysis, atomic structures, gas dynamics, and chemical reactions, among others. Several fascinating mathematical features of wavelets, such as efficient multiscale decompositions and localization qualities in physical and wave-number spaces, encouraged their use in the numerical solution of differential equations. There are various numerical methods for solving differential equations, including the homotopy perturbation method, differential transform method, and adomian decomposition approach. Wavelet-based numerical methods provide an intriguing alternative to the previously described numerical methodologies.

Fibonacci wavelets are an emerging tool in the applied and computational sciences for determining effective numerical solutions. For analyzing solutions with discontinuities and sharp edges, the proposed technique performs efficiently. Sabermahani et al. [15] presented a way to implement a numerical method to solve problems involving optimum control and time-varying delay systems. The Fibonacci operational matrix method has been implemented by Shiralashetti and Lamani [16]. Kurt et al. [17] have used the Fibonacci collocation scheme to solve high-order linear Fredholm integro-differential-difference equations with mixed conditions. A system of nonlinear differential equations with the Fibonacci wavelet method was worked out for numerical solutions by Alkan et al. [18]. Chemical kinetic problems and carbon dioxide absorption into phenyl glycidyl ether have been studied and numerically computed using the Fibonacci wavelet operational matrix by Manohara and Kumbinarasaiah [19].

Karkera [20] employed the uniform Haar wavelet approach to investigate the laminar boundary-layer flow of a viscous incompressible fluid across a moving wedge. Vidya Shree et al. [21] used the Hermite Wavelet technique in an inclined channel to solve coupled nonlinear ordinary differential equations and investigate entropy formation in the MHD Casson fluid flow model. Shiralashetti et al. [22] employed the Muntz-Legendre wavelet operational matrix to study how heat radiation affects the boundary-layer flow of a natural convective vertical plate immersed in a saturated porous material. Heydari and Bavi [23] modified the wavelet to study the behavior of magnetohydrodynamic flow in a channel with a low porosity wall. Kumbinarsaiah and M. Preetham [24] investigated viscous fluid flow and estimated skin friction coefficients using the Bernoulli wavelet approach, which was compared to other wavelet methods for validation.

The current study uses a Fibonacci wavelet technique (FWA) to evaluate how nutrition affects athlete performance. The Fibonacci wavelet operational matrix was employed for the first time to tackle a sports-related problem. Nonlinear coupled equations resulting from basic equations are solved using FWA. Graphical analysis may be used to investigate the influence of all underlying controlling factors. The current study indicates that the wavelet theory may be used to analyze sports difficulties. FWA solutions are satisfactory when compared to previously available numerical solutions. The integral operational matrix corresponding to Fibonacci wavelets is originally built using Chen and Hsiao's [25] approach. The investigated problem is then turned into a collection of algebraic equations using integration operational matrices. These equations can then be solved using any standard approach, including the Newton method.

2 Problem Formulation

The mathematical model considered takes metabolism dynamics, nutrient absorption, and exercise effect into account on the athlete's performance. This is a simplified example that is considered for demonstrative purposes and certain assumptions are made while studying the impact of nutrients intake on the Athlete's performance. The main source of energy athletes depend on carbohydrates, proteins, and fats which are shown in Fig. 1. Actual models would be far more sophisticated to study hence following assumptions are made during the study:

- The model assumes a constant rate of nutrient intake.
- Nutrient absorption rates are higher than the nutrient utilization rates.
- Parameter values are taken arbitrarily for demonstrative purposes.

Fig. 1 Nutrients intake by Athlete

We have considered a model which can be described as a system of differential equations which is as shown below [26].

- Nutrient absorption and metabolism

$$\frac{dC}{dt} = K_{C_1} I_C(t) - K_{C_2} C(t) - \frac{K_{C_3} F(t) C(t)}{K_{C_4} + C(t)} \quad (1)$$

$$\frac{dP}{dt} = K_{P_1} I_P(t) - K_{P_2} P(t) - \frac{K_{P_3} F(t) P(t)}{K_{C_4} + P(t)} \quad (2)$$

$$\frac{dF}{dt} = K_{F_1} I_F(t) - K_{F_2} F(t) - \frac{K_{F_3} C(t) F(t)}{K_{F_4} + F(t)} \quad (3)$$

$$\frac{dM}{dt} = K_{M_1}C(t) + K_{M_2}P(t) + K_{M_3}F(t) - K_{M_4}E(t) - K_{M_5}M(t) \tag{4}$$

- Exercise-induced athletic performance

$$\frac{dA}{dt} = K_A M(t)(1 - e^{-K_E E(t)}) \tag{5}$$

Variables Defined Are as Follows:

$C(t)$—Carbohydrate intake rate at time t.
$P(t)$—Protein intake rate at time t.
$F(t)$—Fat intake rate at time t.
$M(t)$—Metabolic rate at time t.
$E(t)$—Excercise intensity at time t.
$A(t)$—Performance level at time t.
$I_C(t), I_P(t), I_F(t)$—rates at which carbohydrates, proteins, and fats are ingested at the time t.

The Parameters Defined Are as Follows:

$K_{C_1}, K_{P_1}, K_{F_1}$—Carbohydrate, proteins, and fat absorption rate;
$K_{C_2}, K_{P_2}, K_{F_2}$—Carbohydrate, proteins, and fat utilization rate;
$K_{C_3}, K_{P_3}, K_{F_3}$—Interaction between nutrients affecting nutrient utilization;
$K_{C_4}, K_{P_4}, K_{F_4}$—Saturation constants for nutrient utilization;
$K_{M_1}, K_{M_2}, K_{M_3}$—Conversion rate of nutrients to metabolic rate; K_{M_4}—Energy. expenditure due to exercise; K_{M_5}—Metabolic decay rate; K_A—Performance sensitivity.
of metabolic rate; K_E—Exercise effect rate on performance.

3 Numerical Implementation

Fibonacci wavelet is defined over the interval $[0, \delta_l]$ as follows:

$$\tau_{g,d}(\delta) = \begin{cases} 2^{\frac{k-1}{2}}\overline{F}_d(\frac{2^{k-1}}{\delta_l}\delta - g + 1), & \text{if} \frac{-g+1}{2^{k-1}}\delta_l \le \delta \le \frac{-g+2}{2^{k-1}}\delta_l, \\ 0, & \text{otherwise.} \end{cases} \tag{6}$$

with $\overline{F}_d(\delta) = \frac{\widehat{F}_{d(\delta)}}{\sqrt{w}}$ and $w = \int_0^1 (\widehat{F}_d(\delta))^2 d\delta$

where $d = 0, 1, 2, 3, \ldots\ldots(D-1)$, $g = 1, 2, 3, \ldots\ldots, 2^{k-1}$ and $\widehat{F}_d(\delta)$ is Fibonacci polynomial of order d.

Function approximation: Consider square-integrable function $y(\delta)$ expanded in terms of Fibonacci wavelet basis functions:

$$y(\delta) \approx \sum_{n=1}^{\infty} \sum_{m=0}^{\infty} G_{g,d} \tau_{g,d}(\delta), \tag{7}$$

on truncating $y(\delta)$ is

$$y(\delta) \approx \sum_{n=1}^{2^{k-1}} \sum_{m=0}^{M-1} G_{g,d} \tau_{g,d}(\delta) = G^T V(\delta), \tag{8}$$

where G and $V(\delta)$ are $2^{k-1}D \times 1$ matrices, $G^T = [G_{1,0}, \ldots\ldots, G_{1,D-1}, G_{2,0}, \ldots\ldots, G_{2,D-1}, \ldots. G_{2^{k-1},D-1}]$ and $V(\delta) = [\tau_{1,0}, \ldots \tau_{1,D-1}, \tau_{2,0}, \ldots, \tau_{2,D-1}, \ldots, \tau_{2^{k-1},D-1}]$.

As shown below, at $k = 2$ and $D = 7$ we retrieve the Fibonacci wavelet basis.

$$\tau_{2,0} = \sqrt{2}$$

$$\tau_{2,1} = \frac{\sqrt{2}\sqrt{3}}{2}\delta,$$

$$\tau_{2,2} = \frac{\sqrt{2}\sqrt{7}\sqrt{15}}{14}\left(\frac{\delta^2}{4} + 1\right),$$

$$\tau_{2,3} = \frac{\sqrt{2}\sqrt{105}\sqrt{239}}{239}\delta\left(\frac{\delta^2}{8} + 2\right),$$

$$\tau_{2,4} = \frac{3\sqrt{2}\sqrt{35}\sqrt{1943}}{1943}\left(\frac{\delta^4}{16} + 3\frac{\delta^2}{4} + 1\right),$$

$$\tau_{2,5} = \frac{3\sqrt{2}\sqrt{385}\sqrt{2582}}{330496}\delta(\delta^4 + 16\delta^2 + 48),$$

$$\tau_{2,6} = \frac{3\sqrt{2}\sqrt{5005}\sqrt{1268209}}{1268209}\left(\frac{\delta^6}{64} + \frac{5\delta^4}{16} + \frac{3\delta^2}{2} + 1\right),$$

$$\tau_{2,7} = \frac{3\sqrt{2}\sqrt{5005}\sqrt{2827883}}{361969024}\delta(\delta^6 + 24\delta^4 + 160\delta^2 + 256),$$

where $\tau_6(\delta) = [\tau_{1,0}, \tau_{1,1}, \tau_{1,2}, \tau_{1,3}, \tau_{1,4}, \tau_{1,5}]^T$.

Operational matrix of integration: Integrate $\tau_{g,d}(\delta)$ with respect to δ between 0 and δ and represent it as a linear combination of Fibonacci wavelet basis.

$$\int_0^\delta \tau_{2,0}(\delta)d\delta = \left[0\frac{1351}{1170}0000\right]K_6(\delta),$$

$$\int_0^\delta \tau_{2,1}(\delta)d\delta = \left[\frac{-1351}{780}0\frac{1001}{423}000\right]K_6(\delta),$$

$$\int_0^\delta \tau_{2,2}(\delta)d\delta = \left[0\frac{282}{1001}0\frac{1331}{1808}00\right]K_6(\delta),$$

$$\int_0^\delta \tau_{2,3}(\delta)d\delta = \left[\frac{-517}{780}0\frac{681}{1504}0\frac{549}{667}0\right]K_6(\delta),$$

$$\int_0^\delta \tau_{2,4}(\delta)d\delta = \left[000\frac{641}{2638}0\frac{441}{793}\right]K_6(\delta),$$

$$\int_0^\delta \tau_{2,5}(\delta)d\delta = \left[\frac{-708}{3667}000\frac{357}{1489}0\right]K_6(\delta) + \left(\frac{733}{1431})\tau_{1,6}(\delta\right),$$

This implies

$$\int_0^t \tau(\delta)d\delta = H_{6\times 6}\tau_6(t) + \widehat{V}_6(t)$$

where

$$H = \begin{bmatrix} 0 & \frac{1351}{1170} & 0 & 0 & 0 & 0 \\ \frac{-1351}{780} & 0 & \frac{1001}{423} & 0 & 0 & 0 \\ 0 & \frac{282}{1001} & 0 & \frac{1331}{1808} & 0 & 0 \\ \frac{-517}{780} & 0 & \frac{681}{1504} & 0 & 0 & \frac{549}{667} \\ 0 & 0 & 0 & \frac{641}{2638} & 0 & \frac{441}{793} \\ \frac{-708}{3667} & 0 & 0 & 0 & \frac{357}{1489} & 0 \end{bmatrix} \text{and} \widehat{V}_6(\delta) = \begin{bmatrix} 0 \\ 0 \\ 0 \\ 0 \\ 0 \\ \frac{733}{1431}\tau_{1,6}(\delta) \end{bmatrix},$$

Similarly, on repeated integration, we get higher order Operational integration matrix.

Method of Solution.

Let us assume that

$$\frac{dC}{dt} = G^T V(\delta) \tag{9}$$

where $G^T = [g_1^c, g_2^c, g_3^c, g_4^c, g_5^c]$.

Integrating Eq. (9) with respect to δ between the limits 0 to δ

$$C = C_0 + G^T[\widehat{V}(\delta) + HV(\delta)] \tag{10}$$

Let us assume that

$$\frac{dP}{dt} = G^T V(\delta) \tag{11}$$

where $G^T = [g_1^p, g_2^p, g_3^p, g_4^p, g_5^p]$.

Integrating Eq. (11) with respect to δ between the limits 0 to δ

$$P = P_0 + G^T[\widehat{V}(\delta) + HV(\delta)] \tag{12}$$

Let us assume that

$$\frac{dF}{dt} = G^T V(\delta) \tag{13}$$

where $G^T = [g_1^f, g_2^f, g_3^f, g_4^f, g_5^f]$.

Integrating Eq. (13) with respect to δ between the limits 0 to δ

$$F = F_0 + G^T[\widehat{V}(\delta) + HV(\delta)] \tag{14}$$

Let us assume that

$$\frac{dM}{dt} = G^T V(\delta) \tag{15}$$

where $G^T = [g_1^m, g_2^m, g_3^m, g_4^m, g_5^m]$.

Integrating Eq. (15) with respect to δ between the limits 0 to δ

$$M = M_0 + G^T[\widehat{V}(\delta) + HV(\delta)] \tag{16}$$

Let us assume that

$$\frac{dA}{dt} = G^T V(\delta) \tag{17}$$

where $G^T = [g_1^a, g_2^a, g_3^a, g_4^a, g_5^a]$

Integrating Eq. (17) with respect to δ between the limits 0 to δ

$$A = A_0 + G^T[\widehat{V}(\delta) + HV(\delta)] \tag{18}$$

Now substituting the eqns. (9)-(18) in eqns. (1)-(5) which leads to a system of nonlinear.

Equations. Determine the unknown Fibonacci wavelet coefficients by collocating these.

Equations for the given collocation points.

Here collocation points are limit points of sequence: $\delta_i = \frac{2i-1}{2N}$ where $i = 1, 2, 3, \ldots\ldots$

Substitute the obtained coefficient values in Eqs.(9)–(18) to get numerical solutions to equations (1)–(5).

4 Results and Discussion

Figures 2, 3, 4 shows the impact of carbohydrate absorption rate on carbohydrate intake rate, metabolic rate, and performance of the athlete. With increase in carbohydrate absorption rate by the human body makes higher carbohydrates influx into the system therefore there is an increased intake rate. The metabolic rate is expected to increase when there are more carbohydrates available. This is due to the fact that carbohydrates are the body's main energy source and that consuming more of them can boost energy for metabolic functions. An athlete's performance has a positive effect as metabolic rate directly influences the athlete's performance.

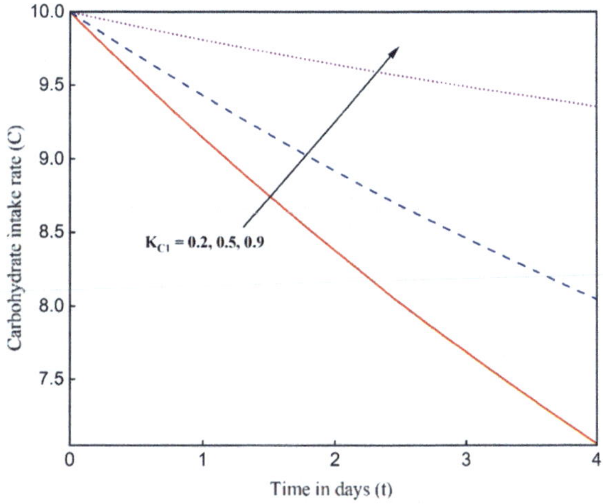

Fig. 2 Impact of carbohydrate absorption rate on carbohydrate intake rate

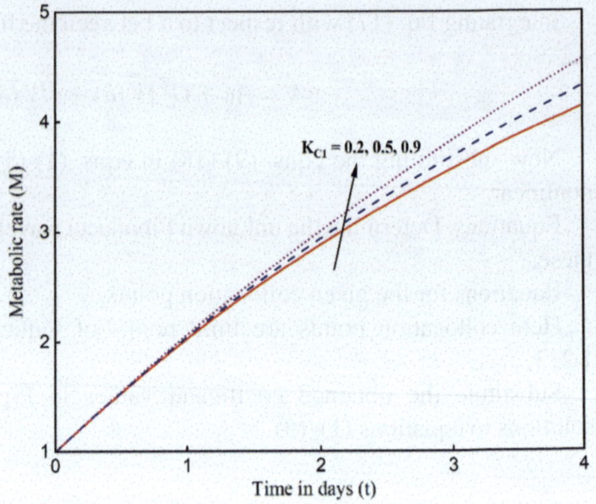

Fig. 3 Impact of carbohydrate absorption rate on metabolic rate

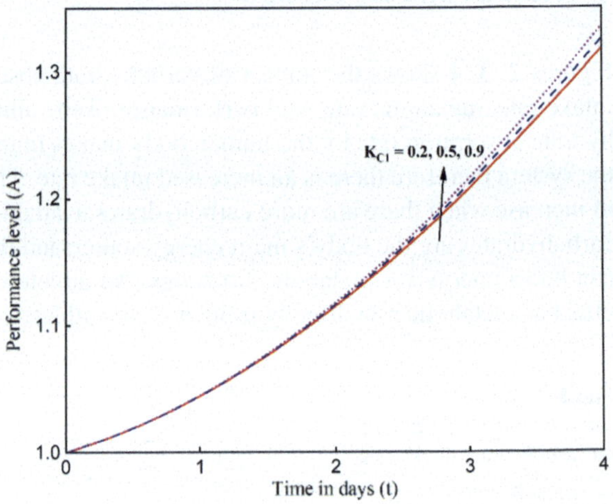

Fig. 4 Impact of carbohydrate absorption rate on athlete's performance

Figures 5, 6, 7, 8 shows the impact of carbohydrate utilization rate on carbohydrate intake rate, fat intake rate, metabolic rate, and performance of the athlete. It is known that the carbohydrate utilization rate is the rate at which carbohydrates are utilized by the body. This rate also shows how quickly carbohydrates are used up for metabolic and energy-producing processes. The body uses carbs more quickly than it absorbs them from the diet when the carbohydrate utilization rate (K_{C2}) is greater than the carbohydrate absorption rate (K_{C1}), causing the carbohydrate reserve to decline. As the body is utilizing it at an extremely fast rate it becomes excess to the body after complete absorption it gets stored as fat. Therefore with greater the value of

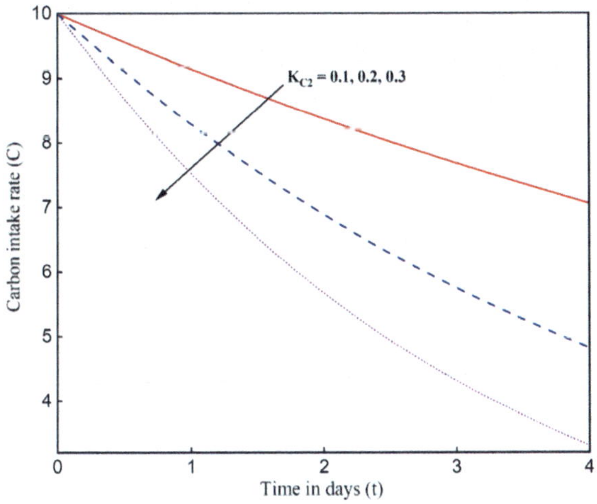

Fig. 5 Impact of carbohydrate utilization rate on carbohydrate intake rate

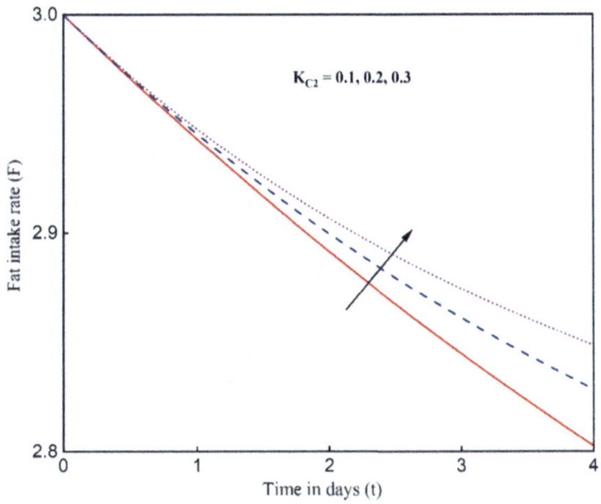

Fig. 6 Impact of carbohydrate utilization rate on fat intake rate

carbohydrate utilization rate fat intake rate increases. With increase in fat content decreases the metabolic rate thereby decreasing the performance of athletes.

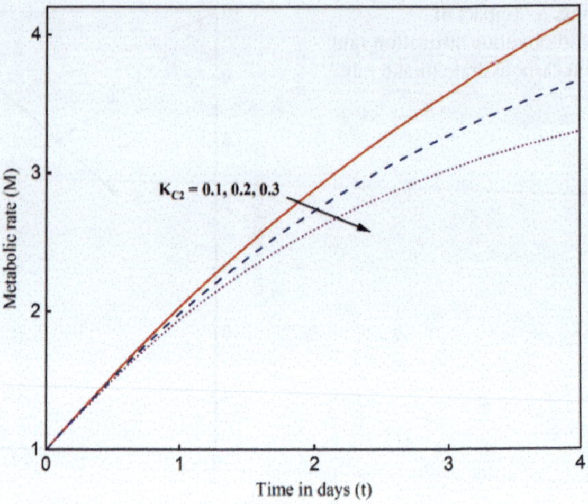

Fig. 7 Impact of carbohydrate utilization rate on metabolic rate

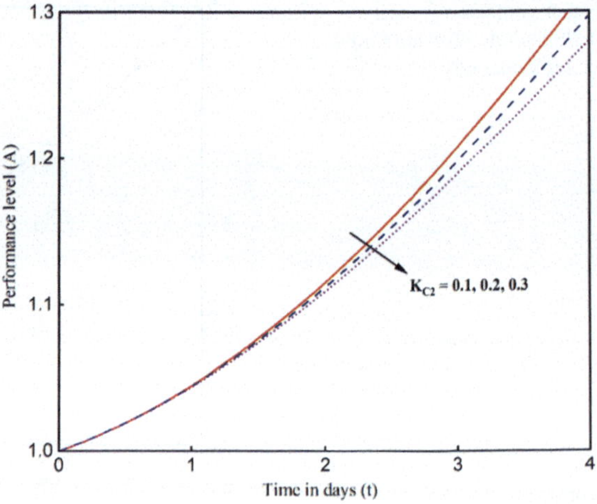

Fig. 8 Impact of carbohydrate absorption rate on athlete's performance

Figure 9a and b shows the impact of nutrient interaction and saturation on carbohydrate intake rate. An interaction factor in the model takes into consideration the potential impact of fats on the consumption of carbohydrates. The rate of utilization of carbohydrates can be slowed down when fats are included in the diet and the interaction coefficient (K_{C3}) is greater than zero. This may help the rate at which carbohydrates are consumed to gradually decrease, particularly if fats make up a large portion of the diet. Thus decreasing the carbohydrate intake rate. A saturation effect is introduced by including a saturation constant (K_{C4}) in the denominator. The term $\frac{K_{C_3}F(t)C(t)}{K_{C_4}+C(t)}$ is more prominent which leads to the decline of carbohydrate intake rate.

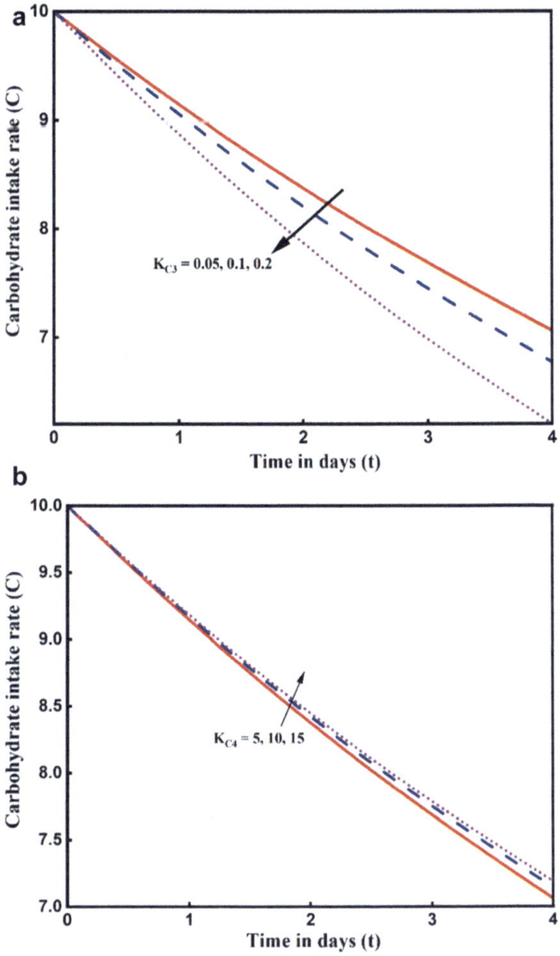

Fig. 9 Impact of **a** nutrient interaction and **b** saturation on carbohydrate intake rate

Figure 10a and b indicates the impact of carbohydrate metabolic conversion and metabolic decay rate on metabolic rate. Metabolic rate increases with growth of metabolic conversion rate. Here in particular for the study, carbohydrate metabolic conversion rate has been considered. Metabolic rate increases because the nutrients are converted to energy and this energy which in turn increases the metabolic rate thus the performance of the athlete increases. Depletion of metabolism occurs due to physical activities such as doing exercise. Therefore the metabolic decay rate decreases the metabolic rate which can be observed from the figure.

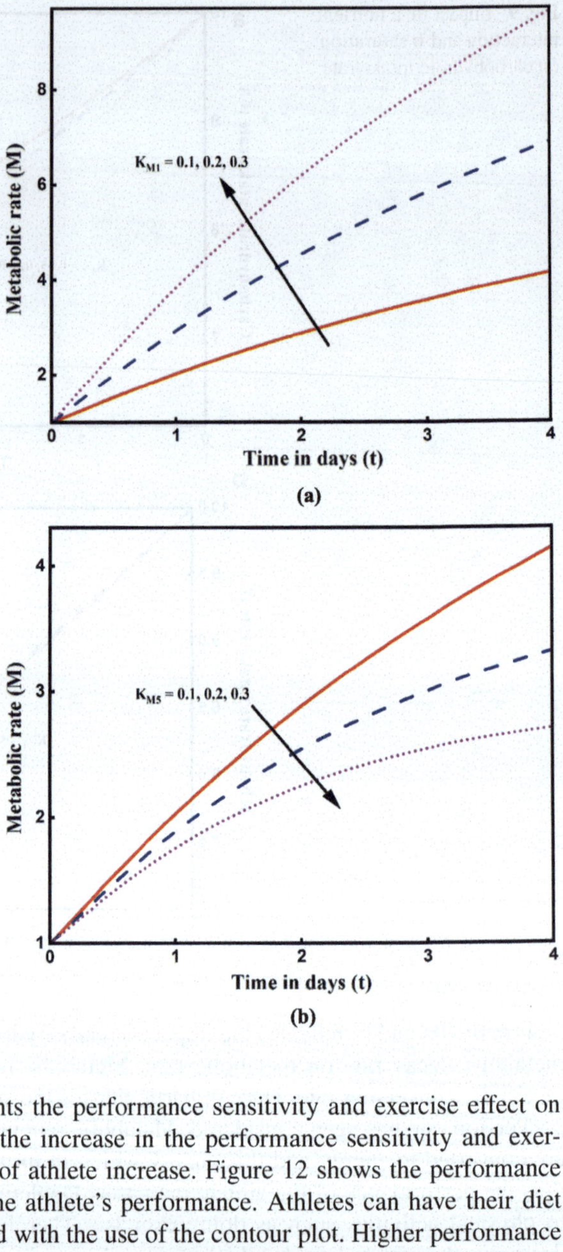

Fig. 10 Impact of **a** carbohydrate metabolic rate conversion and **b** metabolic decay rate on metabolic rate

Figure 11a and b represents the performance sensitivity and exercise effect on athlete's performance. With the increase in the performance sensitivity and exercise effect, the performance of athlete increase. Figure 12 shows the performance saturation contour plot for the athlete's performance. Athletes can have their diet and training plans customized with the use of the contour plot. Higher performance sensitivity values may help athletes become more sensitive to performance, whereas lower K_A values may help athletes reach their peak performance phase sooner.

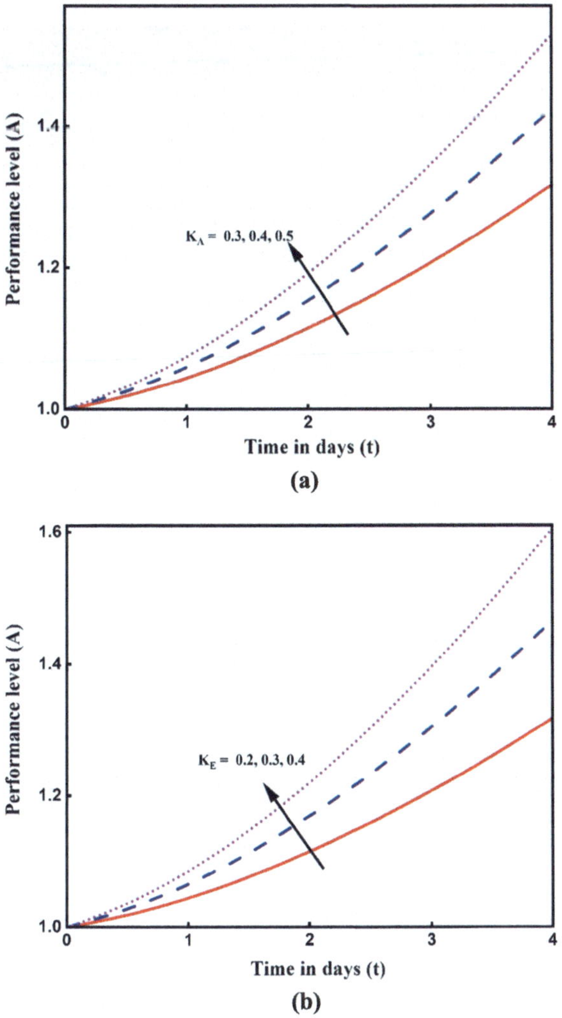

Fig. 11 Impact of **a** performance sensitivity and **b** exercise effect on athlete's performance

5 Concluding Remarks

FWA can solve nonlinear ordinary differential equations. The obtained results demonstrate FWA's capability of solving sports-related issues. In relative to existing numerical solutions in the research, FWA solutions are fairly excellent. This approach may be adopted to solve differential and coupled differential equations with higher order derivatives. In this analysis, the numerical results of the impact of Athlete's performance have been analyzed. The governing nutrient and performance equations are solved using the Fibonacci wavelet method. A model like this can serve

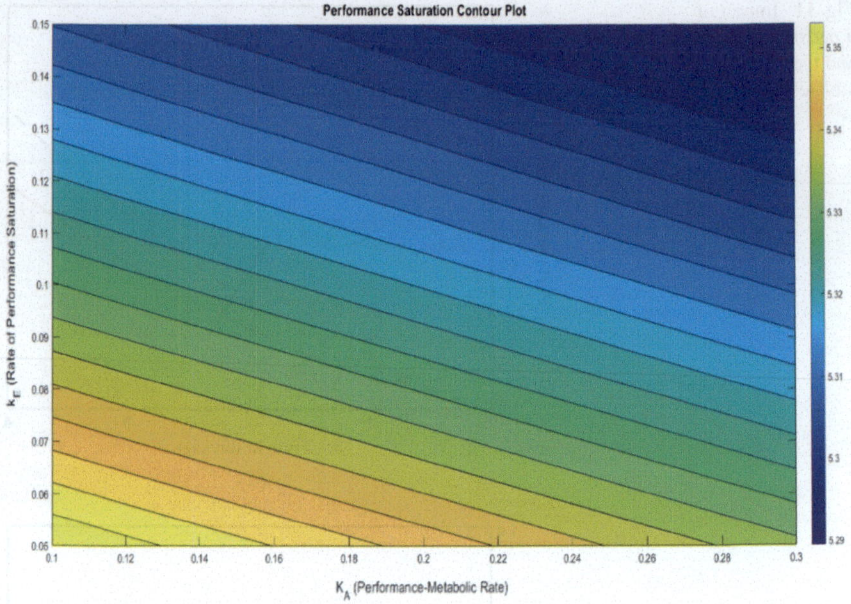

Fig. 12 Performance saturation contour plot

as a starting point for more complex and realistic simulations. Performation saturation graphs can guide decisions about optimizing training and nutrient strategies. The contour plot helps the trainer and dietician to train the athlete to give a better performance.

References

1. Tietjen-Smith T, Mercer J (2008) The female athlete triad: a statement of the problem. 2(1), Accessed: Jan. 07, 2024. Available: https://www.researchgate.net/publication/242142990
2. Volleyball/edited by Jonathan Reeser, Roald Bahr., Volleyball, 2003
3. Swinburn B, ER-TA journal of clinical nutrition, and undefined 1993, Energy balance or fat balance?, Elsevier, Accessed: Jan. 07, 2024. Available: https://www.sciencedirect.com/science/article/pii/S0002916523312309
4. Tatsuta W et al (2009) American dietetic association and Dietitians of Canada: joint position statement nutrition and athletic performance. Med Sci Sports Exerc 41(4):709–731. https://doi.org/10.7600/JSPFSM.61.427
5. Mangels AR, Messina V, Melina V (2003) Position of the American dietetic association and dietitians of Canada: vegetarian diets. J Am Diet Assoc, 103(6), pp 748–766, Accessed: Jan. 07, 2024.
6. Papadopoulou SD (2015) Impact of energy intake and balance on the athletic performance and health of top female volleyball athletes. J RomIan Sport Med Soc, XI, pp 2477–2481
7. Papadopoulou SK, Papadopoulou SD, Gallos GK (2002) Macro- and Micro-nutrient Intake of Adolescent Greek Female Volleyball Players. Int J Sport Nutr Exerc Metab 12(1):73–80. https://doi.org/10.1123/IJSNEM.12.1.73

8. Rico-Sanz J, Zehnder M, Buchli R, Dambach M, Boutellier U (1999) Muscle glycogen degradation during simulation of a fatiguing soccer match in elite soccer players examined noninvasively by 13C-MRS. Med Sci Sports Exerc 31(11):1587–1593. https://doi.org/10.1097/00005768-199911000-00014
9. Barr SI, Costill DL, Fink WJ (1991) Fluid replacement during prolonged exercise: effects of water, saline, or no fluid. Med Sci Sports Exerc, 23(7), pp 811–817, Accessed: Jan. 07, 2024. Available: https://europepmc.org/article/med/1921673
10. Cian C, Barraud PA, Melin B, Raphel C (2001) Effects of fluid ingestion on cognitive function after heat stress or exercise-induced dehydration. Int J Psychophysiol 42(3):243–251. https://doi.org/10.1016/S0167-8760(01)00142-8
11. Burke LM (2001) Nutritional needs for exercise in the heat. Comp Biochem Physiol A Mol Integr Physiol 128(4):735–748. https://doi.org/10.1016/S1095-6433(01)00279-3
12. O'Brien MJ, Viguie CA, Mazzeo RS, Brooks GA (1993) Carbohydrate dependence during marathon running. Med Sci Sports Exerc 25(9):1009–1017. https://doi.org/10.1249/00005768-199309000-00007
13. AC of S Medicine (2009) American dietetic association and Dietitians of Canada : joint position statement nutrition and athletic performance. Med Sci Sports Exerc, 41, pp 709–731, Accessed: Jan. 07, 2024. Available: https://cir.nii.ac.jp/crid/1573387450607751680
14. Hawley JA, Dennis SC, Noakes TD (1992) Oxidation of carbohydrate ingested during prolonged endurance exercise. Sport Med: Int J Appl Med Sci Sport Exerc 14(1):27–42. https://doi.org/10.2165/00007256-199214010-00003/METRICS
15. Sabermahani S, Ordokhani Y, Yousefi SA (2020) Fibonacci wavelets and their applications for solving two classes of time-varying delay problems. Optim Control Appl Methods 41(2):395–416. https://doi.org/10.1002/OCA.2549
16. Shiralashetti SC, Lamani L (2020) Fibonacci wavelet based numerical method for the solution of nonlinear Stratonovich Volterra integral equations. Sci Afr 10:e00594. https://doi.org/10.1016/J.SCIAF.2020.E00594
17. Kurt A, Yalçnbaş S, Sezer M (2013) Fibonacci collocation method for solving high-order linear fredholm integro-differential-difference equations. Int J Math Math Sci, 2013, https://doi.org/10.1155/2013/486013
18. Çakmak M, Alkan S (2021) Fibonacci collocation method for solving a class of systems of nonlinear differential equations. New Trends Math Sci 9(4):11–24. https://doi.org/10.20852/ntmsci.2021.454
19. Manohara G, Kumbinarasaiah S (2023) Fibonacci wavelets operational matrix approach for solving chemistry problems. J Umm Al-Qura Univ Appl Sci 2023 9:4, 9(4), pp 393–410, https://doi.org/10.1007/S43994-023-00046-5
20. Karkera H, Katagi NN, Kudenatti RB (2022) Frontiers in Heat and Mass Transfer analysis of laminar boundary-layer flow over a moving wedge using a uniform haar wavelet method, 18, p 41, https://doi.org/10.5098/hmt.18.41
21. Vidya SR, Patil MB, Kumbinarasaiah S (2023) Entropy generation on an MHD Casson fluid flow in an inclined channel with a permeable walls through Hermite wavelet method. Results Control Optim 12:100261. https://doi.org/10.1016/J.RICO.2023.100261
22. Shiralashetti SC, lowast, Harishkumar ESIH nbsp (2022) Muntz—Legendre wavelet operational matrix method to compute a numerical solution for thermal radiation effect on natural convection boundary layer flow past a vertical plate embedded in a saturated porous medium. Indian J Sci Technol, 15(48), pp. 2777–2790, https://doi.org/10.17485/IJST/V15I48.1619
23. Heydari MH, Bavi O (2022) An optimization method based on the Legendre wavelets for 3D rotating, squeezing and stretching magnetohydrodymanic flow in a channel with porous wall. Eng Comput 38(3):2583–2592. https://doi.org/10.1007/S00366-021-01421-8/FIGURES/7
24. Kumbinarasaiah S, Preetham MP (2022) Applications of the Bernoulli wavelet collocation method in the analysis of MHD boundary layer flow of a viscous fluid. J Umm Al-Qura Univ Appl Sci 2022 9:1, 9(1), pp 1–14, https://doi.org/10.1007/S43994-022-00013-6
25. Chen CF, Hsiao CH (1997) Haar wavelet method for solving lumped and distributed-parameter systems. IEE proceedings: control theory and applications 144(1):87–93. https://doi.org/10.1049/IP-CTA:19970702

26. Kralles E (2024) Digital commons @brockport mathematical modeling of fish populations in lake ontario using differential equations mathematical modeling of fish populations in lake ontario using differential equations a senior honors thesis submitted in partial fulfillment of the requirements for graduation in the honors college. Accessed: Jan. 07, 2024. Available: https://digitalcommons.brockport.edu/honors

Real-Time Insights and Dynamic Performance Visualizations for Unprecedented Impact in Ball Badminton

M. Rohith Gowda, Aditi N. Pai, C. N. Vishwanatha, M. P. Akash, G. H. Ranganath, and H. N. Bhaskar

Abstract The absence of advanced analytical tools has been a major obstacle in the world of ball badminton, a traditional sport. The lack of real-time data and valuable insights has long impeded players, coaches, and spectators. This paper introduces an innovative system aimed at transforming the ball badminton experience, enhancing player performance, engaging spectators, and providing valuable tools for coaches and fans through real-time insights and dynamic performance visualizations. This approach involves developing a system, preferably an application, to track the movement of the player and measure position, acceleration, and speed using computer vision and predicting tactics to assist during training, tracking the short trajectory of the ball to assist during practices for improvement, displaying live statistics such as player stats (points won, errors, aces), match progress (scoreboard), and historical data (head-to-head records) to provide context and analysis during the match, besides assisting the umpire in making decisions. Benefits to the viewers are the display of real-time statistics, fan engagement through quizzes on the sport, and result prediction. Coaches can use it to make strategic decisions, players to review their performance, and fans to gain deeper insights into the match. Additionally, it employs a

M. Rohith Gowda · A. N. Pai · M. P. Akash
Department of Computer Science and Engineering, R. V. College of Engineering, Bengaluru, India
e-mail: rohithgowdam.cs21@rvce.edu.in

A. N. Pai
e-mail: aditinpai.cs21@rvce.edu.in

M. P. Akash
e-mail: mpakash.cs21@rvce.edu.in

C. N. Vishwanatha (✉)
Department of Physical Education and Sports, R. V. College of Engineering, Bengaluru, India
e-mail: dirped@rvce.edu.in

G. H. Ranganath
Department of Physical Education and Sports, SJB Institute of Technology, Bengaluru, India

H. N. Bhaskar
Department of Physical Education and Sports, SBRRMFGC, Mysuru, India

S. Bhattacharyya and D. James (eds.), *Recent Trends in Sports Engineering*, Lecture Notes in Mechanical Engineering, https://doi.org/10.1007/978-981-97-8297-0_20

markerless motion capture system, which relies on deep learning for human position estimation and computer vision techniques. To transform pixel data into court coordinates and enable the measurement of parameters like distance covered, court positioning, and average player speeds for ball badminton players, the methodology relies on the inverse perspective mapping method. One of the algorithms applied in this context is the Lucas-Kanade optical flow algorithm, which aids in player tracking and position estimation, contributing to the overall effectiveness of the technology. The technology will track player movements and tactical insights with great precision. The accuracy of the system's visuals and data is intended to considerably help to a better comprehension of the game. We aim to improve player performance, captivate spectators, and provide vital tools for players, coaches, and fans by providing real-time insights and dynamic performance visualizations.

Keywords Ball badminton · Computer vision · Motion capture · Deep learning

1 Introduction

Within the world of sports, the age-old sport of ball badminton has long been in need of technological advancement to improve player performance, attract passionate spectators, and provide coaches and fans with up-to-date information and dynamic visualizations, due to the non-availability of data as well as analytical tools. This work presents a novel system that is created as an online application using state-of-the-art tools like OpenCV and YOLO for computer vision as well as cutting-edge technologies like React JS, Python-Flask API, and Firebase real-time database. This cutting-edge system offers a comprehensive solution for players, coaches, umpires, and spectators, with the goal of revolutionizing the game of ball badminton. It gives users the ability to take on roles that are customized to their requirements and includes tools like player posture estimate, ball tracking, speed tracking, and court position tracking to aid players and coaches in tactical analysis and performance assessment, to effectively predict the best positions to be maintained in future matches. While viewers can make use of real-time information on scores and win probability percentages based on current as well as historical data, umpires can effectively handle match-related data. The combination of deep learning, computer vision, and markerless motion capture methods guarantees its accuracy in tracking player movements and tactical insights. Because of the system's precision, we should be able to comprehend the game better, which should improve player performance and increase spectator interest. In this paper, we introduce this game-changing technology that, in our opinion, will bring ball badminton into a new era by providing a novel viewpoint and pushing the game to unprecedented heights.

2 Major Research Findings

The suggested human tracking solution for embedded surveillance in [1] is based on the Lucas-Kanade (LK) optical flow algorithm and uses tracked human object bounding boxes that are regularly adjusted to detect human objects and fix tracking problems caused by buildup or missing items. Additionally, it uses a combination of LK and human detector data to address occlusion. The YOLOv2 object detector paradigm is utilized to provide fast and reliable processing. Experiments on this approach demonstrate that, in comparison to other cutting-edge technologies, the method is fast and dependable for addressing occlusions.

By using the former on the data acquired with the latter, the paper [2] compares markerless and marker-based motion capture. Three-dimensional markerless joint kinematics were estimated and compared with conventional marker-based techniques using the measurement of the root mean square distance between joint rotations. The joint angle estimate performance of markerless and marker-based systems was assessed using the same anatomical frames of reference by both methods. By applying the same wand calibration and lens distortion correction to both systems, differences in camera calibration and synchronization were eliminated. While estimations for the hip joint were found to be inaccurate, the knee flexion–extension angle and sagittal plane kinematics were calculated more accurately than on the other planes.

An overview of marker-free human motion analysis for the past ten years is given in the study [3]. The three key elements—target application, complexity, and baseline methods used in generic human motion analysis systems—are the main focus of this investigation. It offers a revolutionary hierarchical classification of computer vision frameworks and directs users to pertinent literature according to the difficulties of each particular scenario. A review of the state-of-the-art in terms of datasets and quantitative outcomes is given along with their real-world applications.

The goal of the paper [4] is to provide a comprehensive, category-organized overview of all published research studies on ball tracking. The paper also examines the methods employed, their efficacy, drawbacks, and appropriateness for a certain sport. The summary of tracking systems based on Kalman filter, particle filter, trajectories, data association, graphs, and objective function optimization is provided by comparing the sport, tracking technique, results, shortcomings, strengths, and also the datasets used in tracking. This review is the ultimate guide to determining which technique is suitable for a sport.

A semi-real-time vehicle tracking technique is presented in the study [5] to use traffic cam footage to calculate the speed of the moving cars. This technique is implemented using MATLAB and Simulink's optical flow (Lucas-Kanade and Horn-Schunck). After the noise in that scene is reduced, the item is identified by segmentation. An object's velocity is determined by measuring its distance traveled across a sequence of frames in relation to the frame rate at which the video was captured. The performance of the algorithms based on average angular error and PSNR is analyzed and compared.

The publication [6] lists the limitations of single-branch networks for posture estimation. It proposes a top-down approach to single-target posture assessment by combining graph convolutional neural networks with multi-branch self-calibrating networks. Human body pose detection and estimation are the two areas of research that are recognized. Coordinate regression-based and heatmap test-based methods are the two used for pose estimation. The high-resolution feature map output from HRNet is utilized for deconvolution to increase the accuracy of single-target pose estimation recognition, hence increasing the accuracy of the heatmap test.

In order to gather pertinent data, the paper [7] analyzes image-processing technologies and path-tracking algorithms using computer vision. It also investigates the parameters that influence badminton path and ball speed transformation. Using sensors, the path-tracking algorithm predicts the rotation angle, ball speed, and athlete's body information during badminton movement. Visual field tracking and target dynamic tracking are utilized to record the moving target's position. The data involves examining the angle of each limb and the racket's speed during movement by establishing distinct placement locations for various motions, such as touching hair, pushing, picking, and hooking a ball. The lowest point of the ball served by player A in the first three serves is higher than that of player B, indicating that there is a relationship between the badminton trajectory's highest and lowest points and the ball's speed despite the fact that both are influenced by a variety of circumstances.

The goal of the study [8] is to remove the obstacles that robotic technologies face when trying to recognize the ball in sports. The parameters this system generates during ball detection include the robot's distance from the ball, the ball's angle of direction, and whether or not the ball is colored. This problem is solved by first converting the RGB image that the camera takes into HSV, then applying color filtering and finishing with the discover contour method. The system tracked ball objects with 92.87% accuracy at the ball's distance from the camera and 88.19% accuracy at the ball's angle of direction, according to the study's findings.

A unique heatmap-free method for joint identification is provided in the paper [9] with a 2D multi-person pose estimate in a photo based on the popular YOLO object recognition framework along with YOLO-position. The framework is used to optimize the object keypoint similarity (OKS) metric after the model has been trained from scratch. By incorporating the best aspects of both top-down and bottom-up approaches, the proposed model overcomes the limitations of not being trainable by learning to simultaneously identify bounding boxes for many persons and their respective 2D poses in a single forward pass. The proposed method eliminates the requirement for post-processing of bottom-up techniques by automatically grouping discovered important points into a skeleton. YOLO-pose exceeds all prior bottom-up algorithms in a single forward pass without a flip test, scoring 90.2% AP50 on COCO validation and 90.3% AP50 on the test-dev set, setting new state-of-the-art results.

The primary objective of the study [10] is to design and evaluate a proposed system for tracking an individual's motions and mobility in various lighting conditions. Using the Independent T-Test statistical technique, the Canny Edge Detector and Lucas-Kanade Optical Flow were utilized to compare the motion detection processing time of the suggested system in three different light situations. The results showed that the t-statistical values were 3.23 in Sunny and Dark lighting, 11.31 in Dim and Dark lighting, and 3.23 in Sunny and Dim lighting. The researchers came to the conclusion that despite a small fluctuation in the lighting, the system can track people's movements.

The investigation of several badminton motions, such as smashing strategies, service recognition, and swing trajectory, is the main topic of the research [11]. It also explores the effect of air resistance on shuttlecock flying and shuttle trajectory analysis. The relationship between racket speed and shuttlecock speed during a smash is examined using cutting-edge techniques including sensing systems with accelerometers and acoustic sensors. They discuss how forehand and backhand smashes differ in terms of forearm, wrist, and upper arm movements. The work uses SVM classifiers and spatial distribution variables to identify swing action in badminton films. These studies add to a thorough knowledge of badminton kinematics and offer insightful information to researchers, coaches, and players.

In article [12], the evaluation focuses on the simulation precision of the pyramid Lucas-Kanade (LK) optical flow algorithm under three distinct environments: urban areas, flat terrains, and mountainous regions. The simulation results signify the LK algorithm's capacity to maintain remarkable accuracy even in the presence of substantial motion, and the unmanned aerial vehicle's (UAV) flight path is determined based on the algorithm's accuracy within each environment, ensuring the optical flow sensor's reliability. In an urban setting with UAV speeds ranging from 12 to 15 m/s, the optimal altitude falls within the range of 80–150 m.

The paper [13] provides an overview of human body pose estimation and its applications, particularly in the context of sign language recognition. The authors discuss various aspects, including the types of input data (such as color-based and static/dynamic images) and different coordinate-based pose estimation approaches. They introduce existing architectures and techniques for human pose estimation, highlighting their applications and limitations. The paper presents an experimental study using MediaPipe and a random forest classifier to recognize isolated sign language gestures, with varying success rates for different signs.

The study [14] focuses on object tracking in sports action recognition videos, specifically addressing the difficulties associated with player tracking, like sudden changes in trajectory and occlusions by assessing three tracking techniques: the simple online and real-time tracking (SORT) method, the Hungarian algorithm-based approach, and the improved SORT that makes use of deep association features (Deep SORT). The YOLOv2 object detector and a bespoke handball dataset are used to evaluate the effectiveness of various techniques. While Deep SORT and other tracking-by-detection techniques show promise in reducing occlusion problems, player re-identification is a challenge.

The paper [15] gives an overview of motion estimation and optical flow. The benefits and drawbacks of the approach are described. The idea of feature tracking—which identifies trackable features—is also covered. Eigenvalues are used in this process. The possible uses of the Lucas-Kanade approach in authentication systems are mentioned in the paper's conclusion. This review of the literature presents important ideas in optical flow and motion estimation and emphasizes the applicability of the Lucas-Kanade approach to feature tracking and computer vision.

The use of artificial neural networks (ANNs) for sports outcome prediction is reviewed in the research [16], with a focus on the significance of precise predictions for different stakeholders. It looks at earlier studies that used ANNs to analyze data from various sports and sources, emphasizing feature selection and data preprocessing to improve accuracy. The SRP-CRISP-DM framework, a systematic method with six essential elements for handling sports prediction, is introduced in this research. The framework emphasizes the importance of feature selection and model evaluation, stresses the necessity of maintaining data order, and uses a round-by-round split for multi-round sports. For those who are interested in utilizing ANNs to predict the results of sporting events, it offers insightful information.

The study [17] uses historical season data analysis to tackle the problem of English Premier League (EPL) soccer match prediction. To provide predictions, the study uses machine learning methods such as Random Forest, SVM, and Naïve Bayes. After hyperparameter adjustment, the SVM method is shown to be the most accurate, with a prediction accuracy of 67%. The study emphasizes the significance of pertinent characteristics like attacking and defensive strength and makes suggestions for future work that could increase forecast accuracy, like adding management records and sentiment analysis.

In the research [18], an effective real-time object detection model called YOLOv3 is presented. Its robust performance is highlighted, particularly when it comes to recognizing objects that have partial alignment at an IoU of 0.5. The authors affirm the efficacy of YOLOv3's existing formulation while briefly mentioning alternate strategies that were investigated throughout development. The article also addresses ethical issues surrounding computer vision technology and its possible uses in a variety of fields, highlighting the duty that researchers have to take society implications into account when doing their study. YOLOv3 is a noteworthy advancement in computer vision, providing competitive and useful object identification capabilities.

3 Proposed Approach

The proposed approach revolves around the development of a web application using ReactJS for frontend, Python-Flask API for the backend, and Firebase real-time database to store real-time data during the match, using other open-source technological tools like OpenCV and the YOLOv3 model pre-trained on Coco dataset. The application is designed with different features and options available for different user roles—coach, umpire, or fan (public). Coaches and umpires who use it for analytical purposes and performance improvements need to authenticate using their credentials before accessing the tools in the web application. This feature is used to prevent the general public or fans from accessing unnecessary analytical data and provide them with the best experience by keeping them engaged in the game.

The first component provides all the necessary tools to the coaches to keep track of each of the players, their relative positions on the court, and their body movements and tactics. This comprises four major options—tracking the speed of five players in the team and displaying continuously on the frames, tracking the ball movement across the court for performance analysis and strategic insights, accurate pose estimation of every player for assessment, injury prevention, and gameplay enhancement, and finally court boundary detection and position tracking of five players for strategic planning, player positioning optimization, and tactical decision-making in real-time, enhancing their ability to refine team performance and game strategies. The second component is designed for the umpire for the management of match details and statistics, by updating the data as and when required during the match. The match number, team players' names, next hand, the score of each team can be simultaneously updated in the application by the umpires and finally, the finishing of the match can be marked to save the data into the database in real time. They can also make use of the ball-tracking feature to know when the ball is out of the boundary, which helps them in making quick precise decisions during the match. The third component is meant exclusively for fan engagement, such as checking live updates of the match and knowing the winning probability based on scores as well as the old and recent historical data alongside the scores. An option to attempt quizzes during the match is also provided by linking to an external site where they can get into the game and enjoy the match.

3.1 Features of the Application

See Fig. 1.

3.2 Pose Estimation Using Markerless Motion Capture

For markerless motion capture systems, we use the YOLO-pose variant of the YOLOv8 model. This model is pre-trained on the coco dataset and can display human poses as skeletal lines joined at specific points identified by the main features of the body such as the head, neck, shoulders, hips, knees, ankles, elbows, and wrists. As it can track the movements based on body features and not using physical markers, it can provide accurate results on the pose of the players as well as distinguish player movements from other objects assumed to be moving due to the motion of the camera. For every frame of the video, the YOLOv8 inference is performed on the current model initialized using pre-trained weights. The time taken to process the frame is recorded, and the Frames Per Second (FPS) is calculated to measure the system's performance as fps = 1/(end_time − start_time), which can be displayed on the window for further information. These detected objects are visually annotated on

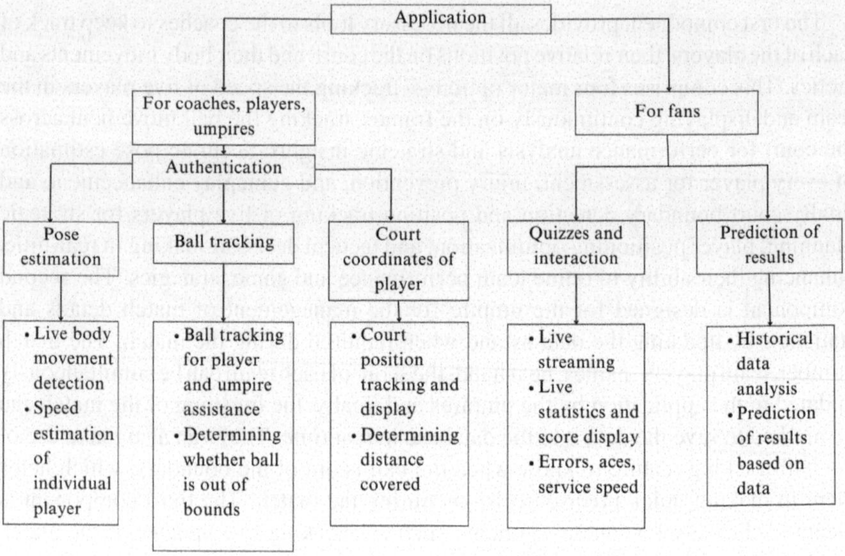

Fig. 1 A workflow of the proposed web application showing major features and options available

the current frame to render the final video with accurate pose estimation. An optional Imageio writer may be used to save the file as per the requirements (Fig. 2).

Markerless motion capture is employed in this case as it does not pose many constraints on the appearance of the object. Although marker-based systems also offer efficient pose estimation techniques, they require precise measurements of markers and their calibration, which is not necessary while implementing markerless systems. They tend to provide the best results when multi-person capture is to be performed.

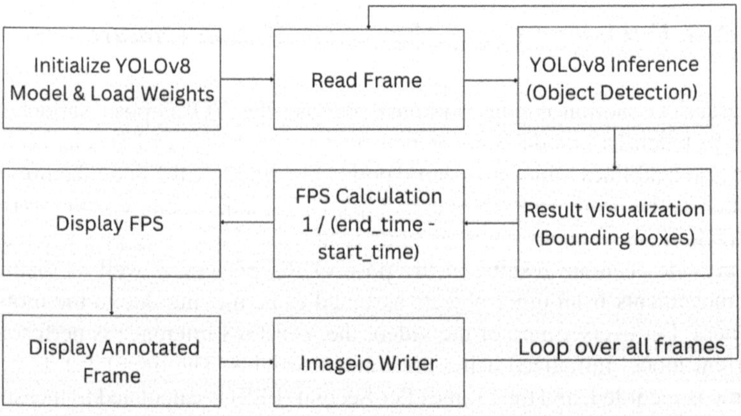

Fig. 2 Procedural flowchart for implementing pose estimation based on markerless motion capture

3.3 Speed and Pose Estimation Using Lucas-Kanade Algorithm

Marker-based motion capture systems identify certain physical markers that move in successive frames or image sequences, through the movement of pixels. The Lucas-Kanade method is an optical flow estimating algorithm. It is employed in order to determine the optical flow vectors, also known as velocity vectors, of a group of pixels or image patches. It makes the assumption that there is little motion between frames and that the motion can be described by a straightforward model, frequently a linear model. The markers may consist of the image's corners, edges, or other distinguishing aspects. To estimate motion, this technique relies on variations in pixel intensity between frames and calculates the optical flow vectors using the gradients or variations in pixel values. As per the comparison, results in [5] the Lucas-Kanade algorithm gives significantly lesser angular errors compared to other algorithms such as Horn-Schunck, hence the proposed algorithm uses this technique for speed estimation.

To identify and track physical markers in the proposed method, the Shi-Tomasi corner detection algorithm has been used along with Lucas-Kanade for optical flow. For every frame of the video, features of interest are found using Shi-Tomasi corner detection. For the proposed implementation, the maximum number of corners to be tracked is set to 100. The minimum possible Euclidean distance between any pair of corners is kept at 7. For every frame of the video, preprocessing is done by converting into grayscale followed by optical flow calculation using cv2.calcOpticalFlowPyrLK, a function that uses the Lucas-Kanade algorithm to track the movement of the feature points from the previous frame to the current frame. The result includes the new positions of the tracked points and status information indicating whether each point was successfully tracked. Good points are selected based on the status information. The last five tracks (i.e., the paths of the tracked feature points over time) are kept track of. For the first five tracks, the speed of each tracked point is calculated and displayed on the frame. The speed is estimated based on the distance the point has moved between frames, which provides a good approximation of it for five players at a time being tracked in the frame. The tracked curves are drawn on a mask by generating random colors for each point to be distinguishable, and for convenience purposes, the track lines are cleared after a specific duration by clearing the mask (Fig. 3).

3.4 Ball Tracking

The proposed approach for ball tracking makes use of features such as the yellow color, boundaries, and shape of the ball to detect the ball in every frame using contour detection. One of the purposes of using this method is that it also helps filter out small, irrelevant details or noise in the image, focusing on the significant objects or regions.

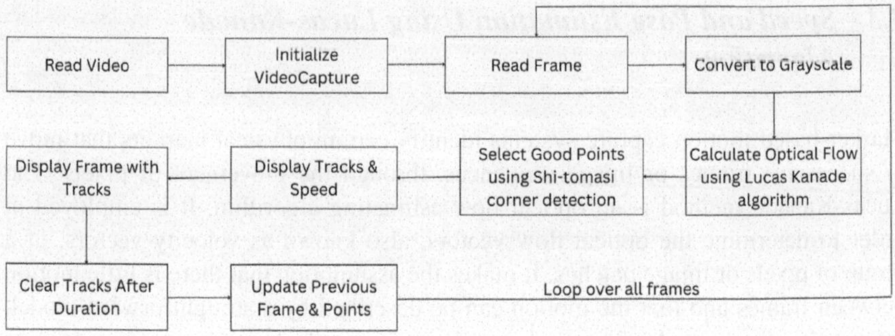

Fig. 3 Execution flowchart for speed estimation using optical flow

Every frame of the video is first converted to HSV color space for better color detection. To detect the color of the ball specifically, a lower and upper threshold is specified to isolate the yellow color of the ball using a mask. Contours are then found in the mask using the findContours() method and for each detected contour, the center and radius of the minimum enclosing circle is calculated. An object is considered detected if the radius is not equal to zero. If an object is detected, it draws a red dot at the center of the object on the frame and saves the frame as an image to display the original frame with the red dot indicating the center of the detected object.

The choice of this algorithm for ball tracking in our application is underpinned by its effectiveness in detecting and monitoring dynamic objects in video frames. The algorithm's capacity to isolate and track the distinctive yellow color of the ball, combined with its real-time processing capability, aligns with our specific requirements for real-time object tracking. Furthermore, contour detection, a core component of this algorithm, provides a reliable means of identifying the object's shape and boundaries, making it an ideal choice for our ball-tracking scenario. Its simplicity and efficiency are particularly well-suited to applications where rapid and accurate object tracking is essential (Fig. 4).

3.5 *Court Boundary Detection and Player Position Estimation*

For detecting the players and estimating their position in the court, the YOLOv3 object detection model is used, which is pre-trained on the Coco dataset. The YOLO model is configured with pre-trained weights and the configuration file. This is used to detect five players, draw bounding boxes around them and finally estimate their coordinates with respect to the court and draw them as points in a small-scaled display of the court on the top left corner of the video being displayed, so that their

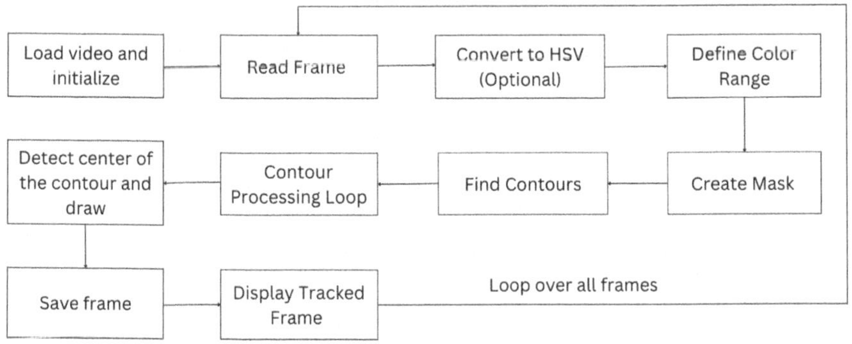

Fig. 4 Tracking the ball using contour detection

movements out of the boundaries can be tracked and their relative positions can be estimated by the coaches, for better monitoring and training purposes (Fig. 5).

The algorithm initially detects players in the frame using YOLOv3 using a confidence threshold of 99% in every frame of the video. To exclude unnecessary detections, the number of players detected to be considered for further processing is considered to be five, which is also the reason for the high specified value of the confidence threshold. The coordinate of the player is measured to be the center of the bounding box that is drawn for every detected player, and this data is stored for further computation in a list. To remove any redundant detections as well as to ensure that only the most confident and distinct player detections are detected, non-maximum suppression is applied. Further, the bounding boxes are drawn for the players considered and then the upper and lower threshold values for detecting the white court lines are specified. Using these values, contours are detected in the

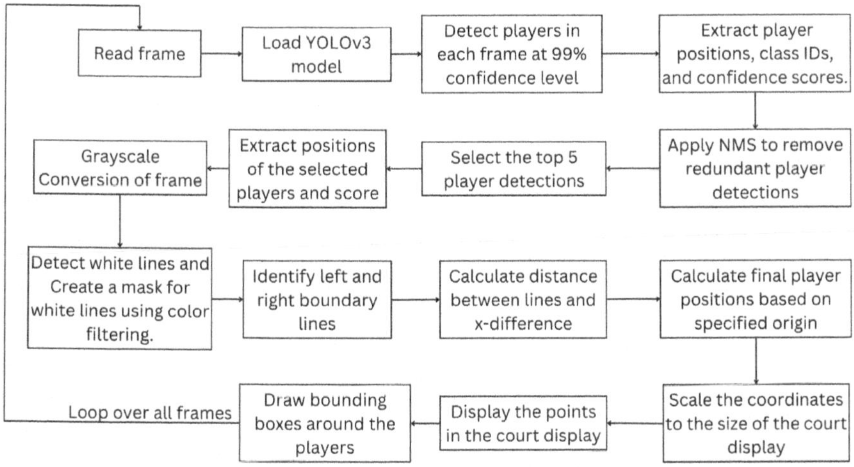

Fig. 5 Algorithm to detect court boundaries and player position estimation

frame after creating masks, and the detected lines are displayed on the frame. The lower leftmost and rightmost points in the detected contours are found, which is in turn used to calculate the x-distance between these lines. This distance is taken as the estimate or approximation of the width of the court visible in the frame. As per the conventions, the dimensional length of the ball badminton court is taken to be twice its width. The dimensions of the displayed small-scale court boundary image on the window are known, so these coordinates are scaled to those dimensions, and taking the bottom right corner of this box as the origin, these coordinates are drawn as points. In every frame, this shows the positions of the players on the court at any time and hence is useful to track the relative position of the players with good accuracy and also to determine out-of-boundary conditions if the threshold is precisely fine tuned. Finally, the bounding boxes and points are displayed on the frame using graphic overlay techniques.

Although this algorithm does not take camera calibration into consideration, it is built keeping in mind only the scaling of the displayed video, which fits well to almost every distance of capture that is clear enough to be processed using computer vision and estimates player positions accurately.

3.6 Predictive Historical Analysis

This part of the application predicts the winning probability of each team giving equal weightage to both the most recent 5 matches played and all previous matches. To access and modify the match records, the algorithm first loads the dataset into a Pandas DataFrame. It gathers every distinct team name from previous games. When the input of the names of two teams is given (as per the ongoing match), the validity of the input is tested. Following this preliminary validation only matches involving the designated teams are included in the Dataframe. The algorithm involves computing a number of statistics, including the total number of games each team has participated in, the total number of games between the two chosen teams, and the number of recent victories. The expected odds of each side winning the next game are determined using these facts by calculating the ratio. For two distinct teams X and Y, the following formula is used to calculate the probability of win,

$$\text{X\_win\_probability} = (0.5 * \text{matches\_won\_by\_X/total\_number\_of\_matches\_between\_X} \\ \text{and Y} + 0.5 * \text{matches\_won\_by\_X\_in\_recent\_five/5}) * 100.$$

If changes are needed, they are made to make sure probabilities are not extreme values (0 or 100) as any prediction cannot guarantee a winning probability of 1 for any team. In case of invalid or inconsistent data, the algorithm outputs the winning probabilities as 50% for each team. The function outputs the anticipated odds of each team prevailing at the end. This technique is quite simple, mostly consisting of data filtering and elementary statistical computations to generate forecasts based on past

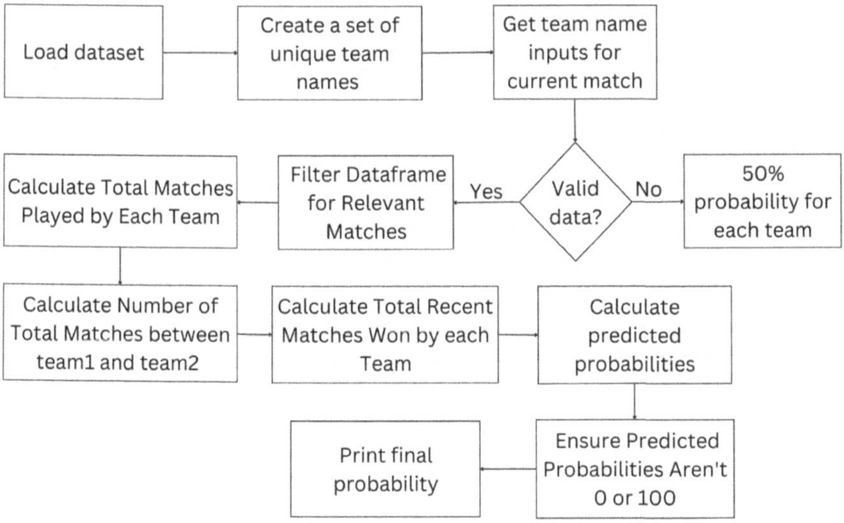

Fig. 6 Algorithm for winner prediction using historical analysis

performance data. With sophisticated datasets having good amounts of data namely the scores of two teams, team names, and the winner, and by numbering the matches according to the order played, this algorithm can be applied to predict the winner of the match with great accuracy (Fig. 6).

4 Purpose

Ball badminton enthusiasts, players, coaches, and umpires will all benefit from the proposed web application, as it attempts to revolutionize the tennis experience with three key aspects. The first function is centered on position estimation, allowing users to capture and analyze tennis shots, making it a useful resource for both players and coaches. The second component of the application delves into speed estimation and ball tracking, which helps umpires make accurate calls and provides viewers with real-time analytics. It also increases fan participation through online quizzes, making tennis viewing more engaging and exciting. The third component enables historical analysis and forecast of results, providing insights into the trends in the sport. This multimodal software is intended to increase fan interaction, improve umpiring accuracy, and provide useful data to players and analysts, ultimately bringing the ball badminton community closer together. This study aims to introduce and propose the technology of this novel application, emphasizing its potential societal significance and impact on promoting the popularity of the sport while providing analytical tools for enhancement.

5 Results

The resulting web application offers a comprehensive solution to tackle the problems faced in the sport of ball badminton. The figures below show the practical implementation of all the algorithms discussed in this paper. In Fig. 7, an update of the current score and match details by the umpire while playing the match is shown, where the umpire can also manage the hand table and click finish to store the data once the match is finished. Figure 8 shows the results for real-time statistics displayed for the fans along with the results of the winner prediction algorithm (based on current and historical analysis) to display the win probability for each team. The method works well even for datasets that are not very sophisticated, as verified while measuring the accuracy of prediction using a sample dataset created for the purpose (Fig. 8).

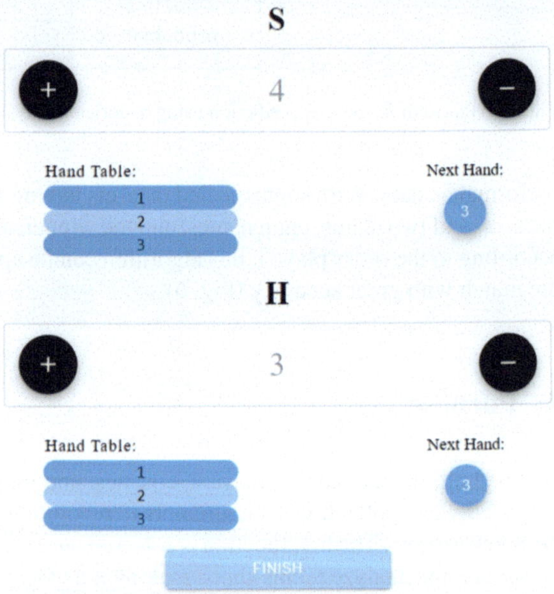

Fig. 7 Real-time statistics updated by the umpire for each team

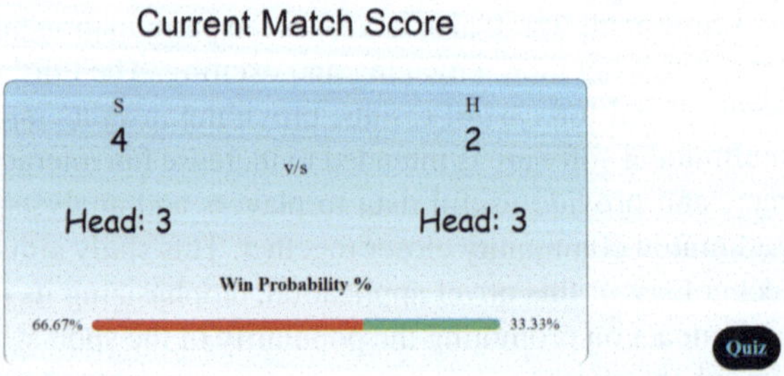

Fig. 8 Live scores and win probability display for fan engagement

Fig. 9 Player detection using YOLOv3 and their positions displayed on the top left as points inside the court boundaries

The accurate results for player position estimation and court boundary detection is shown in Fig. 9. The algorithm tracks five players on one side of the court and after calculating their coordinates with respect to the court boundaries, displays their positions as yellow points inside a small-scaled display of the red court on the top left. The players are tracked only if the detection accuracy is more than 99% (confidence threshold), which provides great precision for estimating coordinates and prevents the audience from being detected as players. YOLOv3 achieves a mAP of 51.9 at an IoU of 0.5 on the COCO dataset. In Fig. 10 the tracking of the ball is shown, as a red dot on the ball detected using contour detection. The algorithm achieves excellent accuracy, often in the range of 80–100%, in the best-case conditions, where the yellow object is consistently colored, well-illuminated, and distinct from the background. However, a number of factors affect it, including variations in illumination, how an object appears, noisy backgrounds, and the dynamics of the object's motion. Accuracy can reach 60–80% in somewhat difficult situations, when the yellow item may sometimes go lost in the backdrop or encounter tracking issues.

Figure 11 shows an implementation of the pose estimation algorithm discussed in Sect. 3.2. The detection accuracy of every player is visible for every bounding box, and every player clearly visible in the frame for a single team is tracked along with his body movements with a minimum of 75% accuracy. For YOLOv8n-pose, the mean Average Precision for poses when the IoU threshold ranges from 50 to 95%, is 50.4% and the mean Average Precision for poses at an IoU threshold of 50% is 80.1% as per [19].

In Fig. 12 the speed of three detected players in the frame is calculated by the algorithm in Sect. 3.3, although the proposed technique can detect five players simultaneously. The optical flow lines are shown around the player's legs, which indicates

Fig. 10 Ball tracking by contour detection as a red dot on the ball

Fig. 11 Markerless motion capture (pose estimation) using YOLOv8

the tracking of markers by the Lucas-Kanade algorithm. Optical flow-based tracking can achieve relatively high tracking accuracy under ideal conditions (e.g., well-lit, stable camera, distinct feature points), with errors often occurring within a few pixels. In these kinds of situations, the accuracy rate may reach 95% or more. The accuracy might drastically drop in more difficult circumstances, such as occlusions, fast motions, or poor feature selection. In such cases, accuracy percentages could be as low as 60% or as high as 80%.

Fig. 12 Optical flow-based speed estimation for five players in the frame

However, the results of the implementation vary greatly based on external factors such as the quality of the video and the position of the recording. For example, court boundary detection works better if the whole court lies within the frame of the video. Other parameters like the position of the camera should also be taken into account since this plays an important role in detection, estimation, and calculations based on it. Multiple cameras are required to track teams on either side of the court, for pose tracking, speed estimation, and player position estimation. The background in the video matters significantly, as techniques like contour detection and court boundary lines require more clarity on the visibility of the boundary lines clearly with respect to the background. A major issue faced throughout the research is the non-availability of official datasets for prediction using historical data. Inconsistent and invalid data can lead to the failure of the proposed prediction algorithm and thus wrong results. Hence, these results can be improved to a great extent if the availability of sufficient quality resources is ensured.

6 Research Implications

In this paper, a complete practical implementation of all the algorithms proposed has been discussed as the main aim of this paper is to provide the methodology for a fully functional analytical and fan engagement website for different users catering to their needs. Also, the proposed approach tends to be simpler as compared to other algorithms proposed in the references and is useful even in cases where the camera calibration or angle of recording is not taken into account. A significant feature of the proposed methodology is authentication to some categories of users, thereby ensuring confidentiality of the analyzed results, as well as better control over the match details updation so that any unauthorized access to such activities is prevented. The research proposed in [1] uses YOLOv2 for improvement in object detection and processing. However, our approach gives the best results for speed estimation without the use of any other additionally integrated techniques for object detection. The algorithm is based only on the tracking of markers, which provides simplicity as well as the required accuracy for the purpose. The research by Ceseracciu et al. [2] compared markerless and marker-based motion capture. Our proposed system takes a step further by integrating both markerless and marker-based approaches. The proposed system effectively uses both techniques, according to which application it is best suited for, markerless motion capture for pose estimation and marker-based motion capture for speed estimation.

Kaushik et al. [3] provided a review of marker-free computer vision for human motion analysis. While they explored the potential of markerless motion capture, our proposed approach implements it practically. The system uses YOLO-pose, a variant of YOLOv8, to perform multi-person pose estimation. This overcomes some of the limitations of marker-based systems by providing accurate pose estimation without the need for physical markers. Our proposed system also aligns with the findings of Kamble and Keskar [4] by offering real-time ball tracking. By using color filtering and contour detection to track the ball, the system ensures that the ball's movement is accurately captured. This overcomes the limitations of some existing systems that might struggle with tracking fast-moving objects like a ball. Duan et al. [6] explored motion capture for sporting events, which aligns with the core concept of our system. Our approach utilizes graph convolutional neural networks and single-target pose estimation algorithms, which enhance motion capture accuracy. It overcomes some limitations of traditional methods that might struggle with complex sports motions.

7 Originality

The proposed web application for ball badminton offers a unique and highly specialized solution, drawing from various cutting-edge technologies and research findings. The originality of the proposed system lies in the fact that although many such analytical systems based on computer vision and deep learning have been presented for other sports, this can be very unique in the field of research in ball badminton. There are not many technologies that are designed specifically for ball badminton, and so this approach presents a novel state-of-the-art system that integrates every aspect of analytics for the sport. The utilization of the Lucas-Kanade optical flow algorithm for human tracking, as reported in [1], ensures precise detection of player movements, addressing the challenge of occlusion handling, which is vital in ball badminton where players frequently cross paths. The use of contour detection to track the ball is especially significant, as it not only identifies the ball based on color but also its shape and boundaries, filtering out irrelevant details in the video frames. This is in line with the findings from [14], where object tracking in sports action recognition videos is explored, emphasizing the importance of player tracking and the challenges it poses, which are effectively addressed in the proposed application. The integration of YOLOv3 for player detection and court boundary estimation provides a level of accuracy in position tracking that is indispensable in ball badminton, ensuring that coaches have access to real time, highly precise data for tactical analysis and player performance assessment. The predictive historical analysis component offers a comprehensive approach to assessing winning probabilities, taking into account both recent performance and overall historical data, a feature that adds significant value to the application for fans and analysts alike. This does not impose the overhead of using machine learning models for the purpose of prediction which give less accurate results.

Also, this system employs certain algorithms already proposed by previous research as mentioned above for speed estimation, player detection, and player pose tracking. However, the algorithms designed for predicting the winner based on historical data, and player position estimation in the court are novel besides being simple as well as accurate compared to other algorithms discussed in the related research section.

8 Conclusion

We aim to create inclusivity throughout the sport by offering technology-driven solutions to ball badminton. This approach promises to connect enthusiasts, coaches, and players, providing an inclusive community that can exchange ideas, methods, and expertise on a hitherto impossible scale. It has the potential to make ball badminton more accessible by removing geographical barriers, thereby creating a global audience. It has the potential to be a valuable teaching tool, assisting coaches in developing and polishing players' talents. Furthermore, the prognostic insights and historical analyses enrich the viewing experience for spectators, promoting a stronger connection to the sport and its players. This technology has the potential to not only revolutionize the sport but also to inspire comparable advancements in other traditional sports, bridging the gap between tradition and innovation in the world of ball badminton. Since there are not many resources available in the technological world that focus on ball badminton, this tool can prove to be a valuable resource in this regard, as it integrates most of the aspects of training and viewer engagement. This ground-breaking technology exemplifies how innovation may bridge the gap between tradition and cutting-edge improvements in sports, ushering in a new era of engagement and interest in ball badminton.

9 Future Work

Exciting potential awaits further effort in this subject. The capabilities of the system can be refined and expanded. Continuous R&D can improve player tracking precision, allowing for even more intricate tactical insights. Furthermore, the introduction of augmented reality (AR) and virtual reality (VR) components has the potential to transform the viewing experience by allowing fans to immerse themselves in the game, analyze statistics, and even virtually engage in training sessions. Using modern biomechanical analysis could provide players with in-depth insights into their technique and aid in injury prevention. Also, the introduction of a virtual coach powered by machine learning would provide tailored coaching tips and real-time tactical assistance, transforming player training. The system might be expanded to accommodate ball badminton at all levels, from grassroots to professional, bringing up new opportunities for skill improvement and talent detection. Collaborations with sporting organizations and federations may result in the inclusion of this system in official events and competitions, establishing it as a standard component in ball badminton. As technology advances, the system's efficiency can be increased to require less hardware, making it more accessible to a wider demographic. Finally, future work on this topic will involve not only enhancing existing technologies but also investigating novel uses and opportunities to make ball badminton more accessible, interesting, and enlightening.

References

1. Dao Vu Q, Chung S-T (2017) Real-time robust human tracking based on Lucas-Kanade optical flow and deep detection for embedded surveillance. In: 2017 8th international conference of information and communication technology for embedded systems (IC-ICTES), Chonburi, Thailand, pp 1–6. https://doi.org/10.1109/ICTEmSys.2017.7958761
2. Ceseracciu E, Sawacha Z, Cobelli C (2014) Comparison of markerless and marker-based motion capture technologies through simultaneous data collection during gait: proof of concept. PLoS One 9(3):e87640
3. Kaushik V, Bhugra S, Gupta A, Lall B (2023) Marker-free computer vision for human motion analysis: a review. In: 2023 IEEE international conference on multimedia and expo workshops (ICMEW), Brisbane, Australia, pp 212–217
4. Kamble P, Keskar A, Bhurchandi K (2019) Ball tracking in sports: a survey. Artif Intell Rev 52. https://doi.org/10.1007/s10462-017-9582-2
5. Patel E, Shukla D (2013) Comparison of optical flow algorithms for speed determination of moving objects. Int J Comput Appl 63:32–37. https://doi.org/10.5120/10465-5180
6. Duan C, Hu B, Liu W, Song J (2023) Motion capture for sporting events based on graph convolutional neural networks and single target pose estimation algorithms. Appl Sci 13:7611
7. Sensors J (2023) Retracted: badminton path tracking algorithm based on computer vision and ball speed analysis. J Sens 2023:1–1. https://doi.org/10.1155/2023/9814728
8. Rizal AM, Taufiqurrohman M, Subur J (2020) Implementation of ball tracking systems in image processing using the find contour method. J Electr Electron Eng-UMSIDA 4(2)
9. Maji D, Nagori S, Mathew M, Poddar D (2022) YOLO-pose: enhancing YOLO for multi person pose estimation using object keypoint similarity loss

10. Leonida KL, Sevilla KV, Manlises CO (2022) A motion-based tracking system using the Lucas-Kanade optical flow method. In: 2022 14th international conference on computer and automation engineering (ICCAE), Brisbane, Australia, pp 86–90. https://doi.org/10.1109/ICCAE55086.2022.9762423
11. Tan DYW, Ting HY, Lau SBY (2016) A review on badminton motion analysis. In: 2016 international conference on robotics, automation and sciences (ICORAS), Melaka, Malaysia, pp 1–4. https://doi.org/10.1109/ICORAS.2016.7872604
12. Xiaochen L et al (2018) Evaluation of Lucas-Kanade based optical flow algorithm. In: 2018 IEEE CSAA Guidance, Navigation and Control Conference (CGNCC), Xiamen, China, pp 1–6. https://doi.org/10.1109/GNCC42960.2018.9018982
13. Amrutha K, Prabu P, Paulose J (2021) Human body pose estimation and applications. In: 2021 innovations in power and advanced computing technologies (i-PACT), Kuala Lumpur, Malaysia, pp 1–6. https://doi.org/10.1109/i-PACT52855.2021.9696513
14. Buric M, Ivasic-Kos M, Pobar M (2019) Player tracking in sports videos. In: 2019 IEEE international conference on cloud computing technology and science (CloudCom), Sydney, NSW, Australia, pp 334–340. https://doi.org/10.1109/CloudCom.2019.00058
15. Patel D, Upadhyay S (2013) Optical flow measurement using Lucas Kanade method. Int J Comput Appl (0975–8887) 61(10)
16. Bunker RP, Thabtah F (2019) A machine learning framework for sport result prediction. Appl Comput Inform 15(1)
17. Ajgaonkar Y, Patil A, Bhoyar K, Shah J (2021) Prediction of winning team using machine learning. Int J Eng Res Technol (IJERT) NTASU–2020, 09(03)
18. Redmon J (2018) YOLOV3: an incremental improvement, April 8, 2018. https://arxiv.org/abs/1804.02767
19. Ultralytics docs. https://docs.ultralytics.com/tasks/pose/#models. Accessed 3 Nov 2023

The Impact of Surface Roughness Angle on Cricket Ball

Arun Jalan, Pratibha Sinha, and Kundrapu Ayyappa Swamy

Abstract For many years, perhaps even from the inception of the game itself, cricket players and spectators have been drawn to the aerodynamic properties of a cricket ball. We have compared the lift force and drag with observations. The Detached Eddy Simulation model (DES) has been used to provide a 2D numerical simulation of a cricket ball. DES was effectively used to simulate the cricket ball's lift and drag forces at various ball speeds, seam angles, and roughness heights. This study focuses on the computational fluid dynamic simulation of the cricket ball in the flow field. The flow field at different roughness angles 45° to 90° with a gradient of 15°, roughness heights in a range of 0.1 to 1 mm, and three bowling velocities of 32.77,35.55, and 45.55 m/s have been computationally analyzed. Aerodynamic forces were obtained using numerical analysis, while boundary layer separation sites were obtained through analysis. The aerodynamics of a cricket ball were proven by comparing the boundary layer separation width and asymmetry to the drag and lift forces.

Keywords Cricket ball · Aerodynamics · Surface roughness angle · Lift force · Drag force

1 Introduction

Arguably, one of the earliest games ever invented is cricket. The ball is the main object in the cricket game. A cricket ball's motion and, thus, its flight are greatly influenced by the related aerodynamics since the ball must be pushed through the air like a three-dimensional body. A cricket ball's aerodynamic qualities can be significantly influenced by the prominence of the seam, the surface roughness of the ball during play, and the bowler's launch attitude. The ball's surface roughness and seam angle can cause an uneven airflow around the ball as a result of pressure variations, which can cause the ball to deviate from its planned flight route. We refer to this possible

A. Jalan · P. Sinha (✉) · K. A. Swamy
Birla Institute of Technology and Science, Pilani, Pilani Campus, RJ 333031, India
e-mail: p20190445@pilani.bits-pilani.ac.in

S. Bhattacharyya and D. James (eds.), *Recent Trends in Sports Engineering*, Lecture Notes in Mechanical Engineering, https://doi.org/10.1007/978-981-97-8297-0_21

flying variation as a swing. Thus, comprehending the employed ball's aerodynamic characteristics is critical. In cricket, spin bowlers commonly use this type of "spin swing" to produce the "Magnus effect." The primary and quarter seams of a swinging cricket ball play a significant role in its aerodynamics [1].

More recently, thorough experimental studies reported by Barton [2], Bentley et al. [3], and Mehta et al. [4] revealed the amplitude of the side force that generates swing and the parameters regulating it. Mehta et al. [5] revisited the previous research based on cricket ball swing. Penrose et al. [6] described a preliminary computational fluid dynamics investigation of cricket ball swings. Mehta [6] outlined a few myths and misconceptions about the aerodynamics of balls used in cricket. A cricket ball's flight can also be influenced by atmospheric conditions, which Binnie examined analytically in great depth [7]. Another study reports on tests conducted to directly quantify the aerodynamic forces of spinning and stationary cricket balls for various seam angles and seam orientations concerning the free stream.

When a bowler releases a cricket ball into the air, aerodynamics is a major factor in its flight. The underlying aerodynamic principles that underpin a cricket ball's nonlinear flight, or "swing," have been the subject of several studies. These concepts were discovered some years ago. Several experimental studies on cricket ball swings have been carried out in the last 20 years, revealing the number of swings that can be achieved and the factors that influence them. The ball's trajectory can be altered with each delivery by adjusting the line, length, tempo, or—most subtly—by "swinging" the ball in the air and causing it to drift sideways. Cricket fans have long been enthralled with this movement but rarely comprehend the fundamental principles that lead to the cricket ball swinging. The turbulent boundary layer splits off from the laminar layer due to its increased energy later (moving back along the ball surface), creating a pressure differential that generates a side force.

Previous works of literature have shown the different parameters that affect the ball's swing, considering different roughness heights, speed of the ball, etc. Still, one more parameter varies throughout the cricket game: the roughness angle of the ball is not considered in the analysis. In this work, the variation of the roughness angle on a cricket ball's swing, as well as the ball's various speeds and roughness heights, are simulated using fluid dynamics computation.

2 Simulation Methodologies, Modeling, and Governing Equations

To determine the cricket ball's aerodynamic characteristics numerically, simulations have been performed in ANSYS Fluent. Figure 1 shows the flow over a cricket ball and the roughness angle.

With the availability and increased numerical simulation power, it is possible to investigate the flow over a ball more precisely. The commercially available ANSYS FLUENT module is used to discretize governing equations as it can handle complex

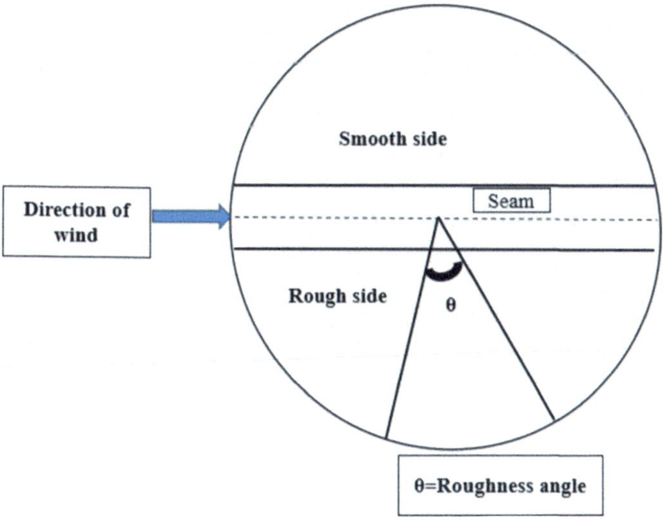

Fig. 1 Representing the flow over a cricket ball and roughness angle

equations. The 74 mm diameter sphere representing the cricket ball was modeled with a different roughness angle for each dimension and placed in a rectangular duct as shown in Fig. 2.

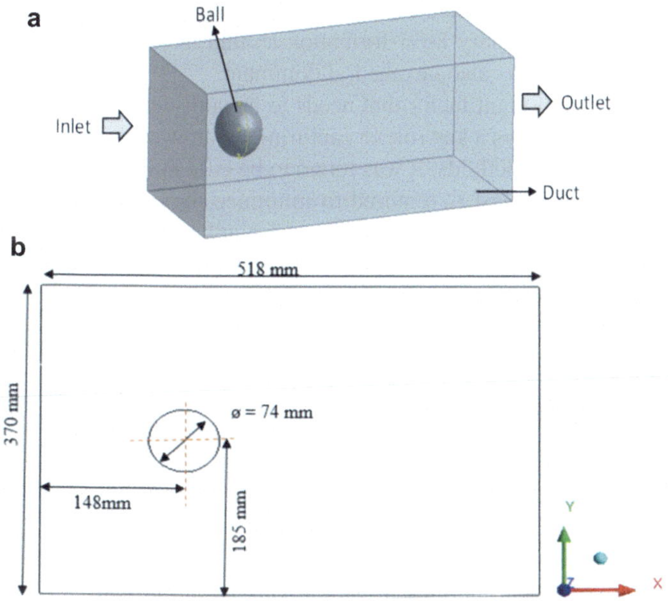

Fig. 2 **a** 3D concept model of cricket ball in ANSYS **b** 2D geometry of cricket ball in ANSYS

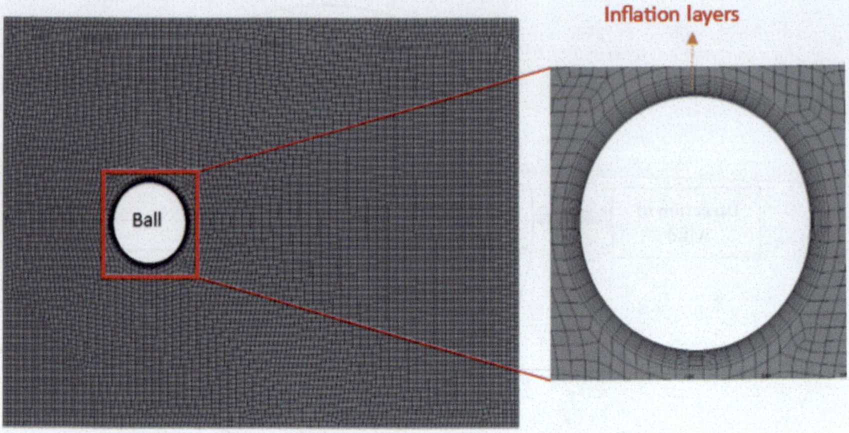

Fig. 3 Hexahedral-dominated mesh domain and inflation layer over the edge of cricket ball

A 2D schematic model is designed for numerical work to minimize and limit computational power, as shown in Fig. 2b. For the present numerical analysis, a two-dimensional fluid domain having a height of 370 mm and a length of 518 mm is used. The ball under investigation is kept at a distance of 148 mm in the horizontal direction and 185 mm in the vertical direction from the inlet. The ball domain is neglected as we are not focusing on the internal pressure or deformation of the ball. The fluid domain around the ball is considered for analysis of aerodynamics. A hexahedral-dominated mesh domain is generated, shown in Fig. 3. An inflation layer is generated to understand better boundary layer formation around the ball and boundary layer separation points, which is also hexahedral dominant.

There is a very important factor that needs to be considered during the inflation layer is 'y + ', which plays a key role in capturing the flow separation point. After a series of trial-and-error methods, it was found to be 0.01 mm.

A grid independence test is required to minimize the computational time. C_d/ C_L is calculated for constant roughness After the number of elements has increased by more than 11,584, there is no significant change in outputs. The following grid independence test results are shown in Fig. 4.

2.1 Governing Equations

Following are the governing equations for k-ϵ model, which are used for the simulation.

$$\frac{\partial}{\partial t}(\rho k)+\frac{\partial}{\partial x_i}(\rho k u_i)=\frac{\partial}{\partial x_j}\left[\left(\mu+\frac{\mu_t}{\sigma_k}\right)\frac{\partial k}{\partial x_j}\right]+P_k+P_b-\rho\epsilon-Y_M+S_k \quad (1)$$

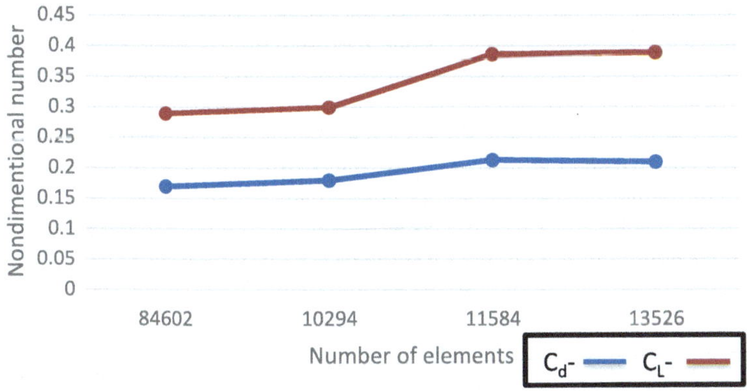

Fig. 4 Grid independency test

$$\frac{\partial}{\partial t}(\rho\varepsilon)+\frac{\partial}{\partial x_i}(\rho\epsilon u_i)=\frac{\partial}{\partial x_j}\left[\left(\mu+\frac{\mu_t}{\sigma_\epsilon}\right)\frac{\partial\epsilon}{\partial x_j}\right]+C_{1\epsilon}\frac{\epsilon}{k}(P_k+C_{3\epsilon}P_b)-C_{2\epsilon}\rho\frac{\epsilon^2}{k}+S_\epsilon \tag{2}$$

ρ = density
k = turbulence kinetic energy.
u = local mean velocity parallel to the wall.
x = Cartesian coordinate in the mean flow direction.
t = time.
μ = dynamic viscosity.
σ = Prandtl number.
i,j = suffices taking values 1, 2, 3
ϵ = rate of dissipation of turbulence kinetic energy.
P_k = generation of turbulence kinetic energy due to the mean velocity gradients
P_b = generation of turbulence kinetic energy due to buoyancy
Y_M = contribution of the fluctuating dilatation in compressible turbulence to the overall dissipation rate
$C_{1\epsilon}$, $C_{2\epsilon}$, $C_{3\epsilon}$ = constants.
S_k and S_ϵ = user-defined source terms

3 Factors Considered for Numerical Analysis

It is understood from the literature that the three major factors that adversely affect the aerodynamics around the ball are roughness height, speed of the ball, and roughness angle. So, the following parameters are considered in our study. Various parameters and the degree of variation are shown in Fig. 5. These parameters are considered when calculating the aerodynamic properties of a cricket ball.

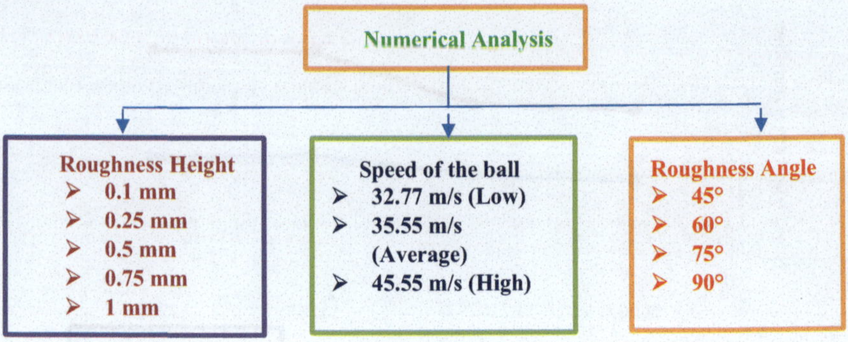

Fig. 5 Parameters considered for numerical analysis

4 Validation

For validating the study, the numerical results of the present investigation are compared with the experimental study of Alam et al. [1]. A very good agreement is observed between the results of the present investigation and with results of Alam et al. [1]. The drag coefficient for the present investigation for different flow velocities is approximately 0.48 ± 0.03.

5 Results and Discussion

To analyze the effects of roughness angles on the cricket ball's swing, simulations have been performed in ANSYS® Fluent. The following are the results of the computational Fluid Dynamics analysis of a cricket ball swing.

5.1 Velocity Contour

Velocity magnitude contours shown in Fig. 6 are plotted using ANSYS® Fluent. It is used for calculating the elemental areas between successive is ovels and adding the products of each area by the mean of its border velocities to determine a stream's discharge.

It is observed from the velocity contour that the velocity gradient on both the shiny and rough sides is high. It is caused due to the turbulence, which is because of the roughness.

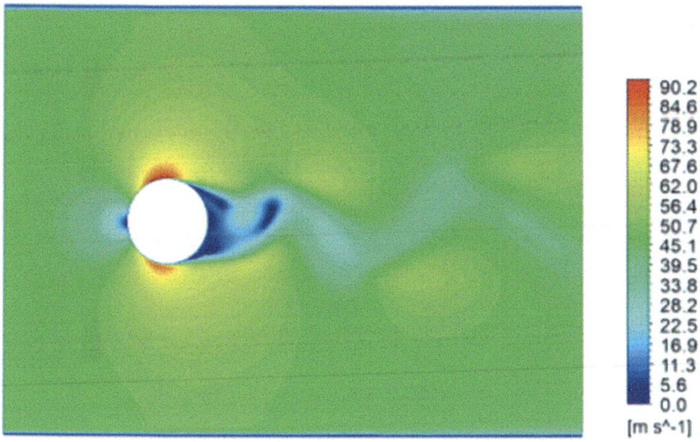

Fig. 6 Velocity contour for x–y plane showing high and low-velocity region because of roughness region

5.2 *Velocity Streamlines*

Velocity streamlines contours are plotted using ANSYS® Fluent and shown in Fig. 7. A streamline is a vector with a magnitude and direction, defined as a line tangential to the instantaneous velocity direction. We could picture the motion of a tiny, indicated fluid element to see this in action.

It is observed from the streamline that the airflow on the shiny surface has a smooth flow. On the other hand, the flow on the rough side is held by the air. Which results in increased inside force and eventually helps in ball swing.

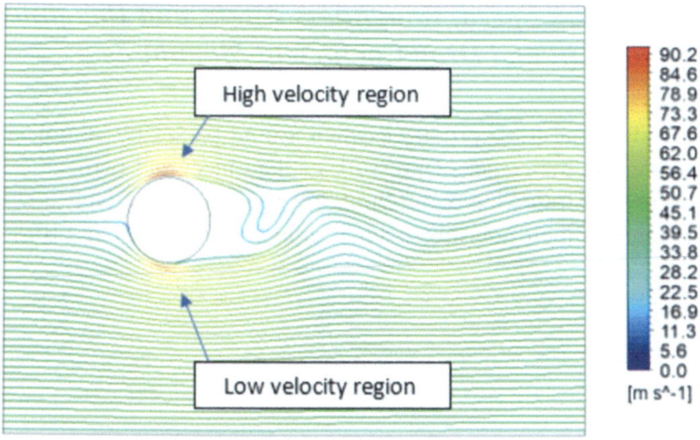

Fig. 7 Velocity streamlines for x–y plane showing asymmetric wake region because of earlier separation on the rough side and delayed separation on the smooth side

Fig. 8 Pressure contour for x–y plane showing asymmetric wake region because of earlier separation on the rough side and delayed separation on the smooth side

5.3 *Pressure Contour*

Pressure Contour plotted using ANSYS® Fluent, It indicates that, due to the middle region's high spinning velocity, the static pressure drops from the wall toward the center.

Similarly, to the velocity gradient, we can observe from Fig. 8, the pressure gradient on the shiny and rough surface. This pressure difference on both sides leads to the ball rolling.

5.4 *Impact of Ball Speed, Roughness Angle, and Roughness Height on Lift and Drag Coefficient Magnitude*

Lift is the aerodynamic force produced when a fluid passes by an object. A lift is defined as acting via the center of pressure of the object and perpendicular to the direction of flow. The following two formulas were used, respectively, to convert the drag and lift force to their non-dimensional drag coefficient C_d and lift coefficient C_L:

$$C_d = \frac{D}{\frac{1}{2}\rho V^2 A} \tag{3}$$

$$C_L = \frac{L}{\frac{1}{2}\rho V^2 A} \tag{4}$$

Here, drag, air density, ball speed, lift force, and projected frontal area of the ball are represented by D, ρ, v, L, and A, respectively. The cricket ball's drag or resistance is represented by the C_d. As a function of Reynolds numbers, C_d and C_L fluctuated according to roughness angle, roughness height, and overall speed. Figures 9 and 10 show the variation of C_d and C_L with different roughness heights (0.1mm, 0.25mm, 0.5 mm, 0.75 mm, and 1mm) at different roughness angles (45°, 60°, 75°, and 90°) for different ball speeds (32.77 m/s, 35.55 m/s, and 45.77 m/s).

The drag force that stops the ball from falling through the air from speeding up too much.

Figure 9 depicts a variation of the drag force's coefficient for different roughness angles. With increased roughness angle C_d increases.

Figure 10 shows the variation of C_L, from Fig; it can be seen that an increment in roughness angle increases the coefficient of lift force, which deflects air downwards by exerting a downward force on the air.

To understand the effect of the surface roughness angle of cricket ball on drag and lift force, simulations have been carried out. Tables 1, 2, and 3 show that when a cricket ball is in flight, it experiences aerodynamic forces like lift and drag force which changes the drag force and lift force. The surface roughness angle is a parameter that is responsible for the change in these aerodynamic forces by affecting the airflow around the ball.

Effect on drag force: Because of the increased turbulence in the boundary layer, a ball with a larger surface roughness angle may encounter more drag, slowing down its flight.

Effect on lift force: The amount of lift produced by a ball with a specific angle of surface roughness varies based on its interaction with the surrounding air.

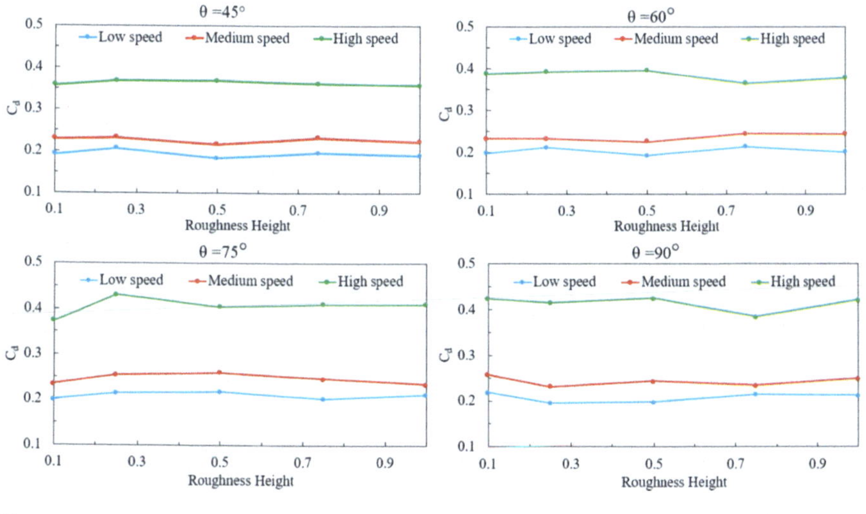

Fig. 9 Depiction of Variation of C_d with different roughness heights at different roughness angles for different ball speeds

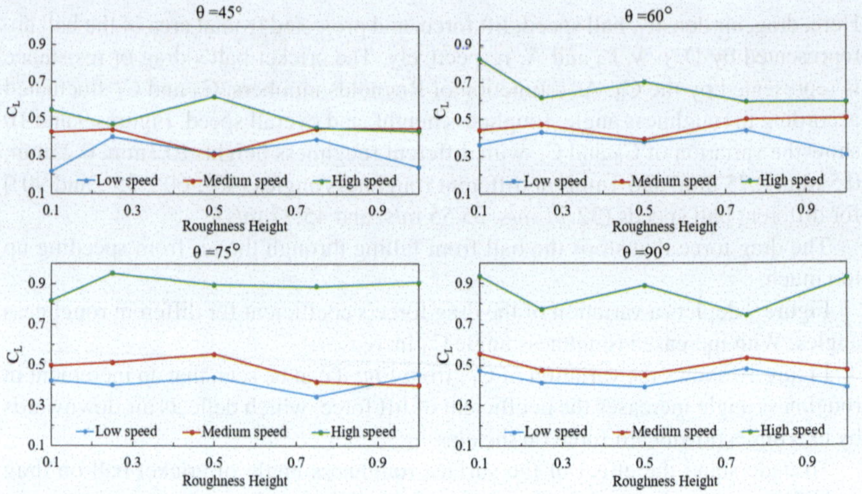

Fig. 10 Variation of C_L with different roughness heights at different roughness angles for different ball speed

Table 1 Drag force and lift force at velocity for 32.77 m/s

	Roughness angle			
	45°	60°	75°	90°
Drag force	3.34062	3.52756	3.73433	3.75917
Lift force	4.2578	5.36894	7.96804	6.8212

Table 2 Drag force and lift force at velocity for 35.55 m/s

	Roughness angle			
	45°	60°	75°	90°
Drag force	3.92458	4.28788	4.15027	4.40731
Lift force	7.88728	8.89435	7.04438	8.34448

Table 3 Drag force and lift force at velocity for 45.55 m/s

	Roughness angle			
	45°	60°	75°	90°
Drag force	6.30149	6.65191	7.24345	7.44768
Lift force	7.67669	10.447	16.0099	16.3397

6 Conclusions

- The aerodynamic behavior of cricket balls significantly varies from shiny to rough ball. The roughness of the surface has a big impact on a cricket ball's lift and drag. A cricket ball's aerodynamic drag is increased by wear and tear.
- Lift and drag experience a non-equal rise in magnitude. Lift grows more quickly than drag, which results in an increase in the lift-to-drag ratio.
- From velocity contour, pressure contour, and velocity streamlines, It is clear from this that, in a laminar state, the boundary layer on the upper surface separates relatively early, but the boundary layer on the rougher bottom side splits later and becomes turbulent. Due to this asymmetry, the ball swings toward the rough side due to a side force.
- The ball's trajectory may also deviate due to variations in the surface roughness angle. It can be difficult for batsmen to precisely forecast the course of the ball due to even little variations in surface roughness, which may result in unpredictable movements during flight. Comprehending these impacts provide substantial perspectives on the ball's behavior while in flight, augmenting both the strategic elements of the game and the players' abilities.

7 Future Recommendation

- There is potential for future research to concentrate on refining the surface properties of cricket balls, with the aim of improving or regulating their aerodynamic performance.
- The combination of data analytics and machine learning has the potential to deepen our comprehension of how different elements, such as the texture of the surface, impact the way cricket balls move through the air.

References

1. Alam F, Hillier D, Xia J, Chowdhury H, Moria H, La Brooy R, Subic A (2010) Aerodynamic of used cricket balls, 17th Australian fluid mechanics conference
2. Barton NG (1982) On the swing of a cricket ball in flight. Proc Roy Soc Lond A 379:109–131
3. Bentley K, Varty P, Proudlove M, Mehta RD (1982) An experimental study of cricket ball swing. ImpAl CollE Aero Tech Note, 82–106
4. Mehta RD, Bentley K, Proudlove M, Varty P (1983) Factors affecting cricket ball swing. Nature 303:787–788
5. Penrose JMT, Hose DR, Trowbridge EA (1996) Cricket ball swing: a preliminary analysis using computational fluid dynamics. In: Haake SJ (Ed) The Engineering of Sport. Balkema AA, Rotterdam, pp 11–19
6. Mehta RD (1985) Aerodynamics of sports balls. Annu Rev Fluid Mech 17:151–189
7. Pahinkar DG, Srinivasan J (2010) Simulation of reverse swing of the cricket ball. Int J Sport Sci Eng 04:053–064

The manufacturer's authorised representative in the EU is Springer Nature Customer Service Centre GmbH, Europaplatz 3, 69115 Heidelberg, Germany. If you have any concerns regarding our products, please contact ProductSafety@springernature.com

Printed and bound by CPI Group (UK) Ltd, Croydon, CR0 4YY
02/10/2026
02227132-0008